Br L Raphael, FSC

2004

Euclid's
Elements

Euclid's
Elements

all thirteen books complete in one volume

The Thomas L. Heath Translation
Dana Densmore, Editor

Green Lion Press
Santa Fe, New Mexico

Manufactured in the United States of America.

Second edition with minor revisions, September 2003.

Published by Green Lion Press,
1611 Camino Cruz Blanca, Santa Fe, New Mexico 87505-0353 USA.
Telephone (505) 983-3675; FAX (505) 989-9314.
www.greenlion.com

Green Lion Press books are printed on acid-free paper. Both softbound and clothbound editions have sewn bindings designed to lie flat and allow heavy use by students and researchers. Clothbound editions meet the guidelines for permanence and durability of the Committee on Production Guidelines for Book Longevity of the Council on Library Resources.

Printed and bound by Sheridan Books, Inc., Ann Arbor, Michigan.

Cover design by Dana Densmore and William H. Donahue with help from Nadine Shea. Adapted from a detail in Raphael's School of Athens, the illustration focuses on a young student being shown a proof by Euclid while friends and mentors offer supportive encouragement.

Cataloging-in-Publication Data:

title of this book: Euclid's Elements
author: Euclid
translation from Greek of: Elements/by Euclid
translated by: Thomas L. Heath

Includes complete unabridged text of all thirteen books of Euclid's Elements in T. L. Heath's translation with minor corrections to text and translation, along with introductions, terminology and biographical notes, bibliography, index and glossary.

ISBN 1-888009-18-7 (cloth binding with dust jacket)

ISBN 1-888009-19-5 (sewn softcover binding)

1. History of Mathematics. 2. Geometry. 3. Classics.
I. Euclid (fl. c. 300 B.C.E.) II. Heath, Thomas L. (1861–1940) III. Title.

QA31.E83 2002

Library of Congress Card Number 2002107461

Table of Contents

Euclid's Elements

The Green Lion's Preface

Euclid's *Elements* has stood for well over two millenia as the exemplar, not just of classical geometry, but of the axiomatic and deductive structure characteristic of all pure mathematics. Its range goes beyond what we think of as geometry, extending to the general theory of proportions, number theory, and an innovative and ingenious treatment of incommensurability.

Euclid speaks to us with a voice as clear and universal as laughter. The simplicity, clarity and elegance of Euclid's proofs and the beautiful sequence of their unfoldings both delight and instruct. They provide a paradigm of what constructing a proof means and an education in how it is done.

Why a New Edition of Euclid?

Students of Euclid have long been needing an unabridged one-volume edition of the *Elements.*

All thirteen books were included in Encyclopaedia Britannica's series, *Great Books of the Western World,* but that generally required purchase of the complete set of volumes. Even if one could acquire an individual volume, the presentation of the text was hardly suited to serious study. The type was tiny and so crammed on the page that there was no space for writing notes. The paper used for the pages was tissue-thin so that any attempt to make notes or mark the diagrams as one worked through the proof resulted in show-through and torn paper. It could be used for reference, which was the primary intent of that series, but it was not suited for anyone actually working through the propositions.

Dover has done us all the service of keeping in print a photo-reproduction of Heath's original Cambridge University Press translation and commentary, in three volumes. We are grateful for this and hope that they will continue to keep it in print. But that edition has several drawbacks for someone whose primary interest is in reading Euclid as opposed to reading Heath.

First there is the inconvenience of having the book divided into three volumes. One must cart them all around or risk not having the

right one in class or at hand. Even at home one finds oneself trying to remember which volume a particular proposition would be in. In class there is a tendency when working on the later books to stop carrying the earlier volumes, with the result that, when a question comes up about an early proposition, Euclid's wording or proof cannot be consulted.

Perhaps even more inconvenient and disruptive of the study of Euclid's actual text is the sheer volume of Heath's historical and scholarly commentary interspersing itself among the propositions. It makes it difficult to find propositions, and to move from one to the next requires carefully paging through extensive commentary looking for where Euclid's actual text resumes. One doesn't get a clean picture of the flow of Euclid's development, which is the great beauty of his work.

We ourselves have experienced these inconveniences as scholars and in the classroom. For these reasons and in response to many requests from other users and teachers of Euclid, Green Lion Press has prepared this one-volume, clean, student-friendly, Euclid-centered edition of the complete text of all thirteen books of *Elements.*

Design Features of this Edition

The design and layout of this Green Lion edition provide ample space around both text and figures, and between propositions, to allow for notes. Substantial, fine quality paper of high opacity has been used for durability under heavy use and to further support the reader's own annotations.

Type size of the proposition numbers, space between propositions, and size and placement of the running heads, make it easy to see which book and proposition one is looking at and to turn quickly and easily to another proposition one may wish to consult.

The typesetting is entirely new for crisp reproduction. Drawings have been newly lettered and largely redrawn. In some few cases, where the original diagram was confusing or hard to read (for example, XII. 16), they have been redrawn differently from Heath's diagrams.

In keeping with Green Lion standard practice, diagrams always appear on the same spread as any text which refers to them, being repeated as often as necessary to maintain this. The reader never needs to be paging back and forth to look at the diagram while following the steps of the proof, as in other editions.

Our extensive index includes the Greek for a large number of the

entries, comprising all the basic geometrical terminology, making it a valuable glossary of Euclid's mathematical language. It is designed to help students find the definitions of crucial terms as well as the chief propositions in which they are used.

Nevertheless, several fundamental Euclidean terms require more extensive introduction than a glossary can provide. For these we have provided a section on Euclidean terminology, where the reader will find discussions of same and compound ratio, *reductio ad absurdum*, terms such as *porism* and *lemma*, and a review of the principal parts of a Euclidean proposition.

The Editor's Preface describes more of the features of this edition.

The Bones

Active users of Euclid will be interested in our companion volume, *The Bones.** This is a pocket reference edition of Euclid's *Elements* with definitions and postulates, and diagrams and enunciations for every proposition, but no proofs. It allows one to quickly find a needed or remembered proposition without having to leaf through so many pages. Sometimes one remembers what the diagram looks like and can most easily find it by scanning the diagrams. Other times one might want to look at the wording of the statements of the propositions to find the required result.

Another important use of *The Bones* is to get an overview of Euclid's system. The succinct presentation of Euclid's propositions, with their diagrams, allows a reader to see the ingenious and elegant structure of *Elements* emerge. *The Bones* includes summaries of the subject matter of each book in the table of contents, along with the locations of the hidden sets of definitions.

May our efforts in producing and presenting these two books make the delicious adventure of Euclid available to many new people, and may it make the study and use of *Elements* by those who already know and love Euclid easier and even more enjoyable.

Dana Densmore
William H. Donahue
for Green Lion Press

* *The Bones: A handy, where-to-find-it pocket reference companion to Euclid's Elements.* Green Lion Press, 2002. 5"x 7", 212 pages. Flexible, waterproof Kivar cover; rounded corners; sewn binding. ISBN 1-888009-21-7.

Editor's Preface

It has been a delight to design and edit and produce this edition of Euclid's wonderful *Elements*. I have appreciated the opportunity to immerse myself in it for an extended period of time and the opportunity to produce an edition with just what I want and need in a Euclid text—both when studying Euclid myself and with students and when calling upon Euclid's propositions in proofs while working with other texts such as Apollonius and Newton.

Why the Heath Translation?

Green Lion Press has chosen to use the Heath translation and to preserve it as he made it, despite the fact that it is now almost a century old, rather than commission a new translation or attempt to modernize the language. Some who have read Euclid with students have said that Heath's language is difficult for students because it is dated (the main offenders seem to be occasions on which he uses the subjunctive). We have never found this to be a problem for our students, nor have our colleagues at St. John's College.

Among those who have had students complain about the language (or who find it difficult themselves), some have attempted to rewrite the text (especially definitions, postulates, and enunciations) in more modern English, or in English language patterns more familiar to the modern ear or more consistent with modern mathematical conventions.

This modernizing of the language has been done without consulting the Greek text, or at least without the scholarly grounding in ancient Greek mathematics that would make such rewriting anything but risky. Especially when done by someone whose field is modern mathematics, there is a real tendency to recast the statement of the definition, postulate, or proposition in the terms of today rather than the terms in which it was understood and expressed in Euclid's time and by Euclid. If one just wants to learn some geometry, that's fine. But if one is looking at this text to know how Euclid thought through and presented this system, it won't do at all.

Thomas L. Heath was a scholar whose depth of knowledge and understanding of the nuances of ancient Greek mathematics was recognized and respected in his time and still stands in our own. It is significant that no present day scholar of ancient Greek mathematics and of Euclid in particular has undertaken to retranslate this work. Heath's translation remains the standard.

Textual Sources

In making textual decisions I used two primary sources, one for Heath's translation and one, that of J.L. Heiberg, for the Greek text.

The Heath Translation

First, for Heath's translation and to consult his commentary, I used the text he originally produced, the one published in his lifetime and presumably with his oversight. This is the Cambridge University Press publication, of which there were two editions, one in 1908 and one in 1926. The Dover edition (1956) is a photo reproduction of the second Cambridge edition.

I consulted both the first and second editions of the Cambridge University Press version where there were questions about the text of the Heath translation. The later Encyclopaedia Britannica edition included some discrepancies from both Cambridge editions. In each case the Britannica emendations seemed erroneous in context, and furthermore each departed from the Greek text, which reads as Heath originally translated (and Heiberg notes no variants in any of the extant Greek manuscripts). In each of these cases, therefore, I followed Heath's original text as published by Cambridge.

The Greek Text

Heath used the first edition of Heiberg's Greek text of the *Elements* and traditional scholia, with Heiberg's Latin footnotes and commentary, published in Leipzig by Teubner 1883–1888. This was the first part of the eight-volume *Euclidis Opera Omnia,* now out of print and virtually unobtainable. I used the six-volume second edition of the *Elements* volumes published in Leipzig by Teubner 1969–1977 and edited by E.S. Stamatis. The Stamatis version is almost identical to the *Elements* part of the first edition, except that the Latin translation is omitted.

When questions about either text or translation arose in the course of preparing this edition, I referred to Heiberg's Greek text and to his discussions of issues raised by difficulties or inconsistencies in the manuscripts.

Nevertheless, although I considered textual issues in making editorial decisions, I generally avoided getting into the intricacies of historical and scholarly cross-referencing and debates about the various manuscripts. The Cambridge/Dover three-volume edition provides Heath's exhaustive discussion of these matters in relation to each proposition. For depth of historical and scholarly considerations, the reader is referred to that work.

Editorial Principles of Emendation

In his translation Heath often put words, sentences, entire sections of proofs, or whole propositions, into brackets. These brackets did not, however, have a uniform signification. Some words or passages were bracketed because Heath (or Heiberg) found them questionable; others because they were not present in any Greek text but were inserted by Heath for clarity. The Green Lion edition follows a consistent practice in use of brackets, as explained below.

My first task in evaluating these passages was to ascertain the meaning of the brackets. Sometimes this took some detective work. Was it Heiberg who objected? On what basis? Was it Heath? On what basis? Was Heath's questioning supported by Heiberg? If Heath doesn't explain why he bracketed a passage, and Heiberg doesn't indicate any problem, and the passage is present in the Greek text, could we guess from the context why Heath might have questioned it?

When brackets indicated a passage regarded as an interpolation, Heath's commentary usually (although not always) explains why he (or Heiberg) questioned the passage. In most cases I also consulted the footnotes and commentary Heiberg provided with his text. Sometimes the passage is regarded by all scholars to be interpolated or corrupt. Some passages are present as marginal notes in particular manuscripts, presumably added by scholiasts. In other cases it is a matter of the passage being suspect because it is missing from some manuscripts although present in others.

In addition to considering the commentary and scholarly apparatus provided by Heath and by Heiberg, I consulted with colleagues working in the field of ancient Greek mathematics through the *Historia Matematica* email discussion forum. I also examined most of the earliest printed editions, both in Greek and in Latin translations. All these investigations have informed my editorial decisions about the text.

My principle in deciding about Heath's bracketed passages and other textual questions has been to consider not only the nature and persuasiveness of the case for a particular passage's spuriousness but also whether the words contribute to Euclid's line of argument. The goal in preparing this edition has been to provide the clearest possible text for understanding Euclid's geometry, while remaining consistent with an authentic reading of the existing manuscripts and with Heath's translation. In each individual case, questionable passages or brackets were included or omitted by determining whether they were more likely to help or to hinder the reader's understanding.

For the most part this edition has kept traditional interpolations (within angle brackets—see below for explanation of bracket conventions in this edition).

I have also kept Heath's numbering of the definitions, postulates, and common notions, even though this numbering represents a bit of recent editorial meddling. (See below for a discussion some of the history of numbering items in these sections.)

In a few places interpolations were omitted where they were confusing because redundant, unnecessary, or incorrect, and where the omission does not affect the traditional numbering of propositions.

For example, an interpolated and unnecessary passage in Book I, Proposition 4, retained within brackets by Heath, has been deleted in the present edition. The passage purports to justify an inference from coincidence of points to coincidence of the lines which join them, whereas the fact really follows from the definition of a straight line and the first Postulate. The passage may carry historical interest, but in a mathematical study its presence is merely confusing. A similar deletion was made, for essentially similar reasons, from Book X, Proposition 26.

In Book IX, the text of Proposition 19 is acknowledged as generally corrupt, and the intact portion of the proof is erroneous. It seems futile to attempt to demonstrate this proposition on the basis of the existing text; better to note the content (given three numbers, to determine whether one can find a fourth proportional) and then either move on or try to construct a proof of one's own. Accordingly, the entirety of this propositon has been removed, retaining only the enunciation and setting-out.

In other places, some words, despite being essential to the line of argument, were bracketed by Heath for technical textual reasons. Most often this was because they were not present in the manuscript Heiberg regarded as the best even though they were present in all other manuscripts. In most such cases, where the passage is essential to the line of argument, this edition simply drops the brackets.

A few of Heath's comments have been summarized in footnotes in this edition. Most of these identify a certain passage as suspect or known to be corrupt. This information could be useful to students who are having trouble interpreting the passage.

Wherever line length permits, the intermediate conclusions in each demonstration have been centered typographically. This helps make the structure of Euclid's chained arguments more evident.

Citations of the form "[id.]" have been removed and replaced with the corresponding proposition number in each case. Readers are

thereby relieved of the necessity to keep paging back to locate the intended reference.

Explanatory footnotes have been added to some of the definitions in Book V. These footnotes give examples of operations such as *componendo, separando, alternando,* which the definitions specify. Citations have also been added to the text in places where these operations are invoked.

Meaning of square [] and angle ⟨ ⟩ brackets

Square brackets [] are used in this book to indicate material which Heath identified as having been supplied by him, adding clarification but not literally present in the Greek text. Square brackets are also used for citations, some of which are traditional, some inserted by Heath, and a few by the Green Lion for this edition. To further emphasize their status as editorial additions, the citations appear in a different type face from the Euclid text.

Angle brackets ⟨ ⟩ are used in this book to indicate text which is regarded by scholars as an interpolation by an early editor, that is, not originally given by Euclid. Some of these presumed interpolations are retained by Heath and by us (within angle brackets) because, having been for so long included in some of the important Greek manuscripts, they have attained quasi-Euclidean status.

Numbering of Definitions, Postulates, and Common Notions

In the earliest printed Greek text, that of Basle 1533, these sections are narratives, not broken into a numbered list, or even into separate sentences. The narrative sections have the function of a general introduction to many of the terms and some of the principles to be used in the book, rather than a formal foundation in the style of modern axiomatic systems. The postulates section included only the first three of those listed in Heiberg's text (and Heath's). A 1550 Greek edition also listed only those three postulates.

Numbering, and the singling out of the terms defined through typographic means such as boldface or italics, were introduced in the Latin translations. Some early printed editions had both Latin and Greek, with the Greek in narrative form and the Latin sporting lists which increasingly often were numbered. These lists varied widely from edition to edition. Each editor seemed to have had his own ideas about what to include. For example, the first printed edition of Euclid, the 1482 Ratdolt Latin translation, had the definitions freely reworded; the definitions, postulates, and common notions divided but not numbered; five postulates could be distinguished but with the first

two of our present canonical five combined and including an added one not in our five; and nine common notions attributed to Euclid plus some extras helpfully added by the commentator. Another early Latin edition featured twenty-two common notions.

The game continues today, with mathematicians criticizing Euclid's shortcomings in the rigor of definitions and axioms and enjoying filling in the missing definitions, postulates, and common notions. The formalizing of a system is, of course, a legitimate project. But the criticism of Euclid may be misplaced. Euclid may not have been trying to cover the foundation exhaustively in these sections. In fact, some present-day scholars doubt that even the oldest of the above-mentioned Greek narrative sections for definitions originated with Euclid. There is some evidence that the definitions we have were introduced by later editors and scholiasts. The same may be true of the postulates and common notions. Of course, if so, Euclid may have had his own formulation of these things in a manuscript that was lost.

Tracing the history and development of these sections is a fascinating project I had to pull back from in order to complete this edition. The investigation involves all the historical and scholarly complications and mysteries and speculations about the Arabic and Greek manuscripts these editors and translators may have used, how much they were introducing their own retellings, and what other influences may have been at work. I mention some of the issues briefly here primarily to warn that there are textual uncertainties in these areas and that it might be prudent to take the received definitions, postulates, and common notions as narrative prefaces which may reconstruct Euclid's usage rather than precisely express his intended foundations.

That he had, at least implicitly, a set of foundations for a tight deductive system is, however, beyond any doubt. The proof of that lies in the fact that these foundations can be reconstructed, as they have been, with formal completeness, by modern mathematicians, and less exhaustively by the editors and scholiasts over the millenia, and as, indeed, they can be by any attentive reader. The primary source of the beauty and delight of reading Euclid is the elegance and perfection of the way the deductive system unfolds.

Acknowledgments

Preparing this new edition was a large project. I am grateful for the help of many people who generously donated their time and talents.

Thanks go first to our wonderful Green Lion interns. Intern Andrea Ruiga did the initial scanning and OCR pass and did preliminary cleanup of the diagrams. Interns Christian Blood, Markus Glodek,

Brendan Norwood, William Whitaker, Louise Guenther, and Kathy Garcia all worked on proofreading and corrections of the OCRed text. Intern Kathleen Kelley helped with proofreading and preparation of the electronic files in the final stages of production.

Kerry Magruder, the Librarian of the History of Science Collections at the University of Oklahoma, was extraordinarily generous and helpful. By happy coincidence, I was at Oklahoma delivering a paper at a History of Science meeting just at the time he had set up a magnificent display of the earliest printed editions of Euclid from their superb rare books collection. He let me immerse myself in these wonderful old books and followed up with slides and photocopies of the sections in which I was particularly interested.

Ron Brashear, Curator of Rare Books in the History of Science and Technology, Dibner Library of the History of Science and Technology of the Smithsonian Institution Libraries, was similarly helpful to me when I made several visits to the Dibner Library to look at their early printed editions and check on certain questions. He also photocopied some very fragile pages of the 1533 Greek edition for me to take away.

I am grateful to many generous colleagues on the *Historia Matematica* e-mail discussion forum for helpful information in response to my queries and for encouragement of this project. I would especially like to acknowledge the assistance of David Fowler, John Harper, Joe Albree, Duncan Melville, David Derbes, Avinoam Mann, Menso Folkerts, Sam Kutler, John Conway, Reviel Netz, Martin Flashman, Heinz Lueneburg, Robert Thomas, Edith Prentice Mendez, John Hannon, Michael Kinyon, Victor Katz, Richard Arthur, and David Zitarelli—and of course many thanks to list owner Julio Gonzalez Cabillon for his tireless work in maintaining this forum.

John S. Steadman of St. John's College was the first to urge us to issue a one-volume unabridged edition, and he encouraged this project throughout its lengthy gestation with his infectious enthusiasm. He enriched the Green Lion index with many extra entries and contributed some helpful supplementary citations.

The Green Lion's new Associate Editor, Howard J. Fisher (who is also a faculty member at St. John's College), made an enormous contribution to this edition. He did most of the page layout implementation (placement of the diagrams and page breaks) according to my editorial and design specifications. He repaired many diagrams and redrew a very large number of them. He helped greatly with the index and terminology sections, and served as a thoughtful consultant on editing decisions. It was a pleasure to continue the fruitful collaboration we began when I edited his Faraday guide, *Faraday's Experimental Researches in Electricity: Guide to a First Reading.*

I thank my husband and Green Lion partner William H. Donahue for his support and assistance throughout the project. He shared supervision of the interns and kept the computers functioning. I called upon his translation abilities on many occasions when I needed to check obscure or subtle questions in the Greek text and on many other occasions when I needed to research and interpret sections in Latin of the Heiberg footnotes and commentary. He is also responsible for the valuable glossary, filling in, looking up, and entering many Greek terms into the index.

The staff of the Meem Library of St. John's College in Santa Fe were patient, cooperative, and helpful in more ways than can be mentioned. Librarians Inga Waite, Laura Cooley, and Heather MacLean were particularly gracious and generous. The Green Lion gives special thanks for the long-term loan of the full set of the Heiberg Greek text and commentary over the extended time of preparation of this new Green Lion edition.

Finally, I thank my students and colleagues at St. John's College, with whom I have shared so much pleasure in reading, discussing, and thinking about Euclid.

 Dana Densmore

About the Author: Euclid

Very little is known about Euclid's life. The *Dictionary of Scientific Biography* begins its long article on Euclid with these words: "Although Euclid is the most celebrated mathematician of all time, whose name became a synonym for geometry until the twentieth century, only two facts of his life are known, and even these are not beyond dispute." These "facts" are inferences or speculations based on references in ancient works. The first is that he lived after Plato (d. 347 B.C.E.) and before Archimedes (b. 287 B.C.E.). The second is that he worked in Alexandria.

He is not the same as Euclid of Megara, a friend of Plato, with whom he is sometimes confused.

Heath says that it is most probable that Euclid (the author of *Elements*) received his mathematical training in Athens from the pupils of Plato because most of the geometers who could have taught him were of that school, and it was in Athens that the mathematicians on whose works Euclid's *Elements* depend had lived and taught. If we are persuaded by this we will place Euclid later than 347 B.C.E.

The argument made for him being older than Archimedes is a reference in Archimedes's work *On the Sphere and the Cylinder* to a proposition of Euclid's *Elements*. However, that reference is regarded now as probably a later interpolation.

Some, therefore, maintain that it is only known with any confidence that he is before (or contemporaneous with) Apollonius (active around 200 B.C.E.). One piece of evidence for this is the same as the evidence for his having worked in Alexandria, namely a reference from Pappus (c. 320 C.E.). Pappus remarks about Apollonius that Apollonius "spent a very long time with the pupils of Euclid at Alexandria, and it was thus that he acquired such a scientific habit of thought." If we are to believe Pappus in this, we must place Euclid earlier than 200 B.C.E.

Heath also concludes from Pappus's remark that Euclid taught and founded a school in Alexandria. But others argue that, although he evidently had pupils in Alexandria, that does not prove that he himself worked there.

In any case, Apollonius himself refers to Euclid in his introductory letter to Book I of the *Conics*. Since Apollonius was born about 262 B.C.E., this would also place Euclid earlier than 200 B.C.E.

So much for what we know about Euclid's life. What about his work?

Heath quotes Proclus (410–485 c.e.) as follows:

> Euclid put together the *Elements*, collecting many of Eudoxus's theorems, perfecting many of Theaetetus's, and also bringing to irrefragable demonstration the things which were only somewhat loosely proved by his predecessors.

The *Dictionary of Scientific Biography* says:

> Euclid's fame rests preeminently upon the *Elements*, which he wrote in thirteen books and which has exercised an influence upon the human mind greater than that of any other work except the Bible. For this reason he became known in antiquity as Ὁ Στοιχειωτής, "the Writer of the Elements" and sometimes simply as Ὁ Γεωμέτρης, "the Geometer". There had been *Elements* written before Euclid—notably by Hippocrates, Leo, and Theudius of Magnesia—but Euclid's work superseded them so completely that they are now known only from Eudemus's references as preserved by Proclus.

Numerous other works have been ascribed to Euclid, in astronomy, optics, and music theory, as well as mathematics. Many of these are no longer extant, or survive in fragments. After the *Elements*, his most significant surviving mathematical work is the *Data* (Δεδομένα), which is evidently designed as an aid in problem solving by analysis. The book *On divisions (of figures)* (περὶ διαιρέσεων βιβλίον), which has survived in an Arabic translation, is about dividing figures into other, unlike, figures. The book of *Porisms* ascribed to Euclid has been lost, but is known through the writings of Pappus. It contained, not porisms such as appear in *Elements*, but, according to Heath, "propositions belonging to the modern theory of transversals and to projective geometry." Euclid is also said to have written a treatise on conic sections, in four books, which was superseded by Apollonius's *Conics*, and has not survived.

The two optical works ascribed to Euclid, *Optics* and *Catoptrics*, inaugurated a long tradition of geometrical optics that lasted until the beginning of the seventeenth century. Further information about these and other Euclidean texts is provided in the bibliography.

What we know of Euclid we know through the brilliant work he left us. This book you have in your hands now offers you the opportunity to experience that work directly and to explore for yourself why "the most celebrated mathematician of all time" has been so highly honored throughout more than two millennia.

About the Translator: Thomas L. Heath

Like the proprietors of Green Lion Press, and, it can safely be said, like Euclid, Thomas Little Heath was an independent scholar.

He was born on October 5, 1861 in Lincolnshire, England. He studied at Trinity College, Cambridge, and received a first class degree in both mathematics and classics from Cambridge University in 1883.

He then embarked on a very successful career as a civil servant, serving first in the Treasury, which he joined in 1884 (being chosen in 1887, after only three years' service, to be Private Secretary to the Permanent Secretary to the Treasury), then serving later as the Comptroller General of the National Debt Office, until he retired in 1926. D'Arcy W. Thompson wrote in an obituary that after his death in 1940 Heath was still remembered by a few old colleagues in the Treasury for his "unfailing kindness and goodness of heart, for quite unusual ability, and for knowing how to write a memo unsurpassed for style and cogency."

Throughout his career in civil service he pursued his independent study as a historian of mathematics and was recognized as one of the world's leading experts in the history of Greek mathematics. His book *Diophantus of Alexandria: A Study in the History of Greek Algebra* was published by Cambridge University Press in 1885. *Apollonius of Perga* (Cambridge) was published in 1896, and his book *The Works of Archimedes* (Cambridge) appeared in 1897.

In 1908 Cambridge University Press published his three-volume work on *Euclid's Elements*, recognized as the standard English version of the text. A second edition of *The Works of Archimedes* appeared in 1912, incorporating a translation of *The Method of Archimedes,* which had been rediscovered by Heiberg in 1906.

His most famous work was the two-volume A *History of Greek Mathematics,* which was published in 1921. A second edition of the three-volume *Euclid's Elements,* revised and updated by Heath, was issued in 1926. *A Manual of Greek Mathematics* (Oxford) appeared in 1931. His last work, *Mathematics in Aristotle,* was published posthumously in 1949 by Oxford University Press.

He received an Sc.D. degree from Cambridge in 1896, the year his work on Apollonius was published. He also received honorary doctorates from Oxford in 1913 and Dublin in 1929, and was made an Honorary Fellow of Trinity College Cambridge in 1920.

Heath was elected a Fellow of the Royal Society in 1912.

Royal honors included the order Companion of the Bath (C.B.),

awarded in 1903, and knighthoods Knight Commander of the Bath (K.C.B.), awarded in 1909, and Knight Commander of the Royal Victorian Order (K.C.V.O.), awarded in 1916.

In addition to all these civil and scholarly accomplishments, his life was rich in other interests and activities. He was an accomplished pianist and lover of music, once in his youth traveling to Vienna just to look at Brahms in the flesh. He loved to travel and took every opportunity for minute exploration. He made a thorough study of some of his favorite works of literature. He had a serious interest in the construction and operation of trains, knowing the different engines and their improvements, and he is said to have known by sound the express engines of all the great European lines along with their schedules.

He married Ada Mary Thomas, a distinguished musician, in 1914. They reared a daughter, Victoria, and a son, Geoffrey. According to a 1936 account by David Eugene Smith, at that time Victoria was studying Jurisprudence in an honors course at Oxford and Geoffrey was about to follow his father to Trinity College Cambridge.

His friend D'Arcy Thompson describes him as an indomitable rock-climber who knew every inch of the Dolomites and who continued rock-climbing till he was over fifty. Throughout his life he was an active mountaineer, with many ascents in the Alps, including at least one first ascent, of Tre Cime di Lavaredo near Cortina. He climbed the Wildspitz in Tirol when he was seventy, fifty years after he had first climbed it.

Heath died of a stroke on March 16, 1940 in Ashtead, Surrey, England, while hard at work on his book about Aristotle's mathematics.

Some Euclidean Terminology

Technical terms in Euclid's language are relatively few; but they do exist, and Euclid does not always offer as much explanation of them as modern readers might wish. The ancient commentators, too, introduced additional formal terminology in their disquisitions on Euclid's work. These terms, and more, were the subject of a massive introductory essay that Heath provided for his original three-volume edition of Euclid. Some of the following material has been adapted from Heath, who in turn quotes extensively from Proclus (410–485 C.E.) and other commentators.

Elements

There are, says Proclus, in the whole of geometry certain leading theorems, bearing to those which follow the relation of a principle, all-pervading, and furnishing proofs of many properties. Such theorems are called by the name of *elements*; and their function may be compared to that of the letters of the alphabet in relation to language, letters being indeed called by the same name in Greek (στοιχεῖα).

The formal divisions of a proposition

Proclus writes:

> Every problem and every theorem, which is complete with all its parts perfect, purports to contain in itself all of the following elements: *enunciation* (πρότασις), *setting-out* (ἔκθεσις), *definition* or *specification* (διορισμός), *construction* or *machinery* (κατα-σκευή), *proof* (ἀπόδειχις), *conclusion* (συμπέρασμα). Now of these the enunciation states what is given and what is that which is sought, the perfect enunciation consisting of both these parts. The setting-out marks off what is given, by itself, and adapts it beforehand for use in the investigation. The definition or διορισμός states separately and makes clear what the particular thing is which is sought. The construction or machinery adds what is wanting to the datum for the purpose of finding what is sought. The proof draws the required inference by reasoning scientifically from acknowledged facts. The conclusion reverts again to the enunciation, confirming what has been demonstrated. These are

all the parts of problems and theorems, but the most essential and those which are found in all are *enunciation, proof, conclusion*. For it is equally necessary to know beforehand what is sought, to prove this by means of the intermediate steps, and to state the proved fact as a conclusion; it is impossible to dispense with any of these three things. The remaining parts are often brought in, but are often left out as serving no purpose.

"Therefore, etc."

Readers will notice that most propositions conclude "Therefore, etc., Q.E.D." The meaning of Q.E.D. is explained below. The words "Therefore, etc." stand for an exact restatement of the enunciation. This restatement, which Proclus called the "conclusion" (συμπέρασμα), was considered formally necessary for a complete demonstration, since the proof itself is always based on setting out of particular lines, circles, etc. and thus must be restated generally.

Of this, Proclus says "[Geometers] pass accordingly to the general in order that we may not conceive the conclusion to be partial. And they are justified in so passing, since they use for the demonstration the particular things set out, not *qua* particulars, but *qua* typical of the rest."

Although the final conclusion is stated in all manuscripts of the Greek text, it was common practice in Latin translations and reworkings either to omit the conclusion entirely (as in the first printed edition of 1482) or to indicate its presence with an abbreviation (as in Clavius's reworking of 1574). More careful versions, such as Commandino's translation of 1572, include the full conclusion.

Heath in his translation followed the practice of abbreviating the conclusion. We are not altering Heath's abbreviations, but we must emphasize that it was Euclid's evident intention to state the conclusion of the propositions in full. Consequently Heath's abbreviation ("Therefore etc.") must always be taken as standing in for that full statement.

Q.E.D. and Q.E.F.

Q.E.D. stands for the Latin *quod erat demonstrandum*, that which was to have been demonstrated, translating the Greek phrase ὅπερ ἔδει δεῖξαι. However, the meaning of the Greek is slightly different: a better translation would be, "precisely what was required to be proved." Q.E.D. is traditionally placed at the end of a proof when that which was specified in the enunciation, and in the setting-out, has been exactly restated as the conclusion of the demonstration. (See above for an explanation of the traditional names for the formal divisions of a proposition.)

Q.E.F. stands for *quod erat faciendum*, that which was to have been done, translating the Greek phrase ὅπερ ἔδει ποιῆσαι. Again, the translation is not quite accurate: "precisely what was required to be done" is closer to the Greek. Q.E.F. is placed at the end of propositions which are constructions.

Porism

Proclus writes:

> The term *porism*...is specially used when from what has been demonstrated some other theorem is revealed at the same time without our propounding it, which theorem has on this very account been called a *porism* (corollary) as being a sort of incidental gain arising from the scientific demonstration.

Lemma

Proclus continues:

> The term lemma is often used of any proposition which is assumed for the construction of something else.

In Euclid's time, the word λῆμμα usually denoted an assumption that was made in order to prove something. Logically, it was the major premise in a syllogism. In later mathematical writings, it came to refer to an *ad hoc* proof adopted only for the purpose of proving a particular proposition, a meaning that seems consistent with the terminology in *Elements*. At least some of the lemmas in *Elements* are thought to be later interpolations aimed at filling gaps in the existing proofs.

Reductio ad absurdum

This is variously called by Aristotle *reductio ad absurdum* (ἡ εἰς τὸ ἀδύνατον ἀπαγωγή) (29b 5, 50a 30), proof *per impossible* (ἡ διὰ τοῦ ἀδυνάτου δεῖξις or ἀπόδειξις) (39b 32, 45a 35), "proof leading to the impossible" (ἡ εἰς τὸ ἀδύνατον ἄγουσα ἀπόδειξις) (85a 10, etc.). It is part of "proof (starting) from a hypothesis, when something impossible results from assuming the contradictory of the original contention, as, for example, when it is proved that the diagonal (of a square) is incommensurable because, if it be assumed commensurable, it will follow that odd (numbers) are equal to even (numbers)." Or again, "proof (leading) to the impossible differs from the direct (δεικτικῆς) in that it assumes what it desires to destroy [namely, the hypothesis of the falsity of the conclusion] and then reduces it to something admittedly false, whereas the direct proof starts from premises admittedly true."

Proclus has the following description of the *reductio ad absurdum*.

> Proofs by *reductio ad absurdum* in every case reach a conclusion manifestly impossible, a conclusion the contradictory of which is admitted. In some cases the conclusions are found to conflict with the common notions, or the postulates, or the hypotheses (from which we started); in others they contradict propositions previously established. ... Every *reductio ad absurdum* assumes what conflicts with the desired result, then, using that as a basis, proceeds until it arrives at an admitted absurdity, and, by thus destroying the hypothesis, establishes the result originally desired.

Figure and boundary

As Heath recounts, Plato observes in the Meno that roundness (στρογγυ-λότης) or *the round* is a "figure," and that *the straight* and many other things are so too; he then inquires what there is common to all of them, in virtue of which we apply the term "figure" to them. His answer is (76 A): "with reference to every figure I say that *that in which the solid terminates is a figure*, or, to put it briefly, *a figure is an extremity of a solid.*"

Instead of calling figure *an extremity* as Plato did, Euclid describes a figure as *that which has* (is contained by) a boundary or boundaries (I. Def. 14). Is the figure then *distinct* from its boundary, or does figure subsume the boundary within itself?

In the case of circles, at least, the commentator An-Nairīzī (d. c. 922) distinguished circumference (περιφέρεια) from circle (κύκλος), arguing that Euclid's definition of circle as "a plane figure" (I. Def. 15) shows that Euclid regarded the circle as the whole surface bounded by the circumference, and not the circumference itself. Heath, however, points out that while Euclid is generally careful to say the "circumference of a circle" when he means the circumference, or an arc, only, there are cases where "circle" *means* "circumference of a circle," for example in III. 10: "A circle does not cut a circle in more points than two."

Same ratio

Euclid defines *same ratio*, or proportion, as follows (V. Def. 5):

> Magnitudes are said to *be in the same ratio*, the first to the second and the third to the fourth, when, if any equimultiples whatever be taken of the first and third, and any equimultiples whatever of the second and fourth, the former equimultiples alike exceed, are alike equal to, or alike fall short of, the latter equimultiples respectively taken in corresponding order.

Although the definition may appear impenetrable, it expresses a plain and

natural insight. The best explanation, says Heath, is one given by Augustus De Morgan (1806–1871), who observes that the definition applies equally to commensurable or incommensurable quantities because no attempt is made to measure one by an aliquot part of another. To elucidate this De Morgan offers an ingenious illustration, the following account of which is adapted from Heath.

Suppose there is a straight colonnade composed of equidistant columns (which we will understand to mean the vertical lines forming the axes of the columns respectively), the first of which is at a distance from a bounding wall equal to the distance between consecutive columns. In front of the colonnade let there be a straight row of equidistant railings (regarded as meaning their axes), the first being at a distance from the bounding wall equal to the distance between consecutive railings. Let the columns be numbered from the wall, and also the railings. We suppose of course that the column distance (say, C) and the railing distance (say, R) are different and that they may bear to each other any ratio, commensurable or incommensurable; i.e., that there need not go any exact number of railings to any exact number of columns.

If the construction be supposed carried on to any extent, a spectator can, by mere inspection, and without measurement, compare C with R to any degree of accuracy. For example, since the 10th railing falls between the 4th and 5th columns, $10R$ is greater than $4C$ and less than $5C$, and therefore R lies between $4/10$ths of C and $5/10$ths of C. To get a more accurate notion, the ten-thousandth railing may be taken; suppose it falls between the 4674th and 4675th columns. Therefore $10,000R$ lies between $4674C$ and $4675C$, or R lies between $4674/10,000$ and $4675/10,000$ of C. There is no limit to the degree of accuracy thus obtainable; and the ratio of A to C is determined when the order of distribution of the railings among the columns is assigned *ad infinitum*; or, in other words, when the position of any given railing can be found, as to the numbers of the columns between which it lies. Any alteration, however small, in the place of the first railing must at last affect the order of distribution. Suppose, e.g., that the first railing is moved from the wall by one part in a thousand of the distance between the columns; then the second railing is pushed forward by $2/1000\,C$, the third by $3/1000\,C$, and so on, so that the railings after the thousandth are pushed forward by more than C; i.e., the order with respect to the columns is disarranged.

Now let it be proposed to make a model of the preceding construction in which c shall be the column distance and r the railing distance. It needs no definition of proportion, nor anything more than the conception which we have of that term prior to definition (and with which we must show the agreement of any definition that we may adopt), to assure us that C must be to R in the same proportion as c to r if the model be truly formed. Nor is it drawing too largely on that conception of proportion to assert that the distribution of the railings among the columns in the model must be everywhere the same as in the original; for example, that the model would be *out of proportion* if its 37th railing fell between the 18th and 19th columns, while the 37th railing of the original fell between the 17th and 18th columns. Thus the dependence of Euclid's definition upon common notions is settled; for the obvious relation between the construction and its model which has just been described contains the collection of conditions, the fulfilment of which, according to Euclid, constitutes proportion. According to Euclid, whenever mC exceeds, equals, or falls short of nR, then mc must exceed, equal, or fall short of nr; and, by the most obvious property of the constructions, according as the mth column comes after, opposite to, or before the nth railing in the original, the mth column must come after, opposite to, or before the nth railing in the correct model.

Thus the test proposed by Euclid is necessary. It is also sufficient. For admitting that, to a given original with a given column-distance in the model, there is one correct model railing distance (which must therefore be that which distributes the railings among the columns as in the original), we have seen that any other railing distance, however slightly different, would at last give a different distribution; that is, the correct distance, and the correct distance only, satisfies all the conditions required by Euclid's definition.

There remains a question relating to the infinite character of the definition. Four magnitudes A, B, C, D are not to be called proportional until it is shown that *every* multiple of A falls in the same intervals among the multiples of B in which the same multiple of C is found among the multiples of D. But how is it possible to compare every one of the infinite number of multiples of A with every one of the multiples of B?

Suppose that the distribution of the railings among the columns should be found to agree in the model and the original as far as the millionth railing. This proves only that the railing distance of the model does not err by the millionth part of the corresponding column distance. We can thus fix limits

to the disproportion, if any, and we may make those limits as small as we please, by carrying on the method of observation; but we cannot obtain an infinite number of cases and so enable ourselves to affirm proportion absolutely. Mathematical methods, however, enable us to avoid the difficulty. We can take *any multiples whatever* and work with them as if they were particular multiples.

Compound ratio

There is only one statement in Euclid's text as we have it indicating what is meant by compound ratio; this is in VI. 23, where, speaking of the sides of parallelograms, he says abruptly "But the ratio of *K* to *M* is compounded of the ratio of *K* to *L* and of that of *L* to *M*." Proposition VIII. 5 constructs an identical argument for numbers, but it is even less explicit as to a definition. Accordingly, *Simson* (1687–1768) gives a definition of compound ratio directly suggested by the statement in VI. 23:

> When there are any number of magnitudes of the same kind, the first is said to have to the last of them the ratio compounded of the ratio which the first has to the second, and of the ratio which the second has to the third, and of the ratio which the third has to the fourth, and so on unto the last magnitude.
>
> For example, if *A*, *B*, *C*, *D*, be four magnitudes of the same kind, the first *A* is said to have to the last *D* the ratio compounded of the ratio of *A* to *B*, and of the ratio of *B* to *C*, and of the ratio of *C* to *D*; or the ratio of *A* to *D* is said to be compounded of the ratios of *A* to *B*, *B* to *C*, and *C* to *D*.
>
> And if *A* has to *B* the same ratio which *E* has to *F*; and *B* to *C* the same ratio that *G* has to *H*; and *C* to *D* the same that *K* has to *L*; then, by this definition, *A* is said to have to *B* the ratio compounded of ratios which are the same with the ratios of *E* to *F*, *G* to *H*, and *K* to *L*: and the same thing is to be understood when it is more briefly expressed, by saying, *A* has to *D* the ratio compounded of the ratios of *E* to *F*, *G* to *H*, and *K* to *L*.
>
> In like manner, the same things being supposed, if *M* has to *N* the same ratio which *A* has to *D*; then, for shortness' sake, *M* is said to have to *N* the ratio compounded of the ratios of *E* to *F*, *G* to *H*, and *K* to *L*.

It is clear that duplicate ratio, triplicate ratio etc. defined in V. Defs. 9 and 10 are merely particular cases of compound ratio, being in fact the ratios compounded of two, three, etc., equal ratios.

Euclid's *Elements*
Book I

Definitions

1. A *point* is that which has no part.

2. A *line* is breadthless length.

3. The extremities of a line are points.

4. A *straight line* is a line which lies evenly with the points on itself.

5. A *surface* is that which has length and breadth only.

6. The extremities of a surface are lines.

7. A *plane surface* is a surface which lies evenly with the straight lines on itself.

8. A *plane angle* is the inclination to one another of two lines in a plane which meet one another and do not lie in a straight line.

9. And when the lines containing the angle are straight, the angle is called *rectilineal*.

10. When a straight line set up on a straight line makes the adjacent angles equal to one another, each of the equal angles is *right*, and the straight line standing on the other is called a *perpendicular* to that on which it stands.

11. An *obtuse angle* is an angle greater than a right angle.

12. An *acute angle* is an angle less than a right angle.

13. A *boundary* is that which is an extremity of anything.

14. A *figure* is that which is contained by any boundary or boundaries.

15. A *circle* is a plane figure contained by one line such that all the straight lines falling upon it from one point among those lying within the figure are equal to one another;

16. And the point is called the *centre* of the circle.

17. A *diameter* of the circle is any straight line drawn through the centre and terminated in both directions by the circumference of the circle, and such a straight line also bisects the circle.

Euclid's definitions, postulates, and common notions—if Euclid is indeed their author—were not numbered, separated, or italicized until translators began to introduce that practice. The Greek text, however, as far back as the 1533 first printed edition, presented the definitions in a running narrative, more as a preface discussing how the terms would be used than as an axiomatic foundation for the propositions to come. We follow Heath's formatting here. —Ed.

18. A *semicircle* is the figure contained by the diameter and the circumference cut off by it. And the centre of the semicircle is the same as that of the circle.

19. *Rectilineal figures* are those which are contained by straight lines, *trilateral* figures being those contained by three, *quadrilateral* those contained by four, and *multilateral* those contained by more than four straight lines.

20. Of trilateral figures, an *equilateral triangle* is that which has its three sides equal, an *isosceles triangle* that which has two of its sides alone equal, and a *scalene triangle* that which has its three sides unequal.

21. Further, of trilateral figures, a *right-angled triangle* is that which has a right angle, an *obtuse-angled triangle* that which has an obtuse angle, and an *acute-angled triangle* that which has its three angles acute.

22. Of quadrilateral figures, a *square* is that which is both equilateral and right-angled; an *oblong* that which is right-angled but not equilateral; a *rhombus* that which is equilateral but not right-angled; and a *rhomboid* that which has its opposite sides and angles equal to one another but is neither equilateral nor right-angled. And let quadrilaterals other than these be called *trapezia*.

23. *Parallel* straight lines are straight lines which, being in the same plane and being produced indefinitely in both directions, do not meet one another in either direction.

Postulates

Let the following be postulated:

1. To draw a straight line from any point to any point.

2. To produce a finite straight line continuously in a straight line.

3. To describe a circle with any centre and distance.

4. That all right angles are equal to one another.

5. That, if a straight line falling on two straight lines make the interior angles on the same side less than two right angles, the two straight lines, if produced indefinitely, meet on that side on which are the angles less than the two right angles.

Common Notions

1. Things which are equal to the same thing are also equal to one another.

2. If equals be added to equals, the wholes are equal.

3. If equals be subtracted from equals, the remainders are equal.

4. Things which coincide with one another are equal to one another.

5. The whole is greater than the part.

Proposition 1

On a given finite straight line to construct an equilateral triangle.

Let AB be the given finite straight line.

Thus it is required to construct an equilateral triangle on the straight line AB.
With centre A and distance AB let the circle BCD be described; [Post. 3]
again, with centre B and distance BA let the circle ACE be described; [Post. 3]
and from the point C, in which the circles cut one another, to the points A, B let the straight lines CA, CB be joined. [Post. 1]

Now, since the point A is the centre of the circle CDB,
AC is equal to AB. [Def. 15]

Again, since the point B is the centre of the circle CAE,
BC is equal to BA. [Def. 15]

But CA was also proved equal to AB;
therefore each of the straight lines CA, CB is equal to AB.

And things which are equal to the same thing are also equal to one another;
[C.N. 1]

therefore CA is also equal to CB.

Therefore the three straight lines CA, AB, BC are equal to one another.

Therefore the triangle ABC is equilateral; and it has been constructed on the given finite straight line AB.

Being what it was required to do.

Proposition 2

To place at a given point [as an extremity][1] *a straight line equal to a given straight line.*

Let A be the given point, and BC the given straight line.

Thus it is required to place at the point A [as an extremity] a straight line equal to the given straight line BC.

From the point A to the point B let the straight line AB be joined; [Post. 1]
and on it let the equilateral triangle DAB be constructed. [I. 1]

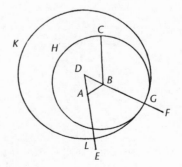

1. Square brackets indicate material which Heath identified as having been supplied by him, adding clarification but not literally present in the Greek text. —Ed.

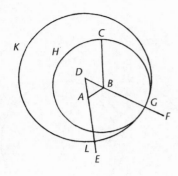

Let the straight lines *AE, BF* be produced in a straight line with *DA, DB*; [Post. 2]

with centre *B* and distance *BC* let the circle *CGH* be described; [Post. 3]

and again, with centre *D* and distance *DG* let the circle *GKL* be described. [Post. 3]

Then, since the point *B* is the centre of the circle *CGH*,

$$BC \text{ is equal to } BG.$$

Again, since the point *D* is the centre of the circle *GKL*,

$$DL \text{ is equal to } DG.$$

And in these *DA* is equal to *DB*;

therefore the remainder *AL* is equal to the remainder *BG*. [C.N. 3]

But *BC* was also proved equal to *BG*;

therefore each of the straight lines *AL, BC* is equal to *BG*.

And things which are equal to the same thing are also equal to one another; [C.N. 1]

therefore *AL* is also equal to *BC*.

Therefore at the given point *A* the straight line *AL* is placed equal to the given straight line *BC*.

Being what it was required to do.

Proposition 3

Given two unequal straight lines, to cut off from the greater a straight line equal to the less.

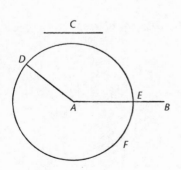

Let *AB, C* be the two given unequal straight lines, and let *AB* be the greater of them.

Thus it is required to cut off from *AB* the greater a straight line equal to *C* the less.

At the point *A* let *AD* be placed equal to the straight line *C*; [I. 2]

and with centre *A* and distance *AD* let the circle *DEF* be described. [Post. 3]

Now, since the point *A* is the centre of the circle *DEF*,

$$AE \text{ is equal to } AD.$$ [Def. 15]

But *C* is also equal to *AD*.

Therefore each of the straight lines *AE, C* is equal to *AD*;

so that *AE* is also equal to *C*. [C.N. 1]

Therefore, given the two straight lines *AB, C*, from *AB* the greater *AE* has been cut off equal to *C* the less.

Being what it was required to do.

Proposition 4

If two triangles have the two sides equal to two sides respectively, and have the angles contained by the equal straight lines equal, they will also have the base equal to the base, the triangle will be equal to the triangle, and the remaining angles will be equal to the remaining angles respectively, namely those which the equal sides subtend.

Let *ABC, DEF* be two triangles having the two sides *AB, AC* equal to the two sides *DE, DF* respectively, namely *AB* to *DE* and *AC* to *DF*, and the angle *BAC* equal to the angle *EDF*.

I say that the base *BC* is also equal to the base *EF*, the triangle *ABC* will be equal to the triangle *DEF*, and the remaining angles will be equal to the remaining angles respectively, namely those which the equal sides subtend, that is, the angle *ABC* to the angle *DEF*, and the angle *ACB* to the angle *DFE*.

For, if the triangle *ABC* be applied to the triangle *DEF*, and if the point *A* be placed on the point *D* and the straight line *AB* on *DE*, then the point *B* will also coincide with *E*, because *AB* is equal to *DE*.

Again, *AB* coinciding with *DE*, the straight line *AC* will also coincide with *DF*, because the angle *BAC* is equal to the angle *EDF*;
hence the point *C* will also coincide with the point *F*, because *AC* is again equal to *DF*.

But *B* also coincided with *E*;
hence the base *BC* will coincide with the base *EF*,
and will be equal to it. [C.N. 4]

Thus the whole triangle *ABC* will coincide with the whole triangle *DEF*,
and will be equal to it. [C.N. 4]

And the remaining angles will also coincide with the remaining angles and will be equal to them,
the angle *ABC* to the angle *DEF*,
and the angle *ACB* to the angle *DFE*. [C.N. 4]

Therefore etc.

Q.E.D.[2]

Proposition 5

In isosceles triangles the angles at the base are equal to one another, and, if the equal straight lines be produced further, the angles under the base will be equal to one another.

2. Q.E.D. stands for the Latin *quod erat demonstrandum*, that which was to have been demonstrated. The use of this and of Q.E.F., *quod erat faciendum*, that which was to have been done, is explained in the section Some Euclidean Terminology. —Ed.

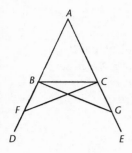

Let *ABC* be an isosceles triangle having the side *AB* equal to the side *AC*;
and let the straight lines *BD*, *CE* be produced further in a straight line with *AB*, *AC*. [Post. 2]

I say that the angle *ABC* is equal to the angle *ACB*, and the angle *CBD* to the angle *BCE*.

Let a point *F* be taken at random on *BD*;
from *AE* the greater let *AG* be cut off equal to *AF* the less; [I. 3]
and let the straight lines *FC*, *GB* be joined. [Post. 1]

Then, since *AF* is equal to *AG* and *AB* to *AC*,
 the two sides *FA*, *AC* are equal to the two sides *GA*, *AB*, respectively;
and they contain a common angle, the angle *FAG*.

Therefore the base *FC* is equal to the base *GB*,
and the triangle *AFC* is equal to the triangle *AGB*,
and the remaining angles will be equal to the remaining angles respectively, namely those which the equal sides subtend,
 that is, the angle *ACF* to the angle *ABG*,
 and the angle *AFC* to the angle *AGB*. [I. 4]

And, since the whole *AF* is equal to the whole *AG*, and in these *AB* is equal to *AC*,
 the remainder *BF* is equal to the remainder *CG*.

But *FC* was also proved equal to *GB*;
 therefore the two sides *BF*, *FC* are equal to the two sides *CG*, *GB* respectively;
and the angle *BFC* is equal to the angle *CGB*,
while the base *BC* is common to them;
therefore the triangle *BFC* is also equal to the triangle *CGB*, and the remaining angles will be equal to the remaining angles respectively, namely those which the equal sides subtend;
 therefore the angle *FBC* is equal to the angle *GCB*,
 and the angle *BCF* to the angle *CBG*.

Accordingly, since the whole angle *ABG* was proved equal to the angle *ACF*, and in these the angle *CBG* is equal to the angle *BCF*,
 the remaining angle *ABC* is equal to the remaining angle *ACB*;
and they are at the base of the triangle *ABC*.

But the angle *FBC* was also proved equal to the angle *GCB*;
and they are under the base.

Therefore etc.

Q.E.D.

Proposition 6

If in a triangle two angles be equal to one another, the sides which subtend the equal angles will also be equal to one another.

Let *ABC* be a triangle having the angle *ABC* equal to the angle *ACB*;
I say that the side *AB* is also equal to the side *AC*.

For, if *AB* is unequal to *AC*, one of them is greater.

Let *AB* be greater;
and from *AB* the greater let *DB* be cut off equal to *AC* the less;
let DC be joined.

Then, since *DB* is equal to *AC*,
and *BC* is common,
the two sides *DB*, *BC* are equal to the two sides *AC*, *CB* respectively;
and the angle *DBC* is equal to the angle *ACB*;
 therefore the base *DC* is equal to the base *AB*,
and the triangle *DBC* will be equal to the triangle *ACB*, the less to the greater:
which is absurd.

Therefore *AB* is not unequal to *AC*;
 it is therefore equal to it.

Therefore etc.

 Q.E.D.

Proposition 7

Given two straight lines constructed on a straight line [from its extremities] and meeting in a point, there cannot be constructed on the same straight line [from its extremities], and on the same side of it, two other straight lines meeting in another point and equal to the former two respectively, namely each to that which has the same extremity with it.

For, if possible, given two straight lines *AC*, *CB* constructed on the straight line *AB* and meeting at the point *C*,
let two other straight lines *AD*, *DB* be constructed on the same straight line *AB*, on the same side of it, meeting in another point *D* and equal to the former two respectively, namely each to that which has the same extremity with it, so that *CA* is equal to *DA* which has the same extremity *A* with it,
and *CB* to *DB* which has the same extremity *B* with it;
and let *CD* be joined.

Then, since *AC* is equal to *AD*,
 the angle *ACD* is also equal to the angle *ADC*; [I. 5]
 therefore the angle *ADC* is greater than the angle *DCB*;
 therefore the angle *CDB* is much greater than the angle *DCB*.

Again, since *CB* is equal to *DB*,
 the angle *CDB* is also equal to the angle *DCB*.

But it was also proved much greater than it:
which is impossible.

Therefore etc.

 Q.E.D.

Proposition 8

If two triangles have the two sides equal to two sides respectively, and have also the base equal to the base, they will also have the angles equal which are contained by the equal straight lines.

Let *ABC, DEF* be two triangles having the two sides *AB, AC* equal to the two sides *DE, DF* respectively, namely *AB* to *DE*, and *AC* to *DF*; and let them have the base *BC* equal to the base *EF*;
I say that the angle *BAC* is also equal to the angle *EDF*.

For, if the triangle *ABC* be applied to the triangle *DEF*, and if the point *B* be placed on the point *E* and the straight line *BC* on *EF*,
 the point *C* will also coincide with *F*,
because *BC* is equal to *EF*.

Then, *BC* coinciding with *EF*,
 BA, AC will also coincide with *ED, DF*;
for, if the base *BC* coincides with the base *EF*, and the sides *BA, AC* do not coincide with *ED, DF* but fall beside them as *EG, GF*,
then, given two straight lines constructed on a straight line [from its extremities] and meeting in a point, there will have been constructed on the same straight line [from its extremities], and on the same side of it, two other straight lines meeting in another point and equal to the former two respectively, namely each to that which has the same extremity with it.

But they cannot be so constructed. [I. 7]

Therefore it is not possible that, if the base *BC* be applied to the base *EF*, the sides *BA, AC* should not coincide with *ED, DF*;
 they will therefore coincide,
so that the angle *BAC* will also coincide with the angle *EDF*,
and will be equal to it.

If therefore etc.

 Q.E.D.

Proposition 9

To bisect a given rectilineal angle.

Let the angle *BAC* be the given rectilineal angle.
Thus it is required to bisect it.

Let a point *D* be taken at random on *AB*;
let *AE* be cut off from *AC* equal to *AD*; [I. 3]
let *DE* be joined,
and on *DE* let the equilateral triangle *DEF* be constructed;
let *AF* be joined.

I say that the angle *BAC* has been bisected by the straight line *AF*.

For, since *AD* is equal to *AE*,
and *AF* is common,
 the two sides *DA*, *AF* are equal to the two sides *EA*, *AF* respectively.

And the base *DF* is equal to the base *EF*;
 therefore the angle *DAF* is equal to the angle *EAF*. [I. 8]
Therefore the given rectilineal angle *BAC* has been bisected by the straight line *AF*.

<div align="right">Q.E.F.</div>

Proposition 10

To bisect a given finite straight line.

Let *AB* be the given finite straight line.

Thus it is required to bisect the finite straight line *AB*.

Let the equilateral triangle *ABC* be constructed on it, [I. 1]
and let the angle *ACB* be bisected by the straight line *CD*;
 [I. 9]
I say that the straight line *AB* has been bisected at the point *D*.

For, since *AC* is equal to *CB*, and *CD* is common,
 the two sides *AC*, *CD* are equal to the two sides *BC*, *CD* respectively;
and the angle *ACD* is equal to the angle *BCD*;
 therefore the base *AD* is equal to the base *BD*. [I. 4]

Therefore the given finite straight line *AB* has been bisected at *D*.

<div align="right">Q.E.F.</div>

Proposition 11

To draw a straight line at right angles to a given straight line from a given point on it.

Let *AB* be the given straight line, and *C* the given point on it.

Thus it is required to draw from the point *C* a straight line at right angles to the straight line *AB*.

Let a point *D* be taken at random on *AC*;
let *CE* be made equal to *CD*; [I. 3]
on *DE* let the equilateral triangle *FDE* be
constructed, [I. 1]
and let *FC* be joined;

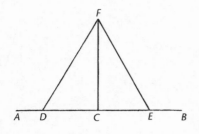

I say that the straight line *FC* has been drawn at right angles to the given straight line *AB* from *C* the given point on it.

For, since *DC* is equal to *CE*, and *CF* is common,
the two sides *DC*, *CF* are equal to the two sides *EC*, *CF* respectively;
and the base *DF* is equal to the base *FE*;

therefore the angle *DCF* is equal to the angle *ECF*; [I. 8]
and they are adjacent angles.

But, when a straight line set up on a straight line makes the adjacent angles equal to one another, each of the equal angles is right; [Def. 10]
therefore each of the angles *DCF*, *FCE* is right.

Therefore the straight line *CF* has been drawn at right angles to the given straight line *AB* from the given point *C* on it.

Q.E.F.

Proposition 12

To a given infinite straight line, from a given point which is not on it, to draw a perpendicular straight line.

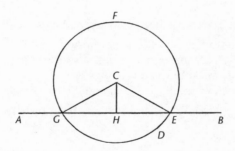

Let *AB* be the given infinite straight line, and *C* the given point which is not on it; thus it is required to draw to the given infinite straight line *AB*, from the given point *C* which is not on it, a perpendicular straight line.

For let a point *D* be taken at random on the other side of the straight line *AB*,
and with centre *C* and distance *CD* let the circle *EFG* be described; [Post. 3]
let the straight line *EG* be bisected at *H*, [I. 10]
and let the straight lines *CG*, *CH*, *CE* be joined. [Post. 1]

I say that *CH* has been drawn perpendicular to the given infinite straight line *AB* from the given point *C* which is not on it.

For, since *GH* is equal to *HE*, and *HC* is common,
the two sides *GH*, *HC* are equal to the two sides *EH*, *HC* respectively;
and the base *CG* is equal to the base *CE*;

therefore the angle *CHG* is equal to the angle *EHC*. [I. 8]

And they are adjacent angles.

But, when a straight line set up on a straight line makes the adjacent angles equal to one another, each of the equal angles is right, and the straight line standing on the other is called a perpendicular to that on which it stands.
[Def. 10]

Therefore *CH* has been drawn perpendicular to the given infinite straight line *AB* from the given point *C* which is not on it.

Q.E.F.

Proposition 13

If a straight line set up on a straight line make angles, it will make either two right angles or angles equal to two right angles.

For let any straight line *AB* set up on the straight line *CD* make the angles *CBA*, *ABD*;
I say that the angles *CBA*, *ABD* are either two right angles or equal to two right angles.

Now, if the angle *CBA* is equal to the angle *ABD*, they are two right angles.
[Def. 10]

But, if not, let *BE* be drawn from the point *B* at right angles to *CD*; [I. 11]
therefore the angles *CBE*, *EBD* are two right angles.

Then, since the angle *CBE* is equal to the two angles *CBA*, *ABE*,
let the angle *EBD* be added to each;
therefore the angles *CBE*, *EBD* are equal to the three angles *CBA*, *ABE*, *EBD*.
[C.N. 2]

Again, since the angle *DBA* is equal to the two angles *DBE*, *EBA*,
let the angle *ABC* be added to each;
therefore the angles *DBA*, *ABC* are equal to the three angles *DBE*, *EBA*, *ABC*.
[C.N. 2]

But the angles *CBE*, *EBD* were also proved equal to the same three angles;
and things which are equal to the same thing are also equal to one another;
[C.N. 1]

therefore the angles *CBE*, *EBD* are also equal to the angles *DBA*, *ABC*.

But the angles *CBE*, *EBD* are two right angles;
therefore the angles *DBA*, *ABC* are also equal to two right angles.

Therefore etc.

Q.E.D.

Proposition 14

If with any straight line, and at a point on it, two straight lines not lying on the same side make the adjacent angles equal to two right angles, the two straight lines will be in a straight line with one another.

For with any straight line *AB*, and at the point *B* on it, let the two straight lines *BC*, *BD* not lying on the same side make the adjacent angles *ABC*, *ABD* equal to two right angles;
I say that *BD* is in a straight line with *CB*.

For, if *BD* is not in a straight line with *BC*, let *BE* be in a straight line with *CB*.

Then, since the straight line *AB* stands on the straight line *CBE*,
the angles *ABC*, *ABE* are equal to two right angles.

[I. 13]

But the angles *ABC*, *ABD* are also equal to two right angles;
therefore the angles *CBA*, *ABE* are equal to the angles *CBA*, *ABD*.

[Post. 4 and C.N. 1]

Let the angle *CBA* be subtracted from each;
therefore the remaining angle *ABE* is equal to the remaining angle *ABD*, [C.N. 3]
the less to the greater: which is impossible.

Therefore *BE* is not in a straight line with *CB*.

Similarly we can prove that neither is any other straight line except *BD*.

Therefore *CB* is in a straight line with *BD*.

Therefore, etc.

Q.E.D.

Proposition 15

If two straight lines cut one another, they make the vertical angles equal to one another.

For let the straight lines *AB*, *CD* cut one another at the point *E*;
I say that the angle *AEC* is equal to the angle *DEB*, and the angle *CEB* to the angle *AED*.

For, since the straight line *AE* stands on the straight line *CD*, making the angles *CEA*, *AED*,
the angles *CEA*, *AED* are equal to two right angles.

[I. 13]

Again, since the straight line *DE* stands on the straight line *AB*, making the angles *AED*, *DEB*,
the angles *AED*, *DEB* are equal to two right angles.

[I. 13]

But the angles *CEA*, *AED* were also proved equal to two right angles;

therefore the angles *CEA, AED* are equal to the angles *AED, DEB*.

[Post. 4 and C.N. 1]

Let the angle *AED* be subtracted from each;

therefore the remaining angle *CEA* is equal to the remaining angle *BED*. [C.N. 3]

Similarly it can be proved that the angles *CEB, DEA* are also equal.

Therefore etc.

Q.E.D.

⟨**PORISM**. From this it is manifest that, if two straight lines cut one another, they will make the angles at the point of section equal to four right angles.⟩[3]

Proposition 16

In any triangle, if one of the sides be produced, the exterior angle is greater than either of the interior and opposite angles.

Let *ABC* be a triangle, and let one side of it *BC* be produced to *D*;
I say that the exterior angle *ACD* is greater than either of the interior and opposite angles *CBA, BAC*.

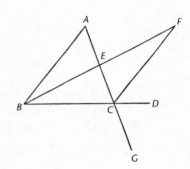

Let *AC* be bisected at *E*, [I. 10]
and let *BE* be joined and produced in a straight line to *F*;
let *EF* be made equal to *BE*, [I. 3]
let *FC* be joined, [Post. 1]
and let *AC* be drawn through to *G*. [Post. 2]

Then, since *AE* is equal to *EC*, and *BE* to *EF*,
the two sides *AE, EB* are equal to the two sides *CE, EF* respectively;
and the angle *AEB* is equal to the angle *FEC*, for they are vertical angles. [I. 15]

Therefore the base *AB* is equal to the base *FC*, and the triangle *ABE* is equal to the triangle *CFE*,
and the remaining angles are equal to the remaining angles respectively, namely those which the equal sides subtend; [I. 4]

therefore the angle *BAE* is equal to the angle *ECF*.

3. Angle brackets are used here and elsewhere in this book to indicate text which is regarded by scholars as an interpolation by an early editor, that is, not originally given by Euclid. Some of these presumed interpolations are retained by Heath and by us (within angle brackets) because, having been for so long included in some of the important Greek manuscripts, they have attained quasi-Euclidean status. The present porism was already present in manuscripts at the time of Proclus (410–485 CE). —Ed.

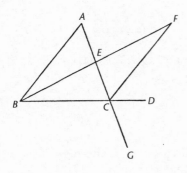

But the angle *ECD* is greater than the angle *ECF*;
[C.N. 5]
therefore the angle *ACD* is greater than the angle *BAE*.

Similarly also, if *BC* be bisected, the angle *BCG*, that is, the angle *ACD* [I. 15], can be proved greater than the angle *ABC* as well.

Therefore etc.

Q.E.D.

Proposition 17

In any triangle two angles taken together in any manner are less than two right angles.

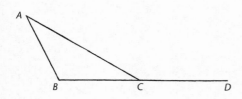

Let *ABC* be a triangle;
I say that two angles of the triangle *ABC* taken together in any manner are less than two right angles.

For let *BC* be produced to *D*. [Post. 2]

Then, since the angle *ACD* is an exterior angle of the triangle *ABC*, it is greater than the interior and opposite angle *ABC*. [I. 16]

Let the angle *ACB* be added to each;
 therefore the angles *ACD*, *ACB* are greater than the angles *ABC*, *BCA*.

But the angles *ACD*, *ACB* are equal to two right angles. [I. 13]

 Therefore the angles *ABC*, *BCA* are less than two right angles.

Similarly we can prove that the angles *BAC*, *ACB* are also less than two right angles, and so are the angles *CAB*, *ABC* as well.

Therefore etc.

Q.E.D.

Proposition 18

In any triangle the greater side subtends the greater angle.

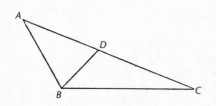

For let *ABC* be a triangle having the side *AC* greater than *AB*;
I say that the angle *ABC* is also greater than the angle *BCA*.

For, since *AC* is greater than *AB*, let *AD* be made equal to *AB* [I. 3], and let *BD* be joined.

Then, since the angle *ADB* is an exterior angle of the triangle *BCD*,

it is greater than the interior and opposite angle *DCB*. [I. 16]

But the angle *ADB* is equal to the angle *ABD,* since the side *AB* is equal to *AD;*
therefore the angle *ABD* is also greater than the angle *ACB;*
therefore the angle *ABC* is much greater than the angle *ACB.*

Therefore etc.

Q.E.D.

Proposition 19

In any triangle the greater angle is subtended by the greater side.

Let *ABC* be a triangle having the angle *ABC* greater than the
angle *BCA;*
I say that the side *AC* is also greater than the side *AB.*

For, if not, *AC* is either equal to *AB* or less.

Now *AC* is not equal to *AB;*
for then the angle *ABC* would also have been equal to the
angle *ACB;* [I. 5]
but it is not;

therefore *AC* is not equal to *AB.*

Neither is *AC* less than *AB,* for then the angle *ABC* would also have been less
than the angle *ACB;* [I. 18]
but it is not;

therefore *AC* is not less than *AB.*

And it was proved that it is not equal either.

Therefore *AC* is greater than *AB.*

Therefore etc.

Q.E.D.

Proposition 20

*In any triangle two sides taken together in any manner are greater than the
remaining one.*

For let *ABC* be a triangle;
I say that in the triangle *ABC* two sides taken together
in any manner are greater than the remaining one,
namely

BA, AC greater than BC,
AB, BC greater than AC,
BC, CA greater than AB.

For let *BA* be drawn through to the point *D,* let *DA* be
made equal to *CA,* and let *DC* be joined.

Then, since *DA* is equal to *AC,*
the angle *ADC* is also equal to the angle *ACD;* [I. 5]

15

therefore the angle *BCD* is greater than the angle *ADC*. [C.N. 5]

And, since *DCB* is a triangle having the angle *BCD* greater than the angle *BDC*, and the greater angle is subtended by the greater side, [I. 19]

therefore *DB* is greater than *BC*.

But *DA* is equal to *AC*;

therefore *BA*, *AC* are greater than *BC*.

Similarly we can prove that *AB*, *BC* are also greater than *CA*, and *BC*, *CA* than *AB*.

Therefore etc.

Q.E.D.

Proposition 21

If on one of the sides of a triangle, from its extremities, there be constructed two straight lines meeting within the triangle, the straight lines so constructed will be less than the remaining two sides of the triangle, but will contain a greater angle.

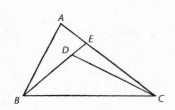

On *BC*, one of the sides of the triangle *ABC*, from its extremities *B*, *C*, let the two straight lines *BD*, *DC* be constructed meeting within the triangle;
I say that *BD*, *DC* are less than the remaining two sides of the triangle *BA*, *AC*, but contain an angle *BDC* greater than the angle *BAC*.

For let *BD* be drawn through to *E*.

Then, since in any triangle two sides are greater than the remaining one, [I. 20]
therefore, in the triangle *ABE*, the two sides *AB*, *AE* are greater than *BE*.

Let *EC* be added to each;

therefore *BA*, *AC* are greater than *BE*, *EC*.

Again, since, in the triangle *CED*, the two sides *CE*, *ED* are greater than *CD*, let *DB* be added to each;

therefore *CE*, *EB* are greater than *CD*, *DB*.

But *BA*, *AC* were proved greater than *BE*, *EC*;

therefore *BA*, *AC* are much greater than *BD*, *DC*.

Again, since in any triangle the exterior angle is greater than the interior and opposite angle, [I. 16]
therefore, in the triangle *CDE*,

the exterior angle *BDC* is greater than the angle *CED*.

For the same reason, moreover, in the triangle *ABE* also,

the exterior angle *CEB* is greater than the angle *BAC*.
But the angle *BDC* was proved greater than the angle *CEB*;

therefore the angle *BDC* is much greater than the angle *BAC*.

Therefore etc.

Q.E.D.

Proposition 22

Out of three straight lines, which are equal to three given straight lines, to construct a triangle: thus it is necessary that two of the straight lines taken together in any manner should be greater than the remaining one.　　[I. 20]

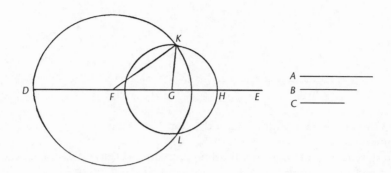

Let the three given straight lines be *A, B, C,* and of these let two taken together in any manner be greater than the remaining one, namely

　　　　　　　　A, B greater than *C,*
　　　　　　　　A, C greater than *B,*
and　　　　　　　*B, C* greater than *A;*

thus it is required to construct a triangle out of straight lines equal to *A, B, C.*

Let there be set out a straight line *DE,* terminated at *D* but of infinite length in the direction of *E,*

and let *DF* be made equal to *A, FG* equal to *B,* and *GH* equal to *C.*　　[I. 3]

With centre *F* and distance *FD* let the circle *DKL* be described;
again, with centre *G* and distance *GH* let the circle *KLH* be described;
and let *KF, KG* be joined;
I say that the triangle *KFG* has been constructed out of three straight lines equal to *A, B, C.*

For, since the point *F* is the centre of the circle *DKL,*

　　　　　　　FD is equal to *FK.*

But *FD* is equal to *A;*

　　　　　　　therefore *KF* is also equal to *A.*

Again, since the point *G* is the centre of the circle *LKH,*

　　　　　　　GH is equal to *GK.*

But *GH* is equal to *C;*

　　　　　　　therefore *KG* is also equal to *C.*

And *FG* is also equal to *B;* therefore the three straight lines *KF, FG, GK* are equal to the three straight lines *A, B, C.*

Therefore out of the three straight lines *KF, FG, GK,* which are equal to the three given straight lines *A, B, C,* the triangle *KFG* has been constructed.

　　　　　　　　　　　　　　　　　　　Q.E.F.

Proposition 23

On a given straight line and at a point on it to construct a rectilineal angle equal to a given rectilineal angle.

Let *AB* be the given straight line, *A* the point on it, and the angle *DCE* the given rectilineal angle;

thus it is required to construct on the given straight line *AB*, and at the point *A* on it, a rectilineal angle equal to the given rectilineal angle *DCE*.

On the straight lines *CD*, *CE* respectively let the points *D*, *E* be taken at random; let *DE* be joined,

and out of three straight lines which are equal to the three straight lines *CD*, *DE*, *CE* let the triangle *AFG* be constructed in such a way that *CD* is equal to *AF*, *CE* to *AG*, and further *DE* to *FG*. **[I. 22]**

Then, since the two sides *DC*, *CE* are equal to the two sides *FA*, *AG* respectively, and the base *DE* is equal to the base *FG*,

 the angle *DCE* is equal to the angle *FAG*. **[I. 8]**

Therefore on the given straight line *AB*, and at the point *A* on it, the rectilineal angle *FAG* has been constructed equal to the given rectilineal angle *DCE*.

 Q.E.F.

Proposition 24

If two triangles have the two sides equal to two sides respectively, but have the one of the angles contained by the equal straight lines greater than the other, they will also have the base greater than the base.

Let *ABC*, *DEF* be two triangles having the two sides *AB*, *AC* equal to the two sides *DE*, *DF* respectively, namely *AB* to *DE*, and *AC* to *DF*, and let the angle at *A* be greater than the angle at *D*;

I say that the base *BC* is also greater than the base *EF*.

For, since the angle *BAC* is greater than the angle *EDF*, let there be constructed, on the straight line *DE*, and at the point *D* on it, the angle *EDG* equal to the angle *BAC*; **[I. 23]**

let *DG* be made equal to either of the two straight lines *AC*, *DF*, and let *EG*, *FG* be joined.

Then, since *AB* is equal to *DE*, and *AC* to *DG*,
 the two sides *BA*, *AC* are equal to the two sides *ED*, *DG*, respectively;
and the angle *BAC* is equal to the angle *EDG*;
 therefore the base *BC* is equal to the base *EG*. [I. 4]

Again, since *DF* is equal to *DG*,
 the angle *DGF* is also equal to the angle *DFG*; [I. 5]
 therefore the angle *DFG* is greater than the angle *EGF*.

 Therefore the angle *EFG* is much greater than the angle *EGF*.

And, since *EFG* is a triangle having the angle *EFG* greater than the angle *EGF*,
and the greater angle is subtended by the greater side, [I. 19]
 the side *EG* is also greater than *EF*.

But *EG* is equal to *BC*.

 Therefore *BC* is also greater than *EF*.

Therefore etc.

 Q.E.D.

Proposition 25

If two triangles have the two sides equal to two sides respectively, but have the base greater than the base, they will also have the one of the angles contained by the equal straight lines greater than the other.

Let *ABC, DEF* be two triangles having the two sides *AB*, *AC* equal to the two sides *DE*, *DF* respectively, namely *AB* to *DE*, and *AC* to *DF*;
and let the base *BC* be greater than the base *EF*;

I say that the angle *BAC* is also greater than the angle *EDF*.

For, if not, it is either equal to it or less.

Now the angle *BAC* is not equal to the angle *EDF*; for then the base *BC* would also have been equal to the base *EF*, [I. 4]
but it is not;
 therefore the angle *BAC* is not equal to the angle *EDF*.

Neither again is the angle *BAC* less than the angle *EDF*;
for then the base *BC* would also have been less than the base *EF*, [I. 24]
but it is not;
 therefore the angle *BAC* is not less than the angle *EDF*.

But it was proved that it is not equal either;
 therefore the angle *BAC* is greater than the angle *EDF*.

Therefore etc.

 Q.E.D.

Proposition 26

If two triangles have the two angles equal to two angles respectively, and one side equal to one side, namely, either the side adjoining the equal angles, or that subtending one of the equal angles, they will also have the remaining sides equal to the remaining sides and the remaining angle to the remaining angle.

Let *ABC*, *DEF* be two triangles having the two angles *ABC*, *BCA* equal to the two angles *DEF*, *EFD* respectively, namely the angle *ABC* to the angle *DEF*, and the angle *BCA* to the angle *EFD*; and let them also have one side equal to one side, first that adjoining the equal angles, namely *BC* to *EF*;

I say that they will also have the remaining sides equal to the remaining sides respectively, namely *AB* to *DE* and *AC* to *DF*, and the remaining angle to the remaining angle, namely the angle *BAC* to the angle *EDF*.

For, if *AB* is unequal to *DE*, one of them is greater.

Let *AB* be greater, and let *BG* be made equal to *DE*; and let *GC* be joined.

Then, since *BG* is equal to *DE*, and *BC* to *EF*,
 the two sides *GB*, *BC* are equal to the two sides *DE*, *EF* respectively;
and the angle *GBC* is equal to the angle *DEF*;
 therefore the base *GC* is equal to the base *DF*,
 and the triangle *GBC* is equal to the triangle *DEF*,
and the remaining angles will be equal to the remaining angles, namely those which the equal sides subtend; [I. 4]
 therefore the angle *GCB* is equal to the angle *DFE*.

But the angle *DFE* is by hypothesis equal to the angle *BCA*;
 therefore the angle *BCG* is equal to the angle *BCA*,
the less to the greater: which is impossible.

Therefore *AB* is not unequal to *DE*,
 and is therefore equal to it.

But *BC* is also equal to *EF*;
 therefore the two sides *AB*, *BC* are equal to the two sides *DE*, *EF* respectively,
and the angle *ABC* is equal to the angle *DEF*;
 therefore the base *AC* is equal to the base *DF*,
 and the remaining angle *BAC* is equal to the remaining angle *EDF*. [I. 4]

Again, let sides subtending equal angles be equal, as *AB* to *DE*;
I say again that the remaining sides will be equal to the remaining sides, namely *AC* to *DF* and *BC* to *EF*, and further the remaining angle *BAC* is equal to the remaining angle *EDF*.

For, if *BC* is unequal to *EF*, one of them is greater.

Let *BC* be greater, if possible, and let *BH* be made equal to *EF*; let *AH* be joined.

Then, since *BH* is equal to *EF*, and *AB* to *DE*,

the two sides *AB, BH* are equal to the two sides *DE, EF* respectively,
and they contain equal angles;
therefore the base *AH* is equal to the base *DF*,
and the triangle *ABH* is equal to the triangle *DEF*,
and the remaining angles will be equal to the remaining angles, namely those
which the equal sides subtend; [I. 4]
therefore the angle *BHA* is equal to the angle *EFD*.

But the angle *EFD* is equal to the angle *BCA*;
therefore, in the triangle *AHC*, the exterior angle *BHA* is equal to the interior and
opposite angle *BCA*: which is impossible. [I. 16]

Therefore *BC* is not unequal to *EF*,
and is therefore equal to it.

But *AB* is also equal to *DE*;
therefore the two sides *AB, BC* are equal to the two sides *DE, EF* respectively,
and they contain equal angles;
therefore the base *AC* is equal to the base *DF*,
the triangle *ABC* equal to the triangle *DEF*,
and the remaining angle *BAC* equal to the remaining angle *EDF*. [I. 4]

Therefore etc.

Q.E.D.

Proposition 27

*If a straight line falling on two straight lines make the alternate angles equal to
one another, the straight lines will be parallel to one another.*

For let the straight line *EF* falling on the
two straight lines *AB, CD* make the
alternate angles *AEF, EFD* equal to one
another;
I say that *AB* is parallel to *CD*.

For, if not, *AB, CD* when produced will
meet either in the direction of *B, D* or
towards *A, C*.

Let them be produced and meet, in the
direction of *B, D*, at *G*.

Then, in the triangle *GEF*,
the exterior angle *AEF* is equal to the interior and opposite angle *EFG*:
which is impossible. [I. 16]

Therefore *AB, CD* when produced will not meet in the direction of *B, D*.

Similarly it can be proved that neither will they meet towards *A, C*.

But straight lines which do not meet in either direction are parallel; [Def. 23]
therefore *AB* is parallel to *CD*.

Therefore etc.

Q.E.D.

Proposition 28

If a straight line falling on two straight lines make the exterior angle equal to the interior and opposite angle on the same side, or the interior angles on the same side equal to two right angles, the straight lines will be parallel to one another.

For let the straight line *EF* falling on the two straight lines *AB*, *CD* make the exterior angle *EGB* equal to the interior and opposite angle *GHD*, or the interior angles on the same side, namely *BGH*, *GHD*, equal to two right angles;
I say that *AB* is parallel to *CD*.

For, since the angle *EGB* is equal to the angle *GHD*, while the angle *EGB* is equal to the angle *AGH*
[I. 15]
the angle *AGH* is also equal to the angle *GHD*;
and they are alternate;
therefore *AB* is parallel to *CD*. [I. 27]

Again, since the angles *BGH*, *GHD* are equal to two right angles, and the angles *AGH*, *BGH* are also equal to two right angles, [I. 13]
the angles *AGH*, *BGH* are equal to the angles *BGH*, *GHD*.

Let the angle *BGH* be subtracted from each;
therefore the remaining angle *AGH* is equal to the remaining angle *GHD*;
and they are alternate;
therefore *AB* is parallel to *CD*. [I. 27]

Therefore etc.

Q.E.D.

Proposition 29

A straight line falling on parallel straight lines makes the alternate angles equal to one another, the exterior angle equal to the interior and opposite angle, and the interior angles on the same side equal to two right angles.

For let the straight line *EF* fall on the parallel straight lines *AB*, *CD*;

I say that it makes the alternate angles *AGH*, *GHD* equal, the exterior angle *EGB* equal to the interior and opposite angle *GHD*, and the interior angles on the same side, namely *BGH*, *GHD*, equal to two right angles.

For, if the angle *AGH* is unequal to the angle *GHD*, one of them is greater.

Let the angle *AGH* be greater.

Let the angle *BGH* be added to each;
therefore the angles *AGH*, *BGH* are greater than the angles *BGH*, *GHD*.

But the angles *AGH*, *BGH* are equal to two right angles; [I. 13]
therefore the angles *BGH*, *GHD* are less than two right angles.

But straight lines produced indefinitely from angles less than two right angles meet;
[Post. 5]
therefore *AB*, *CD*, if produced indefinitely, will meet;
but they do not meet, because they are by hypothesis parallel.

Therefore the angle *AGH* is not unequal to the angle *GHD*,
and is therefore equal to it.

Again, the angle *AGH* is equal to the angle *EGB*; [I. 15]
therefore the angle *EGB* is also equal to the angle *GHD*. [C.N. 1]

Let the angle *BGH* be added to each;
therefore the angles *EGB*, *BGH* are equal to the angles *BGH*, *GHD*. [C.N. 2]

But the angles *EGB*, *BGH* are equal to two right angles; [I. 13]
therefore the angles *BGH*, *GHD* are also equal to two right angles.

Therefore etc.

Q.E.D.

Proposition 30

Straight lines parallel to the same straight line are also parallel to one another.

Let each of the straight lines *AB*, *CD* be parallel to *EF*;
I say that *AB* is also parallel to *CD*.

For let the straight line *GK* fall upon them.

Then, since the straight line *GK* has fallen on the parallel straight lines *AB*, *EF*,
the angle *AGK* is equal to the angle *GHF*. [I. 29]

Again, since the straight line *GK* has fallen on the parallel straight lines *EF*, *CD*,
the angle *GHF* is equal to the angle *GKD*. [I. 29]

But the angle *AGK* was also proved equal to the angle *GHF*;
therefore the angle *AGK* is also equal to the angle *GKD*; [C.N. 1]
and they are alternate.

Therefore *AB* is parallel to *CD*.

Q.E.D.

Proposition 31

Through a given point to draw a straight line parallel to a given straight line.

Let *A* be the given point, and *BC* the given straight line;
thus it is required to draw through the point *A* a straight line parallel to the straight line *BC*.

Let a point *D* be taken at random on *BC*, and let *AD* be joined;
on the straight line *DA*, and at the point *A* on it, let the angle *DAE* be constructed equal to the angle *ADC*; [I. 23]
and let the straight line *AF* be produced in a straight line with *EA*.

Then, since the straight line *AD* falling on the two straight lines *BC*, *EF* has made the alternate angles *EAD*, *ADC* equal to one another,

therefore *EAF* is parallel to *BC*. [I. 27]

Therefore through the given point *A* the straight line *EAF* has been drawn parallel to the given straight line *BC*.

Q.E.F.

Proposition 32

In any triangle, if one of the sides be produced, the exterior angle is equal to the two interior and opposite angles, and the three interior angles of the triangle are equal to two right angles.

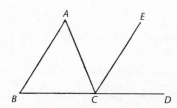

Let *ABC* be a triangle, and let one side of it *BC* be produced to *D*;
I say that the exterior angle *ACD* is equal to the two interior and opposite angles *CAB*, *ABC*, and the three interior angles of the triangle *ABC*, *BCA*, *CAB* are equal to two right angles.

For let *CE* be drawn through the point *C* parallel to the straight line *AB*. [I. 31]

Then, since *AB* is parallel to *CE*, and *AC* has fallen upon them,
the alternate angles *BAC*, *ACE* are equal to one another. [I. 29]

Again, since *AB* is parallel to *CE*, and the straight line *BD* has fallen upon them,
the exterior angle *ECD* is equal to the interior and opposite angle *ABC*. [I. 29]

But the angle *ACE* was also proved equal to the angle *BAC*;
therefore the whole angle *ACD* is equal to the two interior and opposite angles *BAC*, *ABC*.

Let the angle *ACB* be added to each;
therefore the angles *ACD*, *ACB* are equal to the three angles *ABC*, *BCA*, *CAB*.

But the angles *ACD*, *ACB* are equal to two right angles; [I. 13]
therefore the angles *ABC*, *BCA*, *CAB* are also equal to two right angles.

Therefore etc.

Q.E.D.

Proposition 33

The straight lines joining equal and parallel straight lines [at the extremities which are] in the same directions [respectively] are themselves also equal and parallel.

Let *AB*, *CD* be equal and parallel, and let the straight lines *AC*, *BD* join them [at the extremities which are] in the same directions [respectively];
I say that *AC*, *BD* are also equal and parallel.

Let *BC* be joined.

Then, since *AB* is parallel to *CD*, and *BC* has fallen upon them,
the alternate angles *ABC*, *BCD* are equal to one another. [I. 29]

And, since *AB* is equal to *CD*, and *BC* is common,
the two sides *AB*, *BC* are equal to the two sides *DC*, *CB*;
and the angle *ABC* is equal to the angle *BCD*;
therefore the base *AC* is equal to the base *BD*,
and the triangle *ABC* is equal to the triangle *DCB*,
and the remaining angles will be equal to the remaining angles respectively,
namely those which the equal sides subtend; [I. 4]
therefore the angle *ACB* is equal to the angle *CBD*.

And, since the straight line *BC* falling on the two straight lines *AC*, *BD* has made the alternate angles equal to one another,
AC is parallel to *BD*. [I. 27]

And it was also proved equal to it.

Therefore etc.

Q.E.D.

Proposition 34

In parallelogrammic areas the opposite sides and angles are equal to one another, and the diameter bisects the areas.

Let *ACDB* be a parallelogrammic area, and *BC* its diameter;
I say that the opposite sides and angles of the parallelogram *ACDB* are equal to one another, and the diameter *BC* bisects it.

For, since *AB* is parallel to *CD*, and the straight line *BC* has fallen upon them,
the alternate angles *ABC, BCD* are equal to one another. [I. 29]

Again, since *AC* is parallel to *BD*, and *BC* has fallen upon them,
the alternate angles *ACB, CBD* are equal to one another. [I. 29]

Therefore *ABC, DCB* are two triangles having the two angles *ABC, BCA* equal to the two angles *DCB, CBD* respectively, and one side equal to one side, namely that adjoining the equal angles and common to both of them, *BC*;
therefore they will also have the remaining sides equal to the remaining sides respectively, and the remaining angle to the remaining angle; [I. 26]
therefore the side *AB* is equal to *CD*,
and *AC* to *BD*,
and further the angle *BAC* is equal to the angle *CDB*.

And, since the angle *ABC* is equal to the angle *BCD*,
and the angle *CBD* to the angle *ACB*,
the whole angle *ABD* is equal to the whole angle *ACD*. [C.N. 2]

And the angle *BAC* was also proved equal to the angle *CDB*.

Therefore in parallelogrammic areas the opposite sides and angles are equal to one another.

I say, next, that the diameter also bisects the areas.

For, since *AB* is equal to *CD*, and *BC* is common,
the two sides *AB, BC* are equal to the two sides *DC, CB* respectively;
and the angle *ABC* is equal to the angle *BCD*;
therefore the base *AC* is also equal to *DB*,
and the triangle *ABC* is equal to the triangle *DCB*. [I. 4]

Therefore the diameter *BC* bisects the parallelogram *ACDB*.

Q.E.D.

Proposition 35

Parallelograms which are on the same base and in the same parallels are equal to one another.

Let *ABCD, EBCF* be parallelograms on the same base *BC* and in the same parallels *AF, BC*;

I say that *ABCD* is equal to the parallelogram *EBCF*.

For, since *ABCD* is a parallelogram,
$$AD \text{ is equal to } BC.$$ [I. 34]

For the same reason also
$$EF \text{ is equal to } BC,$$
$$\text{so that } AD \text{ is also equal to } EF;$$ [C.N. 1]
and *DE* is common;
$$\text{therefore the whole } AE \text{ is equal to the whole } DF.$$ [C.N. 2]

But *AB* is also equal to *DC*; [I. 34]
therefore the two sides *EA, AB* are equal to the two sides *FD, DC* respectively,
and the angle *FDC* is equal to the angle *EAB*, the exterior to the interior; [1. 29]
$$\text{therefore the base } EB \text{ is equal to the base } FC,$$
$$\text{and the triangle } EAB \text{ will be equal to the triangle } FDC.$$ [I. 4]

Let *DGE* be subtracted from each;
therefore the trapezium *ABGD* which remains is equal to the trapezium *EGCF* which remains. [C.N. 3]

Let the triangle *GBC* be added to each;
therefore the whole parallelogram *ABCD* is equal to the whole parallelogram *EBCF*. [C.N. 2]

Therefore etc.

Q.E.D.

Proposition 36

Parallelograms which are on equal bases and in the same parallels are equal to one another.

Let *ABCD, EFGH* be parallelograms which are on equal bases *BC, FG* and in the same parallels *AH, BG*;
I say that the parallelogram *ABCD* is equal to *EFGH*.

For let *BE, CH* be joined.

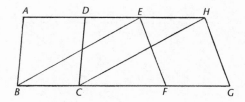

Then, since *BC* is equal to *FG*, while *FG* is equal to *EH*,
<div align="center">*BC* is also equal to *EH*.</div> **[C.N. 1]**

But they are also parallel.

And *EB*, *HC* join them;
but straight lines joining equal and parallel straight lines [at the extremities which are] in the same directions [respectively] are equal and parallel. **[I. 33]**

<div align="center">Therefore *EBCH* is a parallelogram.</div> **[I. 34]**

And it is equal to *ABCD*;
for it has the same base *BC* with it, and is in the same parallels *BC*, *AH* with it. **[I. 35]**

For the same reason also *EFGH* is equal to the same *EBCH*; **[I. 35]**
<div align="center">so that the parallelogram *ABCD* is also equal to *EFGH*.</div> **[C.N. 1]**

Therefore etc.

<div align="right">Q.E.D.</div>

Proposition 37

Triangles which are on the same base and in the same parallels are equal to one another.

Let *ABC*, *DBC* be triangles on the same base *BC* and in the same parallels *AD*, *BC*;
I say that the triangle *ABC* is equal to the triangle *DBC*.

Let *AD* be produced in both directions to *E*, *F*;
through *B* let *BE* be drawn parallel to *CA*, **[I. 31]**
and through *C* let *CF* be drawn parallel to *BD*. **[I. 31]**

Then each of the figures *EBCA*, *DBCF* is a parallelogram;
and they are equal, for they are on the same base *BC* and in the same parallels *BC*, *EF*. **[I. 35]**

Moreover the triangle *ABC* is half of the parallelogram *EBCA*; for the diameter *AB* bisects it. **[I. 34]**

And the triangle *DBC* is half of the parallelogram *DBCF*; for the diameter *DC* bisects it. **[I. 34]**

⟨But the halves of equal things are equal to one another.⟩

<div align="center">Therefore the triangle *ABC* is equal to the triangle *DBC*.</div>

Therefore etc.

<div align="right">Q.E.D.</div>

Proposition 38

Triangles which are on equal bases and in the same parallels are equal to one another.

Let *ABC, DEF* be triangles on equal bases *BC, EF*
and in the same parallels *BF, AD*;
I say that the triangle *ABC* is equal to the
triangle *DEF*.

For let *AD* be produced in both directions to *G, H*;
through *B* let *BG* be drawn parallel to *CA*, [I. 31]
and through *F* let *FH* be drawn parallel to *DE*.

> Then each of the figures *GBCA, DEFH* is a parallelogram;
and *GBCA* is equal to *DEFH*;
for they are on equal bases *BC, EF* and in the same parallels *BF, GH*. [I. 36]

Moreover the triangle *ABC* is half of the parallelogram *GBCA*;
for the diameter *AB* bisects it. [I. 34]

And the triangle *FED* is half of the parallelogram *DEFH*;
for the diameter *DF* bisects it. [I. 34]

⟨But the halves of equal things are equal to one another.⟩

> Therefore the triangle *ABC* is equal to the triangle *DEF*.

Therefore etc.

Q.E.D.

Proposition 39

Equal triangles which are on the same base and on the same side are also in the same parallels.

Let *ABC, DBC* be equal triangles which are on the same
base *BC* and on the same side of it;
⟨I say that they are also in the same parallels.⟩

And ⟨For⟩ let *AD* be joined;
I say that *AD* is parallel to *BC*.

For, if not, let *AE* be drawn through the point *A* parallel to
the straight line *BC*, [I. 31]
and let *EC* be joined.

> Therefore the triangle *ABC* is equal to the triangle *EBC*;
for it is on the same base *BC* with it and in the same parallels. [I. 37]

But *ABC* is equal to *DBC*;
> therefore *DBC* is also equal to *EBC*, [C.N. 1]
the greater to the less: which is impossible.

> Therefore *AE* is not parallel to *BC*.

Similarly we can prove that neither is any other straight line except *AD*;
> therefore *AD* is parallel to *BC*.

Therefore etc.

Q.E.D.

⟨Proposition 40 [4]

Equal triangles which are on equal bases and on the same side are also in the same parallels.

Let *ABC, CDE* be equal triangles on equal bases *BC, CE* and on the same side.

I say that they are also in the same parallels.

For let *AD* be joined;
I say that *AD* is parallel to *BE*.

For, if not, let *AF* be drawn through *A* parallel to *BE*, [I. 31]
and let *FE* be joined.

Therefore the triangle *ABC* is equal to the triangle *FCE*;
for they are on equal bases *BC, CE* and in the same parallels *BE, AF*. [I. 38]

But the triangle *ABC* is equal to the triangle *DCE*;
therefore the triangle *DCE* is also equal to the triangle *FCE*, [C.N. 1]
the greater to the less: which is impossible.

Therefore *AF* is not parallel to *BE*.

Similarly we can prove that neither is any other straight line except *AD*;
therefore *AD* is parallel to *BE*.

Therefore etc.

Q.E.D.⟩

Proposition 41

If a parallelogram have the same base with a triangle and be in the same parallels, the parallelogram is double of the triangle.

For let the parallelogram *ABCD* have the same base *BC* with the triangle *EBC*, and let it be in the same parallels *BC, AE*;
I say that the parallelogram *ABCD* is double of the triangle *BEC*.

For let *AC* be joined.

Then the triangle *ABC* is equal to the triangle *EBC*;
for it is on the same base *BC* with it and in the same parallels *BC, AE*. [I. 37]

But the parallelogram *ABCD* is double of the triangle *ABC*;
for the diameter *AC* bisects it; [I. 34]
so that the parallelogram *ABCD* is also double of the triangle EBC.

Therefore etc.

Q.E.D.

4. The whole of Proposition I.40 is enclosed within angle brackets since, according to Heath, "Heiberg has proved … that this proposition is an interpolation" by an unknown editor.

Proposition 42

To construct, in a given rectilineal angle, a parallelogram equal to given triangle.

Let *ABC* be the given triangle, and *D* the given rectilineal angle;
thus it is required to construct in the rectilineal angle *D* a parallelogram equal to the triangle *ABC*.

Let *BC* be bisected at *E*, and let *AE* be joined;
on the straight line *EC*, and at the point *E* on it, let the angle *CEF* be constructed equal to the angle *D*; [I. 23]
through *A* let *AG* be drawn parallel to *EC*,
and through *C* let *CG* be drawn parallel to *EF*. [I. 31]

Then *FECG* is a parallelogram.

And, since *BE* is equal to *EC*,
the triangle *ABE* is also equal to the triangle *AEC*,
for they are on equal bases *BE*, *EC* and in the same parallels *BC*, *AG*; [I. 38]
therefore the triangle *ABC* is double of the triangle *AEC*.

But the parallelogram *FECG* is also double of the triangle *AEC*, for it has the same base with it and is in the same parallels with it;
therefore the parallelogram *FECG* is equal to the triangle *ABC*.

And it has the angle *CEF* equal to the given angle *D*.

Therefore the parallelogram *FECG* has been constructed equal to the given triangle *ABC*, in the angle *CEF* which is equal to *D*.

Q.E.F.

Proposition 43

In any parallelogram the complements of the parallelograms about the diameter are equal to one another.

Let *ABCD* be a parallelogram, and *AC* its diameter;
and about *AC* let *EH*, *FG* be parallelograms, and *BK*, *KD* the so-called complements;
I say that the complement *BK* is equal to the complement *KD*.

For, since *ABCD* is a parallelogram, and *AC* its diameter,
the triangle *ABC* is equal to the triangle *ACD*. [I. 34]

Again, since *EH* is a parallelogram, and *AK* is its diameter,
the triangle *AEK* is equal to the triangle *AHK*.

For the same reason
the triangle *KFC* is also equal to *KGC*.

Now, since the triangle *AEK* is equal to the triangle *AHK*,
and *KFC* to *KGC*,
the triangle *AEK* together with *KGC* is equal to the triangle *AHK* together with *KFC*. [C.N. 2]

And the whole triangle *ABC* is also equal to the whole *ADC*;

therefore the complement *BK* which remains is equal to the complement *KD* which remains. [C.N. 3]

Therefore etc.

Q.E.D.

Proposition 44

To a given straight line to apply, in a given rectilineal angle, a parallelogram equal to a given triangle.

Let *AB* be the given straight line, *C* the given triangle and *D* the given rectilineal angle;
thus it is required to apply to the given straight line *AB*, in an angle equal to the angle *D*, a parallelogram equal to the given triangle *C*.

Let the parallelogram *BEFG* be constructed equal to the triangle *C*, in the angle *EBG* which is equal to *D*; [I. 42]
let it be placed so that *BE* is in a straight line with *AB*;
let *FG* be drawn through to *H*,
and let *AH* be drawn through *A* parallel to either *BG* or *EF*. [I. 31]

Let *HB* be joined.

Then, since the straight line *HF* falls upon the parallels *AH*, *EF*,
the angles *AHF*, *HFE* are equal to two right angles. [I. 29]

Therefore the angles *BHG*, *GFE* are less than two right angles;
and straight lines produced indefinitely from angles less than two right angles meet; [Post. 5]
therefore *HB*, *FE*, when produced, will meet.

Let them be produced and meet at *K*;
through the point *K* let *KL* be drawn parallel to either *EA* or *FH*, [I. 31]
and let *HA*, *GB* be produced to the points *L*, *M*.

Then *HLKF* is a parallelogram,
HK is its diameter,
and *AG, ME* are parallelograms,
and *LB, BF* the so-called complements, about *HK*;

therefore *LB* is equal to *BF*. [I. 43]

But *BF* is equal to the triangle *C*;

therefore *LB* is also equal to *C*. [C.N. 1]

And, since the angle *GBE* is equal to the angle *ABM*, [I. 15]
while the angle *GBE* is equal to *D*,

the angle *ABM* is also equal to the angle *D*.

Therefore the parallelogram *LB* equal to the given triangle *C* has been applied
to the given straight line *AB*, in the angle *ABM* which is equal to *D*.

Q.E.F.

Proposition 45

To construct, in a given rectilineal angle, a parallelogram equal to a given rectilineal figure.

Let *ABCD* be the given rectilineal figure and *E* the given rectilineal angle;
thus it is required to construct, in the given angle *E*, a parallelogram equal to the
rectilineal figure *ABCD*.

Let *DB* be joined, and let the parallelogram *FH* be constructed equal to the
triangle *ABD*, in the angle *HKF* which is equal to *E*; [I. 42]
let the parallelogram *GM* equal to the triangle *DBC* be applied to the straight
line *GH*, in the angle *GHM* which is equal to *E*. [I. 44]

Then, since the angle *E* is equal to each of the angles *HKF, GHM*,

the angle *HKF* is also equal to the angle *GHM*. [C.N. 1]

Let the angle *KHG* be added to each;

therefore the angles *FKH, KHG* are equal to the angles *KHG, GHM*.

But the angles *FKH, KHG* are equal to two right angles; [I. 29]
therefore the angles *KHG, GHM* are also equal to two right angles.

Thus, with a straight line *GH*, and at the point *H* on it, two straight lines *KH, HM*
not lying on the same side make the adjacent angles equal to two right angles;

therefore *KH* is in a straight line with *HM*. [I. 14]

And, since the straight line *HG* falls upon the parallels *KM, FG*,

the alternate angles *MHG, HGF* are equal to one another. [I. 29]

Let the angle *HGL* be added to each;
therefore the angles *MHG, HGL* are equal to the angles *HGF, HGL*. [C.N. 2]

But the angles *MHG, HGL* are equal to two right angles;
therefore the angles *HGF, HGL* are also equal to two right angles. [C.N. 1]

Therefore *FG* is in a straight line with *GL*. [I. 14]

And, since *FK* is equal and parallel to *HG*, [I. 34]
and *HG* to *ML* also,
KF is also equal and parallel to *ML*; [C.N. 1; I. 30]
and the straight lines *KM, FL* join them [at their extremities];
therefore *KM, FL* are also equal and parallel. [I. 33]

Therefore *KFLM* is a parallelogram.

And, since the triangle *ABD* is equal to the parallelogram *FH*, and *DBC* to *GM*, the whole rectilineal figure *ABCD* is equal to the whole parallelogram *KFLM*.

Therefore the parallelogram *KFLM* has been constructed equal to the given rectilineal figure *ABCD*, in the angle *FKM* which is equal to the given angle *E*.

Q.E.F.

Proposition 46

On a given straight line to describe a square.

Let *AB* be the given straight line;
thus it is required to describe a square on the straight line *AB*.

Let *AC* be drawn at right angles to the straight line *AB* from the point *A* on it, [I. 11]
and let *AD* be made equal to *AB*;
through the point *D* let *DE* be drawn parallel to *AB*,
and through the point *B* let *BE* be drawn parallel to *AD*. [I. 31]

Therefore *ADEB* is a parallelogram;
therefore *AB* is equal to *DE*, and *AD* to *BE*. [I. 34]

But *AB* is equal to *AD*;
therefore the four straight lines *BA, AD, DE, EB* are equal to one another;
therefore the parallelogram *ADEB* is equilateral.

I say next that it is also right-angled.

For, since the straight line *AD* falls upon the parallels *AB, DE*,
the angles *BAD, ADE* are equal to two right angles. [I. 29]

But the angle *BAD* is right;

therefore the angle *ADE* is also right.

And in parallelogrammic areas the opposite sides and angles are equal to one another; [I. 34]

therefore each of the opposite angles *ABE, BED* is also right.

Therefore *ADEB* is right-angled.

And it was also proved equilateral.

Therefore it is a square; and it is described on the straight line *AB*.

Q.E.F.

Proposition 47

In right-angled triangles the square on the side subtending the right angle is equal to the squares on the sides containing the right angle.

Let *ABC* be a right-angled triangle having the angle *BAC* right;
I say that the square on *BC* is equal to the squares on *BA, AC*.

For let there be described on *BC* the square *BDEC*, and on *BA, AC* the squares *GB, HC*;
[I. 46]
through *A* let *AL* be drawn parallel to either *BD* or *CE*,
and let *AD, FC* be joined.

Then, since each of the angles *BAC, BAG* is right,
it follows that with a straight line *BA*, and at the point *A* on it, the two straight lines *AC, AG* not lying on the same side make the adjacent angles equal to two right angles;
 therefore *CA* is in a straight line with *AG*.
[I. 14]

For the same reason *BA* is also in a straight line with *AH*.

And, since the angle *DBC* is equal to the angle *FBA*: for each is right:
let the angle *ABC* be added to each;
 therefore the whole angle *DBA* is equal to the whole angle *FBC*. [C.N. 2]

And, since *DB* is equal to *BC*, and *FB* to *BA*,
 the two sides *AB, BD* are equal to the two sides *FB, BC* respectively;
and the angle *ABD* is equal to the angle *FBC*;
 therefore the base *AD* is equal to the base *FC*,
 and the triangle *ABD* is equal to the triangle *FBC*. [I. 4]

Now the parallelogram *BL* is double of the triangle *ABD*, for they have the same base *BD* and are in the same parallels *BD, AL*. [I. 41]

And the square *GB* is double of the triangle *FBC*, for they again have the same base *FB* and are in the same parallels *FB, GC*. [I. 41]

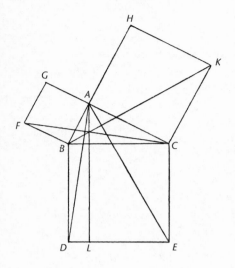

⟨But the doubles of equals are equal to one another.⟩

Therefore the parallelogram *BL* is also equal to the square *GB*.

Similarly, if *AE*, *BK* be joined,
the parallelogram *CL* can also be proved equal to the square *HC*;
therefore the whole square *BDEC* is equal to the two squares *GB*, *HC*. [C.N. 2]

And the square *BDEC* is described on *BC*,
and the squares *GB*, *HC* on *BA*, *AC*.

Therefore the square on the side *BC* is equal to the squares on the sides *BA*, *AC*.

Therefore etc.

Q.E.D.

Proposition 48

If in a triangle the square on one of the sides be equal to the squares on the remaining two sides of the triangle, the angle contained by the remaining two sides of the triangle is right.

For in the triangle *ABC* let the square on one side *BC* be equal to the squares on the sides *BA*, *AC*;
I say that the angle *BAC* is right.

For let *AD* be drawn from the point *A* at right angles to the straight line *AC*,
let *AD* be made equal to *BA*,
and let *DC* be joined.

Since *DA* is equal to *AB*,
the square on *DA* is also equal to the square on *AB*.

Let the square on *AC* be added to each;
therefore the squares on *DA*, *AC* are equal to the squares on *BA*, *AC*.
But the square on *DC* is equal to the squares on *DA*, *AC*, for the angle *DAC* is right;
[I. 47]
and the square on *BC* is equal to the squares on *BA*, *AC*, for this is the hypothesis;
therefore the square on *DC* is equal to the square on *BC*,
so that the side *DC* is also equal to *BC*.

And, since *DA* is equal to *AB*, and *AC* is common,
the two sides *DA*, *AC* are equal to the two sides *BA*, *AC*;
and the base *DC* is equal to the base *BC*;
therefore the angle *DAC* is equal to the angle *BAC*. [I. 8]
But the angle *DAC* is right;
therefore the angle *BAC* is also right.

Therefore etc.

Q.E.D.

Book II

Definitions

1. Any rectangular parallelogram is said to be *contained* by the two straight lines containing the right angle.

2. And in any parallelogrammic area let any one whatever of the parallelograms about its diameter with the two complements be called a *gnomon*.

Proposition 1

If there be two straight lines, and one of them be cut into any number of segments whatever, the rectangle contained by the two straight lines is equal to the rectangles contained by the uncut straight line and each of the segments.

Let *A*, *BC* be two straight lines, and let *BC* be cut at random at the points *D*, *E*;
I say that the rectangle contained by *A*, *BC* is equal to the rectangle contained by *A*, *BD*, that contained by *A*, *DE* and that contained by *A*, *EC*.

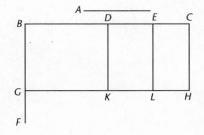

For let *BF* be drawn from *B* at right angles to *BC*;
[I. 11]
let *BG* be made equal to *A*, [I. 3]
through *G* let *GH* be drawn parallel to *BC*, [I. 31]
and through *D*, *E*, *C* let *DK*, *EL*, *CH* be drawn parallel to *BG*.

Then *BH* is equal to *BK*, *DL*, *EH*.

Now *BH* is the rectangle *A*, *BC*, for it is contained by *GB*, *BC*, and *BG* is equal to *A*;
BK is the rectangle *A*, *BD*, for it is contained by *GB*, *BD*, and *BG* is equal to *A*;
and *DL* is the rectangle *A*, *DE*; for *DK*, that is *BG* [I. 34], is equal to *A*.

Similarly also *EH* is the rectangle *A*, *EC*.

Therefore the rectangle *A*, *BC* is equal to the rectangle *A*, *BD*, the rectangle *A*, *DE* and the rectangle *A*, *EC*.

Therefore etc.

Q.E.D.

Proposition 2

If a straight line be cut at random, the rectangles contained by the whole and both of the segments are equal to the square on the whole.

For let the straight line *AB* be cut at random at the point *C*;
I say that the rectangle contained by *AB*, *BC* together with the rectangle contained by *BA*, *AC* is equal to the square on *AB*.

For let the square *ADEB* be described on *AB*, **[I. 46]**
and let *CF* be drawn through *C* parallel to either *AD* or *BE*.
[I. 31]

Then *AE* is equal to *AF*, *CE*.

Now *AE* is the square on *AB*;
AF is the rectangle contained by *BA*, *AC*, for it is contained by *DA*, *AC*, and *AD* is equal to *AB*;
and *CE* is the rectangle *AB*, *BC*, for *BE* is equal to *AB*.

Therefore the rectangle *BA*, *AC* together with the rectangle *AB*, *BC* is equal to the square on *AB*.

Therefore etc.

Q.E.D.

Proposition 3

If a straight line be cut at random, the rectangle contained by the whole and one of the segments is equal to the rectangle contained by the segments and the square on the aforesaid segment.

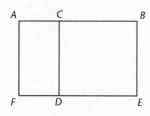

For let the straight line *AB* be cut at random at *C*;
I say that the rectangle contained by *AB*, *BC* is equal to the rectangle contained by *AC*, *CB* together with the square on *BC*.

For let the square *CDEB* be described on *CB*; **[I. 46]**
let *ED* be drawn through to *F*, and through *A* let *AF* be drawn parallel to either *CD* or *BE*. **[I. 31]**

Then *AE* is equal to *AD*, *CE*.

Now *AE* is the rectangle contained by *AB*, *BC*, for it is contained by *AB*, *BE*, and *BE* is equal to *BC*;
AD is the rectangle *AC*, *CB*, for *DC* is equal to *CB*;
and *DB* is the square on *CB*.

Therefore the rectangle contained by *AB*, *BC* is equal to the rectangle contained by *AC*, *CB* together with the square on *BC*.

Therefore etc.

Q.E.D.

Proposition 4

If a straight line be cut at random, the square on the whole is equal to the squares on the segments and twice the rectangle contained by the segments.

For let the straight line *AB* be cut at random at *C*;
I say that the square on *AB* is equal to the squares on *AC*, *CB* and twice the rectangle contained by *AC*, *CB*.

For let the square *ADEB* be described on *AB*, [I. 46]
let *BD* be joined;
through *C* let *CF* be drawn parallel to either *AD* or *EB*,
and through *G* let *HK* be drawn parallel to either *AB* or *DE*. [I. 31]

Then, since *CF* is parallel to *AD*, and *BD* has fallen on them,
the exterior angle *CGB* is equal to the interior and opposite angle *ADB*. [I. 29]

But the angle *ADB* is equal to the angle *ABD*, since the side *BA* is also equal to *AD*; [I. 5]

therefore the angle *CGB* is also equal to the angle *GBC*,
so that the side *BC* is also equal to the side *CG*. [I. 6]

But *CB* is equal to *GK*, and *CG* to *KB*; [I. 34]
therefore *GK* is also equal to *KB*;
therefore *CGKB* is equilateral.

I say next that it is also right-angled.

For, since *CG* is parallel to *BK*,
the angles *KBC*, *GCB* are equal to two right angles. [I. 29]

But the angle *KBC* is right;
therefore the angle *BCG* is also right,
so that the opposite angles *CGK*, *GKB* are also right. [I. 34]

Therefore *CGKB* is right-angled;
and it was also proved equilateral;
therefore it is a square;
and it is described on *CB*.

For the same reason
HF is also a square;
and it is described on *HG*, that is *AC*. [I. 34]

Therefore the squares *HF*, *KC* are the squares on *AC*, *CB*.

Now, since *AG* is equal to *GE*,
and *AG* is the rectangle *AC*, *CB*, for *GC* is equal to *CB*,
therefore *GE* is also equal to the rectangle *AC*, *CB*.

Therefore *AG*, *GE* are equal to twice the rectangle *AC*, *CB*.

But the squares *HF*, *CK* are also the squares on *AC*, *CB*;
therefore the four areas *HF*, *CK*, *AG*, *GE* are equal to the squares on *AC*, *CB* and twice the rectangle contained by *AC*, *CB*.

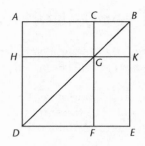

But *HF, CK, AG, GE* are the whole *ADEB,*
which is the square on *AB.*

Therefore the square on *AB* is equal to the squares on *AC, CB* and twice the
rectangle contained by *AC, CB.*

Therefore etc.

Q.E.D.

Proposition 5

*If a straight line be cut into equal and unequal segments, the rectangle contained
by the unequal segments of the whole together with the square on the straight
line between the points of section is equal to the square on the half.*

For let a straight line *AB* be cut into equal segments at *C* and into unequal
segments at *D*;
I say that the rectangle contained by *AD, DB* together with the square on *CD* is
equal to the square on *CB.*

For let the square *CEFB* be described on *CB,* [I. 46]
and let *BE* be joined;
through *D* let *DG* be drawn parallel to either *CE* or *BF,*
through *H* again let *KM* be drawn parallel to either *AB* or *EF,*
and again through *A* let *AK* be drawn parallel to either *CL* or *BM.* [I. 31]

Then, since the complement *CH* is equal to the complement *HF,* [I. 43]
let *DM* be added to each;
 therefore the whole *CM* is equal to the whole *DF.*

But *CM* is equal to *AL,*
since *AC* is also equal to *CB*; [I. 36]
 therefore *AL* is also equal to *DF.*

Let *CH* be added to each;

therefore the whole *AH* is equal to the gnomon *NOP*.

But *AH* is the rectangle *AD, DB,* for *DH* is equal to *DB,*

therefore the gnomon *NOP* is also equal to the rectangle *AD, DB.*

Let *LG,* which is equal to the square on *CD,* be added to each;
therefore the gnomon *NOP* and *LG* are equal to the rectangle contained by *AD, DB* and the square on *CD.*

But the gnomon *NOP* and *LG* are the whole square *CEFB,* which is described on *CB;*
therefore the rectangle contained by *AD, DB* together with the square on *CD* is equal to the square on *CB.*

Therefore etc.

Q.E.D.

Proposition 6

If a straight line be bisected and a straight line be added to it in a straight line, the rectangle contained by the whole with the added straight line and the added straight line together with the square on the half is equal to the square on the straight line made up of the half and the added straight line.

For let a straight line *AB* be bisected at the point *C,*
and let a straight line *BD* be added to it in a straight line;
I say that the rectangle contained by *AD, DB* together with the square on *CB* is equal to the square on *CD.*

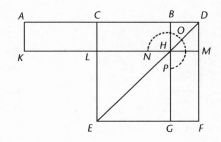

For let the square *CEFD* be described on *CD,* [I. 46]
and let *DE* be joined;
through the point *B* let *BG* be drawn parallel to either *EC* or *DF,*
through the point *H* let *KM* be drawn parallel to either *AB* or *EF,*
and further through *A* let *AK* be drawn parallel to either *CL* or *DM.* [I. 31]

Then, since *AC* is equal to *CB,*

 AL is also equal to *CH.* [I. 36]

But *CH* is equal to *HF.* [I. 43]

 Therefore *AL* is also equal to *HF.*

Let *CM* be added to each;

 therefore the whole *AM* is equal to the gnomon *NOP.*

But *AM* is the rectangle *AD, DB,* for *DM* is equal to *DB;*

 therefore the gnomon *NOP* is also equal to the rectangle *AD, DB.*

Let *LG,* which is equal to the square on *BC,* be added to each;
therefore the rectangle contained by *AD, DB* together with the square on *CB* is equal to the gnomon *NOP* and *LG.*

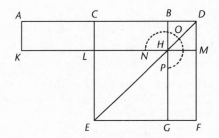

But the gnomon *NOP* and *LG* are the whole square *CEFD*, which is described on *CD*;
therefore the rectangle contained by *AD*, *DB* together with the square on *CB* is equal to the square on *CD*.

Therefore etc.

<div align="right">Q.E.D.</div>

Proposition 7

If a straight line be cut at random, the square on the whole and that on one of the segments both together are equal to twice the rectangle contained by the whole and the said segment and the square on the remaining segment.

For let a straight line *AB* be cut at random at the point *C*; I say that the squares on *AB*, *BC* are equal to twice the rectangle contained by *AB*, *BC* and the square on *CA*.

For let the square *ADEB* be described on *AB*, **[I. 46]** and let the figure be drawn.

Then, since *AG* is equal to *GE*, **[I. 43]** let *CF* be added to each;
therefore the whole *AF* is equal to the whole *CE*.

Therefore *AF*, *CE* are double of *AF*.

But *AF*, *CE* are the gnomon *KLM* and the square *CF*; therefore the gnomon *KLM* and the square *CF* are double of *AF*.

But twice the rectangle *AB*, *BC* is also double of *AF*;
for *BF* is equal to *BC*;
therefore the gnomon *KLM* and the square *CF* are equal to twice the rectangle *AB*, *BC*.

Let *DG*, which is the square on *AC*, be added to each;
therefore the gnomon *KLM* and the squares *BG*, *GD* are equal to twice the rectangle contained by *AB*, *BC* and the square on *AC*.

But the gnomon *KLM* and the squares *BG*, *GD* are the whole *ADEB* and *CF*, which are squares described on *AB*, *BC*;
therefore the squares on *AB*, *BC* are equal to twice the rectangle contained by *AB*, *BC* together with the square on *AC*.

Therefore etc.

<div align="right">Q.E.D.</div>

Proposition 8

If a straight line be cut at random, four times the rectangle contained by the whole and one of the segments together with the square on the remaining segment is equal to the square described on the whole and the aforesaid segment as on one straight line.

For let a straight line *AB* be cut at random at the point *C*;
I say that four times the rectangle contained by *AB*, *BC* together with the square on *AC* is equal to the square described on *AB*, *BC* as on one straight line.

For let [the straight line] *BD* be produced in a straight line with *AB*, and let *BD* be made equal to *CB*;
let the square *AEFD* be described on *AD*, and let the figure be drawn double.

Then, since *CB* is equal to *BD*, while *CB* is equal to *GK*, and *BD* to *KN*,
therefore *GK* is also equal to *KN*.

For the same reason
QR is also equal to *RP*.

And, since *BC* is equal to *BD*, and *GK* to *KN*,
therefore *CK* is also equal to *KD*, and *GR* to *RN*. [I. 36]

But *CK* is equal to *RN*, for they are complements of the parallelogram *CP*; [I. 43]
therefore *KD* is also equal to *GR*;
therefore the four areas *DK, CK, GR, RN* are equal to one another.

Therefore the four are quadruple of *CK*.

Again, since *CB* is equal to *BD*,
while *BD* is equal to *BK*, that is *CG*,
and *CB* is equal to *GK*, that is *GQ*,
therefore *CG* is also equal to *GQ*.

And, since *CG* is equal to *GQ*, and *QR* to *RP*,
AG is also equal to *MQ*, and *QL* to *RF*. [I. 36]

But *MQ* is equal to *QL*, for they are complements of the parallelogram *ML*; [I. 43]
therefore *AG* is also equal to *RF*;
therefore the four areas *AG, MQ, QL, RF* are equal to one another.

Therefore the four are quadruple of *AG*.

But the four areas *CK, KD, GR, RN* were proved to be quadruple of *CK*;
therefore the eight areas, which contain the gnomon *STU*, are quadruple of *AK*.

Now, since *AK* is the rectangle *AB, BD*, for *BK* is equal to *BD*,
therefore four times the rectangle *AB, BD* is quadruple of *AK*.

But the gnomon *STU* was also proved to be quadruple of *AK*;
therefore four times the rectangle *AB, BD* is equal to the gnomon *STU*.

Let *OH*, which is equal to the square on *AC*, be added to each;

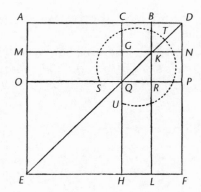

therefore four times the rectangle *AB*, *BD* together with the square on *AC* is equal to the gnomon *STU* and *OH*.

But the gnomon *STU* and *OH* are the whole square *AEFD*,
which is described on *AD*;
therefore four times the rectangle *AB*, *BD* together with the square on *AC* is equal to the square on *AD*.

But *BD* is equal to *BC*;
therefore four times the rectangle contained by *AB*, *BC* together with the square on *AC* is equal to the square on *AD*, that is to the square described on *AB* and *BC* as on one straight line.

Therefore etc.

Q.E.D.

Proposition 9

If a straight line be cut into equal and unequal segments, the squares on the unequal segments of the whole are double of the square on the half and of the square on the straight line between the points of section.

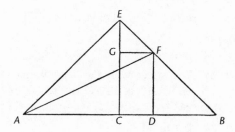

For let a straight line *AB* be cut into equal segments at *C*, and into unequal segments at *D*;
I say that the squares on *AD*, *DB* are double of the squares on *AC*, *CD*.

For let *CE* be drawn from *C* at right angles to *AB*, and let it be made equal to either *AC* or *CB*;
let *EA*, *EB* be joined,
let *DF* be drawn through *D* parallel to *EC*,
and *FG* through *F* parallel to *AB*,

and let *AF* be joined.

Then, since *AC* is equal to *CE*,
 the angle *EAC* is also equal to the angle *AEC*.

And, since the angle at *C* is right,
 the remaining angles *EAC*, *AEC* are equal to one right angle. **[I. 32]**

And they are equal;
 therefore each of the angles *CEA*, *CAE* is half a right angle.

For the same reason
 each of the angles *CEB*, *EBC* is also half a right angle;
 therefore the whole angle *AEB* is right.

And, since the angle *GEF* is half a right angle and the angle *EGF* is right, for it
is equal to the interior and opposite angle *ECB*, **[I. 29]**
 the remaining angle *EFG* is half a right angle; **[I. 32]**
 therefore the angle *GEF* is equal to the angle *EFG*,
 so that the side *EG* is also equal to *GF*. **[I. 6]**

Again, since the angle at *B* is half a right angle,
and the angle *FDB* is right, for it is again equal to the interior and opposite
angle *ECB*, **[I. 29]**
 the remaining angle *BFD* is half a right angle; **[I. 32]**
 therefore the angle at *B* is equal to the angle *DFB*,
 so that the side *FD* is also equal to the side *DB*. **[I. 6]**

Now, since *AC* is equal to *CE*,
 the square on *AC* is also equal to the square on *CE*;
 therefore the squares on *AC*, *CE* are double of the square on *AC*.

But the square on *EA* is equal to the squares on *AC*, *CE*, for the angle *ACE* is
right; **[I. 47]**
 therefore the square on *EA* is double of the square on *AC*.

Again, since *EG* is equal to *GF*,
 the square on *EG* is also equal to the square on *GF*;
 therefore the squares on *EG*, *GF* are double of the square on *GF*.

But the square on *EF* is equal to the squares on *EG*, *GF*;
 therefore the square on *EF* is double of the square on *GF*.

But *GF* is equal to *CD*; **[I. 34]**
 therefore the square on *EF* is double of the square on *CD*.

But the square on *EA* is also double of the square on *AC*;
 therefore the squares on *AE*, *EF* are double of the squares on *AC*, *CD*.

And the square on *AF* is equal to the squares on *AE*, *EF*, for the angle *AEF* is
right; **[I. 47]**
 therefore the square on *AF* is double of the squares on *AC*, *CD*.

But the squares on *AD*, *DF* are equal to the square on *AF*, for the angle at *D* is
right; **[I. 47]**
therefore the squares on *AD*, *DF* are double of the squares on *AC*, *CD*.

And *DF* is equal to *DB*;
therefore the squares on *AD*, *DB* are double of the squares on *AC*, *CD*.

Therefore etc.

 Q.E.D.

Proposition 10

If a straight line be bisected, and a straight line be added to it in a straight line, the square on the whole with the added straight line and the square on the added straight line both together are double of the square on the half and of the square described on the straight line made up of the half and the added straight line as on one straight line.

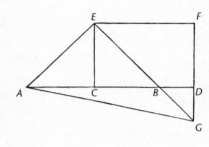

For let a straight line *AB* be bisected at *C*, and let a straight line *BD* be added to it in a straight line; I say that the squares on *AD*, *DB* are double of the squares on *AC*, *CD*.

For let *CE* be drawn from the point *C* at right angles to *AB*, [I. 11]
and let it be made equal to either *AC* or *CB*; [I. 3]
let *EA*, *EB* be joined;
through *E* let *EF* be drawn parallel to *AD*,
and through *D* let *FD* be drawn parallel to *CE*.
[I. 31]

Then, since a straight line *EF* falls on the parallel straight lines *EC*, *FD*,
the angles *CEF*, *EFD* are equal to two right angles; [I. 29]
therefore the angles *FEB*, *EFD* are less than two right angles.

But straight lines produced from angles less than two right angles meet;
[I. Post. 5]
therefore *EB*, *FD*, if produced in the direction *B*, *D*, will meet.

Let them be produced and meet at *G*,
and let *AG* be joined.

Then, since *AC* is equal to *CE*,
the angle *EAC* is also equal to the angle *AEC*; [I. 5]
and the angle at *C* is right;
therefore each of the angles *EAC*, *AEC* is half a right angle. [I. 32]

For the same reason
each of the angles *CEB*, *EBC* is also half a right angle;
therefore the angle *AEB* is right.

And, since the angle *EBC* is half a right angle,
the angle *DBG* is also half a right angle.

But the angle *BDG* is also right,
for it is equal to the angle *DCE*, they being alternate; [I. 29]
therefore the remaining angle *DGB* is half a right angle; [I. 32]
therefore the angle *DGB* is equal to the angle *DBG*,
so that the side *BD* is also equal to the side *GD*. [I. 6]

Again, since the angle *EGF* is half a right angle,
and the angle at *F* is right, for it is equal to the opposite angle, the angle at *C*,
[I. 34]
the remaining angle *FEG* is half a right angle; [I. 32]
therefore the angle *EGF* is equal to the angle *FEG*,
so that the side *GF* is also equal to the side *EF*. [I. 6]

Now, since the square on *EC* is equal to the square on *CA*,
 the squares on *EC, CA* are double of the square on *CA*.

But the square on *EA* is equal to the squares on *EC, CA*; **[I. 47]**
 therefore the square on *EA* is double of the square on *AC*. **[C.N. 1]**

Again, since *FG* is equal to *EF*,
 the square on *FG* is also equal to the square on *FE*;
 therefore the squares on *GF, FE* are double of the square on *EF*.

But the square on *EG* is equal to the squares on *GF, FE*; **[I. 47]**
 therefore the square on *EG* is double of the square on *EF*.

And *EF* is equal to *CD*; **[I. 34]**
 therefore the square on *EG* is double of the square on *CD*.

But the square on *EA* was also proved double of the square on *AC*;
 therefore the squares on *AE, EG* are double of the squares on *AC, CD*.

And the square on *AG* is equal to the squares on *AE, EG*; **[I. 47]**
 therefore the square on *AG* is double of the squares on *AC, CD*.

But the squares on *AD, DG* are equal to the square on *AG*; **[I. 47]**
 therefore the squares on *AD, DG* are double of the squares on *AC, CD*.

And *DG* is equal to *DB*;
 therefore the squares on *AD, DB* are double of the squares on *AC, CD*.

Therefore etc.

 Q.E.D.

Proposition 11

To cut a given straight line so that the rectangle contained by the whole and one of the segments is equal to the square on the remaining segment.

Let *AB* be the given straight line;
thus it is required to cut *AB* so that the rectangle contained by the whole and one of the segments is equal to the square on the remaining segment.

For let the square *ABDC* be described on *AB*; **[I. 46]**
let *AC* be bisected at the point *E*, and let *BE* be joined;
let *CA* be drawn through to *F*, and let *EF* be made equal to *BE*;
let the square *FH* be described on *AF*, and let *GH* be drawn through to *K*.

I say that *AB* has been cut at *H* so as to make the rectangle contained by *AB, BH* equal to the square on *AH*.

For, since the straight line *AC* has been bisected at *E*, and *FA* is added to it,
the rectangle contained by *CF, FA* together with the square on *AE* is equal to the square on *EF*. **[II. 6]**

But *EF* is equal to *EB*;
therefore the rectangle *CF, FA* together with the square on *AE* is equal to the square on *EB*.

But the squares on *BA, AE* are equal to the square on *EB*, for the angle at *A* is right;
 [I. 47]

therefore the rectangle *CF, FA* together with the square on *AE* is equal to the squares on *BA, AE*.

Let the square on *AE* be subtracted from each;
therefore the rectangle *CF, FA* which remains is equal to the square on *AB*.

Now the rectangle *CF, FA* is *FK*, for *AF* is equal to *FG*;
and the square on *AB* is *AD*;
therefore *FK* is equal to *AD*.

Let *AK* be subtracted from each;
therefore *FH* which remains is equal to *HD*.

And *HD* is the rectangle *AB, BH*, for *AB* is equal to *BD*;
and *FH* is the square on *AH*;
therefore the rectangle contained by *AB, BH* is equal to the square on *HA*.

Therefore the given straight line *AB* has been cut at *H* so as to make the rectangle contained by *AB, BH* equal to the square on *HA*.

Q.E.F.

Proposition 12

In obtuse-angled triangles the square on the side subtending the obtuse angle is greater than the squares on the sides containing the obtuse angle by twice the rectangle contained by one of the sides about the obtuse angle, namely that on which the perpendicular falls, and the straight line cut off outside by the perpendicular towards the obtuse angle.

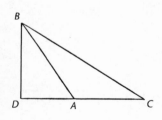

Let *ABC* be an obtuse-angled triangle having the angle *BAC* obtuse, and let *BD* be drawn from the point *B* perpendicular to *CA* produced;
I say that the square on *BC* is greater than the squares on *BA, AC* by twice the rectangle contained by *CA, AD*.

For, since the straight line *CD* has been cut at random at the point *A*, the square on *DC* is equal to the squares on *CA, AD* and twice the rectangle contained by *CA, AD*.

[II. 4]

Let the square on *DB* be added to each;
therefore the squares on *CD, DB* are equal to the squares on *CA, AD, DB* and twice the rectangle *CA, AD*.

But the square on *CB* is equal to the squares on *CD, DB*, for the angle at *D* is right;
[I. 47]
and the square on *AB* is equal to the squares on *AD, DB*;
[I. 47]
therefore the square on *CB* is equal to the squares on *CA, AB* and twice the rectangle contained by *CA, AD*;
so that the square on *CB* is greater than the squares on *CA, AB* by twice the rectangle contained by *CA, AD*.

Therefore etc.

Q.E.D.

Proposition 13

In acute-angled triangles the square on the side subtending the acute angle is less than the squares on the sides containing the acute angle by twice the rectangle contained by one of the sides about the acute angle, namely that on which the perpendicular falls, and the straight line cut off within by the perpendicular towards the acute angle.

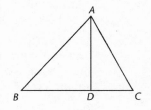

Let *ABC* be an acute-angled triangle having the angle at *B* acute, and let *AD* be drawn from the point *A* perpendicular to *BC*;
I say that the square on *AC* is less than the squares on *CB, BA* by twice the rectangle contained by *CB, BD.*

For, since the straight line *CB* has been cut at random at *D*, the squares on *CB, BD* are equal to twice the rectangle contained by *CB, BD* and the square on *DC.* [II. 7]

Let the square on *DA* be added to each;
therefore the squares on *CB, BD, DA* are equal to twice the rectangle contained by *CB, BD* and the squares on *AD, DC.*

But the square on *AB* is equal to the squares on *BD, DA*, for the angle at *D* is right; [I. 47]
and the square on *AC* is equal to the squares on *AD, DC*;
therefore the squares on *CB, BA* are equal to the square on *AC* and twice the rectangle *CB, BD*,
so that the square on *AC* alone is less than the squares on *CB, BA* by twice the rectangle contained by *CB, BD.*

Therefore etc.

Q.E.D.

Proposition 14

To construct a square equal to a given rectilineal figure.

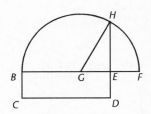

Let *A* be the given rectilineal figure;
thus it is required to construct a square equal to the rectilineal figure *A*.

For let there be constructed the rectangular parallelogram *BD* equal to the rectilineal figure *A*. [I. 45]

Then, if *BE* is equal to *ED*, that which was enjoined will have been done; for a square *BD* has been constructed equal to the rectilineal figure *A*.

But, if not, one of the straight lines *BE, ED* is greater.

Let *BE* be greater, and let it be produced to *F*;
let *EF* be made equal to *ED*, and let *BF* be bisected at *G*.

With centre G and distance one of the straight lines GB, GF let the semicircle BHF be described; let DE be produced to H, and let GH be joined.

Then, since the straight line BF has been cut into equal segments at G, and into unequal segments at E,
the rectangle contained by BE, EF together with the square on EG is equal to the square on GF. [II. 5]

But GF is equal to GH;
therefore the rectangle BE, EF together with the square on GE is equal to the square on GH.

But the squares on HE, EG are equal to the square on GH; [I. 47]
therefore the rectangle BE, EF together with the square on GE is equal to the squares on HE, EG.

Let the square on GE be subtracted from each;
therefore the rectangle contained by BE, EF which remains is equal to the square on EH.

But the rectangle BE, EF is BD, for EF is equal to ED;
therefore the parallelogram BD is equal to the square on HE.

And BD is equal to the rectilineal figure A.

Therefore the rectilineal figure A is also equal to the square which can be described on EH.

Therefore a square, namely that which can be described on EH, has been constructed equal to the given rectilineal figure A.

Q.E.F.

Book III

Definitions

1. *Equal circles* are those the diameters of which are equal, or the radii of which are equal.

2. A straight line is said to *touch a circle* which, meeting the circle and being produced, does not cut the circle.

3. *Circles* are said to *touch one another* which, meeting one another, do not cut one another.

4. In a circle straight lines are said *to be equally distant from the centre* when the perpendiculars drawn to them from the centre are equal.

5. And that straight line is said to be *at a greater distance* on which the greater perpendicular falls.

6. A *segment of a circle* is the figure contained by a straight line and a circumference of a circle.

7. An *angle of a segment* is that contained by a straight line and a circumference of a circle.

8. An *angle in a segment* is the angle which, when a point is taken on the circumference of the segment and straight lines are joined from it to the extremities of the straight line which is the *base of the segment*, is contained by the straight lines so joined.

9. And, when the straight lines containing the angle cut off a circumference, the angle is said to *stand upon* that circumference.

10. A *sector of a circle* is the figure which, when an angle is constructed at the centre of the circle, is contained by the straight lines containing the angle and the circumference cut off by them.

11. *Similar segments of circles* are those which admit equal angles, or in which the angles are equal to one another.

Proposition 1

To find the centre of a given circle.

Let ABC be the given circle;
 thus it is required to find the centre of the circle ABC.

Let a straight line AB be drawn through it at random, and let it be bisected at the point D;
from D let DC be drawn at right angles to AB and let it be drawn through to E; let CE be bisected at F;
I say that F is the centre of the circle ABC.

For suppose it is not, but, if possible, let G be the centre,

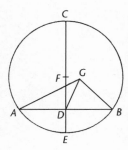

and let *GA*, *GD*, *GB* be joined.

Then, since *AD* is equal to *DB*, and *DG* is common,
the two sides *AD*, *DG* are equal to the two sides *BD*, *DG* respectively;
and the base *GA* is equal to the base *GB*, for they are radii;
therefore the angle *ADG* is equal to the angle *GDB*. [I. 8]

But, when a straight line set up on a straight line makes the adjacent angles equal to one another, each of the equal angles is right; [I. Def. 10]
therefore the angle *GDB* is right.

But the angle *FDB* is also right;
therefore the angle *FDB* is equal to the angle *GDB*,
the greater to the less: which is impossible.

Therefore *G* is not the centre of the circle *ABC*.

Similarly we can prove that neither is any other point except *F*.

Therefore the point *F* is the centre of the circle *ABC*.

Q.E.F.

PORISM. From this it is manifest that, if in a circle a straight line cut a straight line into two equal parts and at right angles, the centre of the circle is on the cutting straight line.

Proposition 2

If on the circumference of a circle two points be taken at random, the straight line joining the points will fall within the circle.

Let *ABC* be a circle, and let two points *A*, *B* be taken at random on its circumference;
I say that the straight line joined from *A* to *B* will fall within the circle.

For suppose it does not, but, if possible, let it fall outside, as *AEB*;
let the centre of the circle *ABC* be taken [III. 1], and let it be *D*;
let *DA*, *DB* be joined, and let *DFE* be drawn through.

Then, since *DA* is equal to *DB*,
the angle *DAE* is also equal to the angle *DBE*. [I. 5]

And, since one side *AEB* of the triangle *DAE* is produced,
the angle *DEB* is greater than the angle *DAE*. [I. 16]

But the angle *DAE* is equal to the angle *DBE*;
therefore the angle *DEB* is greater than the angle *DBE*.

And the greater angle is subtended by the greater side; [I. 19]
therefore *DB* is greater than *DE*.

But *DB* is equal to *DF*;

therefore *DF* is greater than *DE*,
the less than the greater: which is impossible.

Therefore the straight line joined from *A* to *B* will not fall outside the circle.

Similarly we can prove that neither will it fall on the circumference itself;
therefore it will fall within.

Therefore etc.

Q.E.D.

Proposition 3

If in a circle a straight line through the centre bisect a straight line not through the centre, it also cuts it at right angles; and if it cut it at right angles, it also bisects it.

Let *ABC* be a circle, and in it let a straight line *CD* through the centre bisect a straight line *AB* not through the centre at the point *F*;
I say that it also cuts it at right angles.

For let the centre of the circle *ABC* be taken, and let it be *E*; let *EA*, *EB* be joined.

Then, since *AF* is equal to *FB*, and *FE* is common,
two sides are equal to two sides;
and the base *EA* is equal to the base *EB*;
therefore the angle *AFE* is equal to the angle *BFE*. **[I. 8]**

But, when a straight line set up on a straight line makes the adjacent angles equal to one another, each of the equal angles is right; **[I. Def. 10]**
therefore each of the angles *AFE*, *BFE* is right.

Therefore *CD*, which is through the centre, and bisects *AB* which is not through the centre, also cuts it at right angles.

Again, let *CD* cut *AB* at right angles;
I say that it also bisects it, that is, that *AF* is equal to *FB*.

For, with the same construction,
since *EA* is equal to *EB*,
the angle *EAF* is also equal to the angle *EBF*. **[I. 5]**

But the right angle *AFE* is equal to the right angle *BFE*,
therefore *EAF*, *EBF* are two triangles having two angles equal to two angles and one side equal to one side, namely *EF*, which is common to them, and subtends one of the equal angles;
therefore they will also have the remaining sides equal to the remaining sides; **[I. 26]**
therefore *AF* is equal to *FB*.

Therefore etc.

Q.E.D.

Proposition 4

If in a circle two straight lines cut one another which are not through the centre, they do not bisect one another.

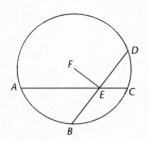

Let *ABCD* be a circle, and in it let the two straight lines *AC, BD,* which are not through the centre, cut one another at *E;* I say that they do not bisect one another.

For, if possible, let them bisect one another, so that *AE* is equal to *EC,* and *BE* to *ED;*
let the centre of the circle *ABCD* be taken [III. 1], and let it be *F;* let *FE* be joined.

Then, since a straight line *FE* through the centre bisects a straight line *AC* not through the centre,
<div align="center">it also cuts it at right angles; [III. 3]</div>
<div align="center">therefore the angle *FEA* is right.</div>

Again, since a straight line *FE* bisects a straight line *BD,*
<div align="center">it also cuts it at right angles; [III. 3]</div>
<div align="center">therefore the angle *FEB* is right.</div>

But the angle *FEA* was also proved right;
<div align="center">therefore the angle *FEA* is equal to the angle *FEB,*</div>
the less to the greater: which is impossible.

Therefore *AC, BD* do not bisect one another.

Therefore etc.

<div align="right">Q.E.D.</div>

Proposition 5

If two circles cut one another, they will not have the same centre.

For let the circles *ABC, CDG* cut one another at the points *B, C;*
I say that they will not have the same centre.

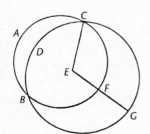

For, if possible, let it be *E;* let *EC* be joined, and let *EFG* be drawn through at random.

Then, since the point *E* is the centre of the circle *ABC,*
<div align="center">*EC* is equal to *EF.* [I. Def. 15]</div>

Again, since the point *E* is the centre of the circle *CDG,*
<div align="center">*EC* is equal to *EG.*</div>

But *EC* was proved equal to *EF* also;
<div align="center">therefore *EF* is also equal to *EG,*</div>
the less to the greater: which is impossible.

Therefore the point *E* is not the centre of the circles *ABC, CDG.*

Therefore etc.

<div align="right">Q.E.D.</div>

Proposition 6

If two circles touch one another, they will not have the same centre.

For let the two circles *ABC, CDE* touch one another at
the point *C*;
I say that they will not have the same centre.

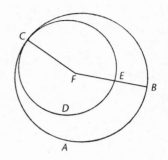

For, if possible, let it be *F*;
let *FC* be joined, and let *FEB* be drawn through at
random.

Then, since the point *F* is the centre of the circle *ABC*,
 FC is equal to *FB*.

Again, since the point *F* is the centre of the circle *CDE*,
 FC is equal to *FE*.

But *FC* was proved equal to *FB*;
 therefore *FE* is also equal to *FB*,
the less to the greater: which is impossible.

Therefore *F* is not the centre of the circles *ABC, CDE*.

Therefore etc.

Q.E.D.

Proposition 7

*If on the diameter of a circle a point be taken which is not the centre of the circle,
and from the point straight lines fall upon the circle, that will be greatest on
which the centre is, the remainder of the same diameter will be least, and of the
rest the nearer to the straight line through the centre is always greater than the
more remote, and only two equal straight lines will fall from the point on the
circle, one on each side of the least straight line.*

Let *ABCD* be a circle, and let *AD* be a diameter of
it; on *AD* let a point *F* be taken which is not the
centre of the circle, let *E* be the centre of the circle,
and from *F* let straight lines *FB, FC, FG* fall upon
the circle *ABCD*;
I say that *FA* is greatest, *FD* is least, and of the rest
FB is greater than *FC*, and *FC* than *FG*.

For let *BE, CE, GE* be joined.

Then, since in any triangle two sides are greater
than the remaining one, [I. 20]
 EB, EF are greater than *BF*.

But *AE* is equal to *BE*;
 therefore *AF* is greater than *BF*.

Again, since *BE* is equal to *CE*, and *FE* is common,
 the two sides *BE, EF* are equal to the two sides *CE, EF*.

But the angle *BEF* is also greater than the angle *CEF*;

therefore the base *BF* is greater than the base *CF*. [I. 24]

For the same reason

CF is also greater than *FG*.

Again, since *GF*, *FE* are greater than *EG*,
and *EG* is equal to *ED*,

GF, *FE* are greater than *ED*.

Let *EF* be subtracted from each;

therefore the remainder *GF* is greater than the remainder *FD*.

Therefore *FA* is greatest, *FD* is least, and *FB* is greater than *FC*, and *FC* than *FG*.

I say also that from the point *F* only two equal straight lines will fall on the circle *ABCD*, one on each side of the least *FD*.

For on the straight line *EF*, and at the point *E* on it, let the angle *FEH* be constructed equal to the angle *GEF* [I. 23], and let *FH* be joined.

Then, since *GE* is equal to *EH*,
and *EF* is common,

the two sides *GE*, *EF* are equal to the two sides *HE*, *EF*;
and the angle *GEF* is equal to the angle *HEF*;

therefore the base *FG* is equal to the base *FH*. [I. 4]

I say again that another straight line equal to *FG* will not fall on the circle from the point *F*.

For, if possible, let *FK* so fall.

Then, since *FK* is equal to *FG*, and *FH* to *FG*,

FK is also equal to *FH*,

the nearer to the straight line through the centre being thus equal to the more remote: which is impossible.

Therefore another straight line equal to *GF* will not fall from the point *F* upon the circle;

therefore only one straight line will so fall.

Therefore etc.

Q.E.D.

Proposition 8

If a point be taken outside a circle and from the point straight lines be drawn through to the circle, one of which is through the centre and the others are drawn at random, then, of the straight lines which fall on the concave circumference, that through the centre is greatest, while of the rest the nearer to that through the centre is always greater than the more remote, but, of the straight lines falling on the convex circumference, that between the point and the diameter is least, while of the rest the nearer to the least is always less than the more remote, and only two equal straight lines will fall on the circle from the point, one on each side of the least.

Let *ABC* be a circle, and let a point *D* be taken outside *ABC*;
let there be drawn through from it straight lines *DA, DE, DF, DC*, and let *DA* be through the centre;
I say that, of the straight lines falling on the concave circumference *AEFC*, the straight line *DA* through the centre is greatest,
while *DE* is greater than *DF* and *DF* than *DC*;
but, of the straight lines falling on the convex circumference *HLKG*, the straight line *DG* between the point and the diameter *AG* is least; and the nearer to the least *DG* is always less than the more remote, namely *DK* than *DL*, and *DL* than *DH*.

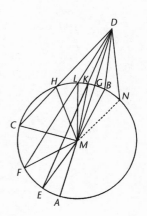

For let the centre of the circle *ABC* be taken, **[III. 1]**
and let it be *M*; let *ME, MF, MC, MK, ML, MH* be joined.

Then, since *AM* is equal to *EM*, let *MD* be added to each;
 therefore *AD* is equal to *EM, MD*.

But *EM, MD* are greater than *ED*;
 therefore *AD* is also greater than *ED*.

Again, since *ME* is equal to *MF*,
and *MD* is common,
 therefore *EM, MD* are equal to *FM, MD*;
and the angle *EMD* is greater than the angle *FMD*;
 therefore the base *ED* is greater than the base *FD*. **[I. 24]**

Similarly we can prove that *FD* is greater than *CD*; therefore *DA* is greatest, while *DE* is greater than *DF*, and *DF* than *DC*.

Next, since *MK, KD* are greater than *MD*, **[I. 20]**
and *MG* is equal to *MK*,
 therefore the remainder *KD* is greater than the remainder *GD*,
 so that *GD* is less than *KD*.

And, since on *MD*, one of the sides of the triangle *MLD*, two straight lines *MK, KD* were constructed meeting within the triangle,
 therefore *MK, KD* are less than *ML, LD*; **[I. 21]**
and *MK* is equal to *ML*;
 therefore the remainder *DK* is less than the remainder *DL*.

Similarly we can prove that *DL* is also less than *DH*;
 therefore *DG* is least, while *DK* is less than *DL*, and *DL* than *DH*.

I say also that only two equal straight lines will fall from the point *D* on the circle, one on each side of the least *DG*.

On the straight line *MD*, and at the point *M* on it, let the angle *DMB* be constructed equal to the angle *KMD*, and let *DB* be joined.

Then, since *MK* is equal to *MB*,
and *MD* is common,
 the two sides *KM, MD* are equal to the two sides *BM, MD* respectively;
and the angle *KMD* is equal to the angle *BMD*;
 therefore the base *DK* is equal to the base *DB*. **[I. 4]**

I say that no other straight line equal to the straight line *DK* will fall on the circle from the point *D*.

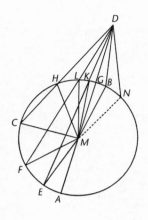

For, if possible, let a straight line so fall, and let it be *DN*.

Then, since *DK* is equal to *DN*,
while *DK* is equal to *DB*,
 DB is also equal to *DN*,
that is, the nearer to the least *DG* equal to the more remote: which was proved impossible.

Therefore no more than two equal straight lines will fall on the circle *ABC* from the point *D*, one on each side of *DG* the least.

Therefore etc.

Q.E.D.

Proposition 9

If a point be taken within a circle, and more than two equal straight lines fall from the point on the circle, the point taken is the centre of the circle.

Let *ABC* be a circle and *D* a point within it, and from *D* let more than two equal straight lines, namely *DA, DB, DC*, fall on the circle *ABC*;
I say that the point *D* is the centre of the circle *ABC*.

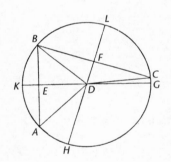

For let *AB, BC* be joined and bisected at the points *E, F*, and let *ED, FD* be joined and drawn through to the points *G, K, H, L*.

Then, since *AE* is equal to *EB*, and *ED* is common,
 the two sides *AE, ED* are equal to the two sides *BE, ED*;
and the base *DA* is equal to the base *DB*;
 therefore the angle *AED* is equal to the angle *BED*.
 [I. 8]

 Therefore each of the angles *AED, BED* is right;
 [I. Def. 10]
 therefore *GK* cuts *AB* into two equal parts and at right angles.

And since, if in a circle a straight line cut a straight line into two equal parts and at right angles, the centre of the circle is on the cutting straight line, [III. 1, Por.]
 the centre of the circle is on *GK*.

For the same reason
 the centre of the circle *ABC* is also on *HL*.

And the straight lines *GK, HL* have no other point common but the point *D*;
 therefore the point *D* is the centre of the circle *ABC*.

Therefore etc.

Q.E.D.

Proposition 10

A circle does not cut a circle at more points than two.

For, if possible, let the circle *ABC* cut the circle *DEF* at more points than two, namely *B, G, F, H*; let *BH, BG* be joined and bisected at the points *K, L*, and from *K, L* let *KC, LM* be drawn at right angles to *BH, BG* and carried through to the points *A, E*.

Then, since in the circle *ABC* a straight line *AC* cuts a straight line *BH* into two equal parts and at right angles,

the centre of the circle *ABC* is on *AC*. [III. 1, Por.]

Again, since in the same circle *ABC* a straight line *NO* cuts a straight line *BG* into two equal parts and at right angles,

the centre of the circle *ABC* is on *NO*.

But it was also proved to be on *AC*, and the straight lines *AC, NO* meet at no point except at *P*;

therefore the point *P* is the centre of the circle *ABC*.

Similarly we can prove that *P* is also the centre of the circle *DEF*;
therefore the two circles *ABC, DEF* which cut one another have the same centre *P*:
which is impossible. [III. 5]

Therefore etc.

Q.E.D.

Proposition 11

If two circles touch one another internally, and their centres be taken, the straight line joining their centres, if it be also produced, will fall on the point of contact of the circles.

For let the two circles *ABC, ADE* touch one another internally at the point *A*, and let the centre *F* of the circle *ABC*, and the centre *G* of *ADE*, be taken;
I say that the straight line joined from *G* to *F* and produced will fall on *A*.

For suppose it does not, but, if possible, let it fall as *FGH*, and let *AF, AG* be joined.

Then, since *AG, GF* are greater than *FA*, that is, than *FH*,
let *FG* be subtracted from each;
therefore the remainder *AG* is greater than the remainder *GH*.

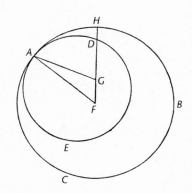

But *AG* is equal to *GD*;
therefore *GD* is also greater than *GH*,
the less than the greater: which is impossible.

Therefore the straight line joined from *F* to *G* will not fall outside;
therefore it will fall at *A* on the point of contact.

Therefore etc.

Q.E.D.

Proposition 12

If two circles touch one another externally, the straight line joining their centres will pass through the point of contact.

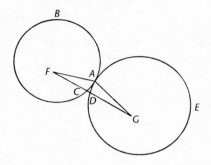

For let the two circles *ABC, ADE* touch one another externally at the point *A,* and let the centre *F* of *ABC,* and the centre *G* of *ADE,* be taken; I say that the straight line joined from *F* to *G* will pass through the point of contact at *A.*

For suppose it does not, but, if possible, let it pass as *FCDG,* and let *AF, AG* be joined.

Then, since the point *F* is the centre of the circle *ABC,*

FA is equal to *FC.*

Again, since the point *G* is the centre of the circle *ADE,*

GA is equal to *GD.*

But *FA* was also proved equal to *FC;*

therefore *FA, AG* are equal to *FC, GD,*

so that the whole *FG* is greater than *FA, AG;*

but it is also less [I. 20]: which is impossible.

Therefore the straight line joined from *F* to *G* will not fail to pass through the point of contact at *A;*

therefore it will pass through it.

Therefore etc.

Q.E.D.

Proposition 13

A circle does not touch a circle at more points than one, whether it touch it internally or externally.

For, if possible, let the circle *ABDC* touch the circle *EBFD,* first internally, at more points than one, namely *D, B.*

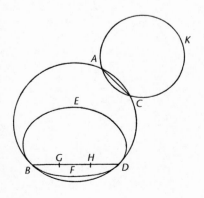

Let the centre *G* of the circle *ABDC,* and the centre *H* of *EBFD,* be taken.

Therefore the straight line joined from *G* to *H* will fall on *B, D.* [III. 11]

Let it so fall, as *BGHD.*

Then, since the point *G* is the centre of the circle *ABCD,*

BG is equal to *GD;*

therefore *BG* is greater than *HD;*

therefore *BH* is much greater than *HD.*

Again, since the point *H* is the centre of the circle *EBFD,*

<div align="center">BH is equal to HD;</div>

but it was also proved much greater than it: which is impossible.

Therefore a circle does not touch a circle internally at more points than one.

I say further that neither does it so touch it externally.

For, if possible, let the circle *ACK* touch the circle *ABDC* at more points than one, namely *A, C,*
and let *AC* be joined.

Then, since on the circumference of each of the circles *ABDC, ACK* two points *A, C* have been taken at random, the straight line joining the points will fall within each circle; [III. 2]
but it fell within the circle *ABCD* and outside *ACK* [III. Def. 3]: which is absurd.

Therefore a circle does not touch a circle externally at more points than one.

And it was proved that neither does it so touch it internally.

Therefore etc.

<div align="right">Q.E.D.</div>

Proposition 14

In a circle equal straight lines are equally distant from the centre, and those which are equally distant from the centre are equal to one another.

Let *ABDC* be a circle, and let *AB, CD* be equal straight lines in it;
I say that *AB, CD* are equally distant from the centre.

For let the centre of the circle *ABDC* be taken [III. 1], and let it be *E*; from *E* let *EF, EG* be drawn perpendicular to *AB, CD*; and let *AE, EC* be joined.

Then, since a straight line *EF* through the centre cuts a straight line *AB* not through the centre at right angles, it also bisects it. [III. 3]

Therefore *AF* is equal to *FB*;
<div align="center">therefore *AB* is double of *AF*.</div>

For the same reason
<div align="center">*CD* is also double of *CG*;</div>
and *AB* is equal to *CD*;
<div align="center">therefore *AF* is also equal to *CG*.</div>

And, since *AE* is equal to *EC*,
<div align="center">the square on *AE* is also equal to the square on *EC*.</div>

But the squares on *AF, EF* are equal to the square on *AE*, for the angle at *F* is right;
and the squares on *EG, GC* are equal to the square on *EC*, for the angle at *G* is right; [I. 47]
<div align="center">therefore the squares on *AF, FE* are equal to the squares on *CG, GE*,</div>
of which the square on *AF* is equal to the square on *CG*, for *AF* is equal to *CG*;
<div align="center">therefore the square on *FE* which remains is equal to the square on *EG*,</div>
<div align="center">therefore *EF* is equal to *EG*.</div>

But in a circle straight lines are said to be equally distant from the centre when the perpendiculars drawn to them from the centre are equal; [III. Def. 4]

therefore *AB*, *CD* are equally distant from the centre.

Next, let the straight lines *AB*, *CD* be equally distant from the centre; that is, let *EF* be equal to *EG*.

I say that *AB* is also equal to *CD*.

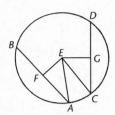

For, with the same construction, we can prove, similarly, that *AB* is double of *AF*, and *CD* of *CG*.

And, since *AE* is equal to *CE*,

the square on *AE* is equal to the square on *CE*.

But the squares on *EF*, *FA* are equal to the square on *AE*, and the squares on *EG*, *GC* equal to the square on *CE*. [I. 47]

Therefore the squares on *EF*, *FA* are equal to the squares on *EG*, *GC*, of which the square on *EF* is equal to the square on *EG*, for *EF* is equal to *EG*;

therefore the square on *AF* which remains is equal to the square on *CG*;

therefore *AF* is equal to *CG*.

And *AB* is double of *AF*, and *CD* double of *CG*;

therefore *AB* is equal to *CD*.

Therefore etc.

Q.E.D.

Proposition 15

Of straight lines in a circle the diameter is greatest, and of the rest the nearer to the centre is always greater than the more remote.

Let *ABCD* be a circle, let *AD* be its diameter and *E* the centre; and let *BC* be nearer to the diameter *AD*, and *FG* more remote; I say that *AD* is greatest and *BC* greater than *FG*.

For from the centre *E* let *EH*, *EK* be drawn perpendicular to *BC*, *FG*.

Then, since *BC* is nearer to the centre and *FG* more remote, *EK* is greater than *EH*. [III. Def. 5]

Let *EL* be made equal to *EH*, through *L* let *LM* be drawn at right angles to *EK* and carried through to *N*, and let *ME*, *EN*, *FE*, *EG* be joined.

Then, since *EH* is equal to *EL*,

BC is also equal to *MN*. [III. 14]

Again, since *AE* is equal to *EM*, and *ED* to *EN*,

AD is equal to *ME*, *EN*.

But *ME*, *EN* are greater than *MN*, [I. 20]
and *MN* is equal to *BC*;

therefore *AD* is greater than *BC*.

And, since the two sides *ME*, *EN* are equal to the two sides *FE*, *EG*,

and the angle *MEN* greater than the angle *FEG*,

therefore the base *MN* is greater than the base *FG*. [I. 24]

But *MN* was proved equal to *BC*.

Therefore the diameter *AD* is greatest and *BC* greater than *FG*.

Therefore etc.

<div align="right">Q.E.D.</div>

Proposition 16

The straight line drawn at right angles to the diameter of a circle from its extremity will fall outside the circle, and into the space between the straight line and the circumference another straight line cannot be interposed; further the angle of the semicircle is greater, and the remaining angle less, than any acute rectilineal angle.

Let *ABC* be a circle about *D* as centre and *AB* as diameter;
I say that the straight line drawn from *A* at right angles to *AB* from its extremity will fall outside the circle.

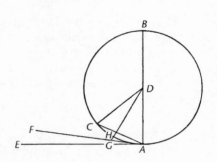

For suppose it does not, but, if possible, let it fall within as *CA*, and let *DC* be joined.

Since *DA* is equal to *DC*,
the angle *DAC* is also equal to the angle *ACD*. [I. 5]

But the angle *DAC* is right;
<div align="center">therefore the angle *ACD* is also right:</div>
thus, in the triangle *ACD*, the two angles *DAC*, *ACD* are equal to two right angles: which is impossible. [I. 17]

Therefore the straight line drawn from the point *A* at right angles to *BA* will not fall within the circle.

Similarly we can prove that neither will it fall on the circumference;
<div align="center">therefore it will fall outside.</div>

Let it fall as *AE*;
I say next that into the space between the straight line *AE* and the circumference *CHA* another straight line cannot be interposed.

For, if possible, let another straight line be so interposed, as *FA*, and let *DG* be drawn from the point *D* perpendicular to *FA*.

Then, since the angle *AGD* is right,
and the angle *DAG* is less than a right angle,
<div align="center">*AD* is greater than *DG*. [I. 19]</div>

But *DA* is equal to *DH*;
<div align="center">therefore *DH* is greater than *DG*,</div>
the less than the greater: which is impossible.

Therefore another straight line cannot be interposed into the space between the straight line and the circumference.

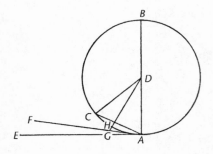

I say further that the angle of the semicircle contained by the straight line *BA* and the circumference *CHA* is greater than any acute rectilineal angle,

and the remaining angle contained by the circumference *CHA* and the straight line *AE* is less than any acute rectilineal angle.

For, if there is any rectilineal angle greater than the angle contained by the straight line *BA* and the circumference *CHA*, and any rectilineal angle less than the angle contained by the circumference *CHA* and the straight line *AE*, then into the space between the circumference and the straight line *AE* a straight line will be interposed such as will make an angle contained by straight lines which is greater than the angle contained by the straight line *BA* and the circumference *CHA*, and another angle contained by straight lines which is less than the angle contained by the circumference *CHA* and the straight line *AE*.

But such a straight line cannot be interposed;
therefore there will not be any acute angle contained by straight lines which is greater than the angle contained by the straight line *BA* and the circumference *CHA*, nor yet any acute angle contained by straight lines which is less than the angle contained by the circumference *CHA* and the straight line *AE*.

Q.E.D.

PORISM. From this it is manifest that the straight line drawn at right angles to the diameter of a circle from its extremity touches the circle.

Proposition 17

From a given point to draw a straight line touching a given circle.

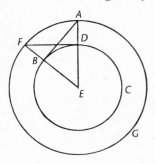

Let *A* be the given point, and *BCD* the given circle;
thus it is required to draw from the point *A* a straight line touching the circle *BCD*.

For let the centre *E* of the circle be taken; [III. 1]
let *AE* be joined, and with centre *E* and distance *EA* let the circle *AFG* be described;
from *D* let *DF* be drawn at right angles to *EA*,
and let *EF*, *AB* be joined;
I say that *AB* has been drawn from the point *A* touching the circle *BCD*.

For, since *E* is the centre of the circles *BCD*, *AFG*,
EA is equal to *EF*, and *ED* to *EB*;
therefore the two sides *AE*, *EB* are equal to the two sides *FE*, *ED*;
and they contain a common angle, the angle at *E*;
therefore the base *DF* is equal to the base *AB*,
and the triangle *DEF* is equal to the triangle *BEA*,
and the remaining angles to the remaining angles; [I. 4]

therefore the angle *EDF* is equal to the angle *EBA*.

But the angle *EDF* is right;
therefore the angle *EBA* is also right.

Now *EB* is a radius;
and the straight line drawn at right angles to the diameter of a circle, from its extremity, touches the circle; [III. 16, Por.]
therefore *AB* touches the circle *BCD*.

Therefore from the given point *A* the straight line *AB* has been drawn touching the circle *BCD*.

Q.E.F.

Proposition 18

If a straight line touch a circle, and a straight line be joined from the centre to the point of contact, the straight line so joined will be perpendicular to the tangent.

For let a straight line *DE* touch the circle *ABC* at the point *C*, let the centre *F* of the circle *ABC* be taken, and let *FC* be joined from *F* to *C*;

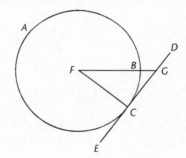

I say that *FC* is perpendicular to *DE*.

For, if not, let *FG* be drawn from *F* perpendicular to *DE*.

Then, since the angle *FGC* is right,
the angle *FCG* is acute; [I. 17]
and the greater angle is subtended by the greater side; [I. 19]
therefore *FC* is greater than *FG*.

But *FC* is equal to *FB*;
therefore *FB* is also greater than *FG*,
the less than the greater: which is impossible.

Therefore *FG* is not perpendicular to *DE*.

Similarly we can prove that neither is any other straight line except *FC*;
therefore *FC* is perpendicular to *DE*.

Therefore etc.

Q.E.D.

Proposition 19

If a straight line touch a circle, and from the point of contact a straight line be drawn at right angles to the tangent, the centre of the circle will be on the straight line so drawn.

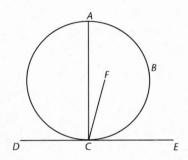

For let a straight line *DE* touch the circle *ABC* at the point *C*, and from *C* let *CA* be drawn at right angles to *DE*; I say that the centre of the circle is on *AC*.

For suppose it is not, but, if possible, let *F* be the centre, and let *CF* be joined.

Since a straight line *DE* touches the circle *ABC*, and *FC* has been joined from the centre to the point of contact,

FC is perpendicular to *DE*; [III. 18]
therefore the angle *FCE* is right.

But the angle *ACE* is also right;
therefore the angle *FCE* is equal to the angle *ACE*,
the less to the greater: which is impossible.

Therefore *F* is not the centre of the circle *ABC*.

Similarly we can prove that neither is any other point except a point on *AC*.

Therefore etc.

Q.E.D.

Proposition 20

In a circle the angle at the centre is double of the angle at the circumference, when the angles have the same circumference as base.

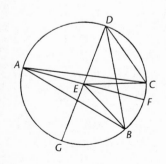

Let *ABC* be a circle, let the angle *BEC* be an angle at its centre, and the angle *BAC* an angle at the circumference, and let them have the same circumference *BC* as base; I say that the angle *BEC* is double of the angle *BAC*.

For let *AE* be joined and drawn to *F*.

Then, since *EA* is equal to *EB*,
the angle *EAB* is also equal to the angle *EBA*; [I. 5]
therefore the angles *EAB*, *EBA* are double of the angle *EAB*.

But the angle *BEF* is equal to the angles *EAB*, *EBA*; [I. 32]
therefore the angle *BEF* is also double of the angle *EAB*.

For the same reason
the angle *FEC* is also double of the angle *EAC*.

Therefore the whole angle *BEC* is double of the whole angle *BAC*.

Again let another straight line be inflected, and let there be another angle *BDC*; let *DE* be joined and produced to *G*.

Similarly then we can prove that the angle *GEC* is double of the angle *EDC*,
of which the angle *GEB* is double of the angle *EDB*;
therefore the angle *BEC* which remains is double of the angle *BDC*.

Therefore etc.

Q.E.D.

Proposition 21

In a circle the angles in the same segment are equal to one another.

Let *ABCD* be a circle, and let the angles *BAD*, *BED* be
angles in the same segment *BAED*;
I say that the angles *BAD*, *BED* are equal to one another.

For let the centre of the circle *ABCD* be taken, and let it be *F*;
let *BF*, *FD* be joined.

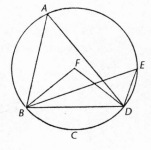

Now, since the angle *BFD* is at the centre,
and the angle *BAD* at the circumference,
and they have the same circumference *BCD* as base,
 therefore the angle *BFD* is double of the angle *BAD*.

[III. 20]

For the same reason
 the angle *BFD* is also double of the angle *BED*;
therefore the angle *BAD* is equal to the angle *BED*.

Therefore etc.

Q.E.D.

Proposition 22

The opposite angles of quadrilaterals in circles are equal to two right angles.

Let *ABCD* be a circle, and let *ABCD* be a quadrilateral in it;
I say that the opposite angles are equal to two right angles.

Let *AC*, *BD* be joined.

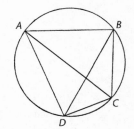

Then, since in any triangle the three angles are equal to two
right angles, [I. 32]
the three angles *CAB*, *ABC*, *BCA* of the triangle *ABC* are
equal to two right angles.

But the angle *CAB* is equal to the angle *BDC*, for they are in
the same segment *BADC*; [III. 21]
and the angle *ACB* is equal to the angle *ADB*, for they are
in the same segment *ADCB*;
 therefore the whole angle *ADC* is equal to the angles *BAC*, *ACB*.

Let the angle *ABC* be added to each;
 therefore the angles *ABC*, *BAC*, *ACB* are equal to the angles *ABC*, *ADC*.

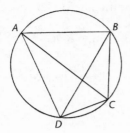

But the angles *ABC, BAC, ACB* are equal to two right angles;
therefore the angles *ABC, ADC* are also equal to two right angles.

Similarly we can prove that the angles *BAD, DCB* are also equal to two right angles.

Therefore etc.

Q.E.D.

Proposition 23

On the same straight line there cannot be constructed two similar and unequal segments of circles on the same side.

For, if possible, on the same straight line *AB* let two similar and unequal segments of circles *ACB*, *ADB* be constructed on the same side;
let *ACD* be drawn through, and let *CB, DB* be joined.

Then, since the segment *ACB* is similar to the segment *ADB*,
and similar segments of circles are those which admit equal angles, **[III. Def. 11]**
 the angle *ACB* is equal to the angle *ADB*,
the exterior to the interior: which is impossible. **[I. 16]**

Therefore etc.

Q.E.D.

Proposition 24

Similar segments of circles on equal straight lines are equal to one another.

For let *AEB, CFD* be similar segments of circles on equal straight lines *AB, CD*;
I say that the segment *AEB* is equal to the segment *CFD*.

For, if the segment *AEB* be applied to *CFD*, and if the point *A* be placed on *C* and the straight line *AB* on *CD*,
the point *B* will also coincide with the point *D*, because *AB* is equal to *CD*;
and, *AB* coinciding with *CD*,
 the segment *AEB* will also coincide with *CFD*.

For, if the straight line *AB* coincide with *CD* but the segment *AEB* do not coincide with *CFD*,
 it will either fall within it, or outside it;

or it will fall awry, as *CGD*, and a circle cuts a circle at more points than two: which is impossible. **[III. 10]**

Therefore, if the straight line *AB* be applied to *CD*, the segment *AEB* will not fail to coincide with *CFD* also;

therefore it will coincide with it and will be equal to it.

Therefore etc.

Q.E.D.

Proposition 25

Given a segment of a circle, to describe the complete circle of which it is a segment.

Let *ABC* be the given segment of a circle;
thus it is required to describe the complete circle belonging to the segment *ABC*, that is, of which it is a segment.

For let *AC* be bisected at *D*, let *DB* be drawn from the point *D* at right angles to *AC*, and let *AB* be joined;
the angle *ABD* is then greater than, equal to, or less than the angle *BAD*.

First let it be greater;
and on the straight line *BA*, and at the point *A* on it, let the angle *BAE* be constructed equal to the angle *ABD*; let *DB* be drawn through to *E*, and let *EC* be joined.

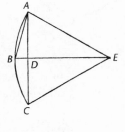

Then, since the angle *ABE* is equal to the angle *BAE*,
the straight line *EB* is also equal to *EA*. **[I. 6]**

And, since *AD* is equal to *DC*, and *DE* is common,
the two sides *AD*, *DE* are equal to the two sides *CD*, *DE* respectively;
and the angle *ADE* is equal to the angle *CDE*, for each is right;
therefore the base *AE* is equal to the base *CE*.

But *AE* was proved equal to *BE*;
therefore *BE* is also equal to *CE*;
therefore the three straight lines *AE*, *EB*, *EC* are equal to one another.

Therefore the circle drawn with centre *E* and distance one of the straight lines *AE*, *EB*, *EC* will also pass through the remaining points and will have been completed. **[III. 9]**

Therefore, given a segment of a circle, the complete circle has been described.

And it is manifest that the segment *ABC* is less than a semicircle, because the centre *E* happens to be outside it.

Similarly, even if the angle *ABD* be equal to the angle *BAD*,
AD being equal to each of the two *BD*, *DC*,
the three straight lines *DA*, *DB*, *DC* will be equal to one another,
D will be the centre of the completed circle,
and *ABC* will clearly be a semicircle.

But, if the angle *ABD* be less than the angle *BAD,* and if we construct, on the straight line *BA* and at the point *A* on it, an angle equal to the angle *ABD,* the centre will fall on *DB* within the segment *ABC,* and the segment *ABC* will clearly be greater than a semicircle.

Therefore, given a segment of a circle, the complete circle has been described.

Q.E.F.

Proposition 26

In equal circles equal angles stand on equal circumferences, whether they stand at the centres or at the circumferences.

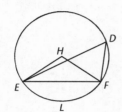

Let *ABC, DEF* be equal circles, and in them let there be equal angles, namely at the centres the angles *BGC, EHF,* and at the circumferences the angles *BAC, EDF;*
I say that the circumference *BKC* is equal to the circumference *ELF.*

For let *BC, EF* be joined.

Now, since the circles *ABC, DEF* are equal,

the radii are equal.

Thus the two straight lines *BG, GC* are equal to the two straight lines *EH, HF;* and the angle at *G* is equal to the angle at *H;*

therefore the base *BC* is equal to the base *EF.* [I. 4]

And, since the angle at *A* is equal to the angle at *D,*

the segment *BAC* is similar to the segment *EDF;* [III. Def. 11]
and they are upon equal straight lines.

But similar segments of circles on equal straight lines are equal to one another;
[III. 24]

therefore the segment *BAC* is equal to *EDF.*

But the whole circle *ABC* is also equal to the whole circle *DEF;*
therefore the circumference *BKC* which remains is equal to the circumference *ELF.*

Therefore etc.

Q.E.D.

Proposition 27

In equal circles angles standing on equal circumferences are equal to one another, whether they stand at the centres or at the circumferences.

For in equal circles *ABC*, *DEF*, on equal circumferences *BC*, *EF*, let the angles *BGC*, *EHF* stand at the centres *G*, *H*, and the angles *BAC*, *EDF* at the circumferences;

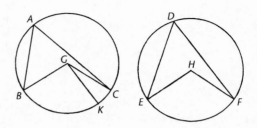

I say that the angle *BGC* is equal to the angle *EHF*,

and the angle *BAC* is equal to the angle *EDF*.

For, if the angle *BGC* is unequal to the angle *EHF*,

one of them is greater.

Let the angle *BGC* be greater: and on the straight line *BG*, and at the point *G* on it, let the angle *BGK* be constructed equal to the angle *EHF*. [I. 23]

Now equal angles stand on equal circumferences, when they are at the centres; [III. 26]

therefore the circumference *BK* is equal to the circumference *EF*.

But *EF* is equal to *BC*;

therefore *BK* is also equal to *BC*, the less to the greater: which is impossible.

Therefore the angle *BGC* is not unequal to the angle *EHF*;

therefore it is equal to it.

And the angle at *A* is half of the angle *BGC*, and the angle at *D* half of the angle *EHF*; [III. 20]

therefore the angle at *A* is also equal to the angle at *D*.

Therefore etc.

 Q.E.D.

Proposition 28

In equal circles equal straight lines cut off equal circumferences, the greater equal to the greater and the less to the less.

Let *ABC*, *DEF* be equal circles, and in the circles let *AB*, *DE* be equal straight lines cutting off *ACB*, *DFE* as greater circumferences and *AGB*, *DHE* as lesser;

I say that the greater circumference *ACB* is equal to the greater circumference *DFE*, and the less circumference *AGB* to *DHE*.

For let the centres *K*, *L* of the circles be taken, and let *AK*, *KB*, *DL*, *LE* be joined.

Now, since the circles are equal,

the radii are also equal;

therefore the two sides *AK, KB* are equal to the two sides *DL, LE*;
and the base *AB* is equal to the base *DE*;

therefore the angle *AKB* is equal to the angle *DLE*. [I. 8]

But equal angles stand on equal circumferences, when they are at the centres; [III. 26]

therefore the circumference *AGB* is equal to *DHE*.

And the whole circle *ABC* is equal to the whole circle *DEF*;
therefore the circumference *ACB* which remains is also equal to the circumference *DFE* which remains.

Therefore etc.

Q.E.D.

Proposition 29

In equal circles equal circumferences are subtended by equal straight lines.

Let *ABC, DEF* be equal circles, and in them let equal circumferences *BGC, EHF* be cut off; and let the straight lines *BC, EF* be joined;
I say that *BC* is equal to *EF*.

For let the centres of the circles be taken, and let them be *K, L*; let *BK, KC, EL, LF* be joined.

Now, since the circumference *BGC* is equal to the circumference *EHF*,
the angle *BKC* is also equal to the angle *ELF*. [III. 27]

And, since the circles *ABC, DEF* are equal,
the radii are also equal;
therefore the two sides *BK, KC* are equal to the two sides *EL, LF*;
and they contain equal angles;

therefore the base *BC* is equal to the base *EF*. [I. 4]

Therefore etc.

Q.E.D.

Proposition 30

To bisect a given circumference.

Let *ADB* be the given circumference;
thus it is required to bisect the circumference *ADB*.

Let *AB* be joined and bisected at *C*; from the point *C* let
CD be drawn at right angles to the straight line *AB*, and
let *AD*, *DB* be joined.

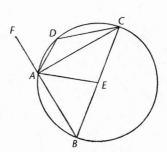

Then, since *AC* is equal to *CB*, and *CD* is common,
the two sides *AC*, *CD* are equal to the two sides *BC*, *CD*;
and the angle *ACD* is equal to the angle *BCD*, for each is right;
therefore the base *AD* is equal to the base *DB*. [I. 4]

But equal straight lines cut off equal circumferences, the greater equal to the
greater, and the less to the less; [III. 28]
and each of the circumferences *AD*, *DB* is less than a semicircle;
therefore the circumference *AD* is equal to the circumference *DB*.

Therefore the given circumference has been bisected at the point *D*.

Q.E.F.

Proposition 31

*In a circle the angle in the semicircle is right, that in a greater segment less than
a right angle, and that in a less segment greater than a right angle; and further
the angle of the greater segment is greater than a right angle, and the angle of
the less segment less than a right angle.*

Let *ABCD* be a circle, let *BC* be its diameter, and *E*
its centre, and let *BA*, *AC*, *AD*, *DC* be joined;
I say that the angle *BAC* in the semicircle *BAC* is
right,
the angle *ABC* in the segment *ABC* greater than the
semicircle is less than a right angle,
and the angle *ADC* in the segment *ADC* less than
the semicircle is greater than a right angle.

Let *AE* be joined, and let *BA* be carried through to *F*.

Then, since *BE* is equal to *EA*,
 the angle *ABE* is also equal to the angle *BAE*. [I. 5]

Again, since *CE* is equal to *EA*,
 the angle *ACE* is also equal to the angle *CAE*. [I. 5]

 Therefore the whole angle *BAC* is equal to the two angles *ABC*, *ACB*.

But the angle *FAC* exterior to the triangle *ABC* is also equal to the two angles
ABC, *ACB*; [I. 32]
 therefore the angle *BAC* is also equal to the angle *FAC*;
 therefore each is right ; [I. Def. 10]
 therefore the angle *BAC* in the semicircle *BAC* is right.

Next, since in the triangle *ABC* the two angles *ABC*, *BAC* are less than two
right angles, [I. 17]
and the angle *BAC* is a right angle,
 the angle *ABC* is less than a right angle;

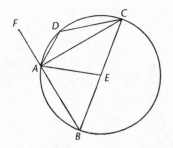

and it is the angle in the segment *ABC* greater than the semicircle.

Next, since *ABCD* is a quadrilateral in a circle, and the opposite angles of quadrilaterals in circles are equal to two right angles, **[III. 22]** while the angle *ABC* is less than a right angle, therefore the angle *ADC* which remains is greater than a right angle; and it is the angle in the segment *ADC* less than the semicircle.

I say further that the angle of the greater segment, namely that contained by the circumference *ABC* and the straight line *AC*, is greater than a right angle; and the angle of the less segment, namely that contained by the circumference *ADC* and the straight line *AC*, is less than a right angle.

This is at once manifest.

For, since the angle contained by the straight lines *BA*, *AC* is right, the angle contained by the circumference *ABC* and the straight line *AC* is greater than a right angle.

Again, since the angle contained by the straight lines *AC*, *AF* is right, the angle contained by the straight line *CA* and the circumference *ADC* is less than a right angle.

Therefore etc.

Q.E.D.

Proposition 32

If a straight line touch a circle, and from the point of contact there be drawn across, in the circle, a straight line cutting the circle, the angles which it makes with the tangent will be equal to the angles in the alternate segments of the circle.

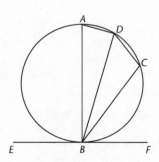

For let a straight line *EF* touch the circle *ABCD* at the point *B*, and from the point *B* let there be drawn across, in the circle *ABCD*, a straight line *BD* cutting it;

I say that the angles which *BD* makes with the tangent *EF* will be equal to the angles in the alternate segments of the circle, that is, that the angle *FBD* is equal to the angle constructed in the segment *BAD*, and the angle *EBD* is equal to the angle constructed in the segment *DCB*.

For let *BA* be drawn from *B* at right angles to *EF*, let a point *C* be taken at random on the circumference *BD*, and let *AD*, *DC*, *CB* be joined.

Then, since a straight line *EF* touches the circle *ABCD* at *B*,

and *BA* has been drawn from the point of contact at right angles to the tangent,
<div align="center">the centre of the circle *ABCD* is on *BA*.</div> [III. 19]

Therefore *BA* is a diameter of the circle *ABCD*;
<div align="center">therefore the angle *ADB*, being an angle in a semicircle, is right.</div> [III. 31]

Therefore the remaining angles *BAD*, *ABD* are equal to one right angle. [I. 32]

But the angle *ABF* is also right;
<div align="center">therefore the angle *ABF* is equal to the angles *BAD*, *ABD*.</div>

Let the angle *ABD* be subtracted from each;
therefore the angle *DBF* which remains is equal to the angle *BAD* in the alternate segment of the circle.

Next, since *ABCD* is a quadrilateral in a circle,
<div align="center">its opposite angles are equal to two right angles.</div> [III. 22]

But the angles *DBF*, *DBE* are also equal to two right angles;
<div align="center">therefore the angles *DBF*, *DBE* are equal to the angles *BAD*, *BCD*,</div>
of which the angle *BAD* was proved equal to the angle *DBF*;
therefore the angle *DBE* which remains is equal to the angle *DCB* in the alternate segment *DCB* of the circle.

Therefore etc.

<div align="right">Q.E.D.</div>

Proposition 33

On a given straight line to describe a segment of a circle admitting an angle equal to a given rectilineal angle.

Let *AB* be the given straight line, and the angle at *C* the given rectilineal angle;
thus it is required to describe on the given straight line *AB* a segment of a circle admitting an angle equal to the angle at *C*.

The angle at *C* is then acute, or right, or obtuse.

First, let it be acute, and, as in the first figure, on the straight line *AB*, and at the point *A*, let the angle *BAD* be constructed equal to the angle at *C*;
therefore the angle *BAD* is also acute.

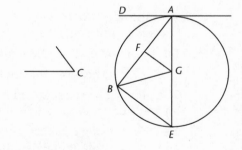

Let *AE* be drawn at right angles to *DA*, let *AB* be bisected at *F*, let *FG* be drawn from the point *F* at right angles to *AB*, and let *GB* be joined.

Then, since *AF* is equal to *FB*,
and *FG* is common,
<div align="center">the two sides *AF*, *FG* are equal to the two sides *BF*, *FG*;</div>
and the angle *AFG* is equal to the angle *BFG*;
<div align="center">therefore the base *AG* is equal to the base *BG*.</div> [I. 4]

Therefore the circle described with centre *G* and distance *GA* will pass through *B* also.

<div align="right">75</div>

Let it be drawn, and let it be *ABE*;
let *EB* be joined.

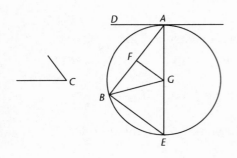

Now, since *AD* is drawn from *A*, the extremity of the diameter *AE*, at right angles to *AE*,

therefore *AD* touches the circle *ABE*.

[III. 16, Por.]

Since then a straight line *AD* touches the circle *ABE*,
and from the point of contact at *A* a straight line *AB* is drawn across in the circle *ABE*,
the angle *DAB* is equal to the angle *AEB* in the alternate segment of the circle. [III. 32]

But the angle *DAB* is equal to the angle at *C*;
therefore the angle at *C* is also equal to the angle *AEB*.

Therefore on the given straight line *AB* the segment *AEB* of a circle has been described admitting the angle *AEB* equal to the given angle, the angle at *C*.

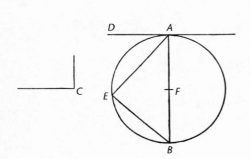

Next let the angle at *C* be right;
and let it be again required to describe on *AB* a segment of a circle admitting an angle equal to the right angle at *C*.

Let the angle *BAD* be constructed equal to the right angle at *C*, as is the case in the second figure;
let *AB* be bisected at *F*, and with centre *F* and distance either *FA* or *FB* let the circle *AEB* be described.

Therefore the straight line *AD* touches the circle *ABE*, because the angle at *A* is right.

[III. 16, Por.]

And the angle *BAD* is equal to the angle in the segment *AEB*, for the latter too is itself a right angle, being an angle in a semicircle. [III. 31]

But the angle *BAD* is also equal to the angle at *C*.

Therefore the angle *AEB* is also equal to the angle at *C*.

Therefore again the segment *AEB* of a circle has been described on *AB* admitting an angle equal to the angle at *C*.

Next, let the angle at *C* be obtuse;
and on the straight line *AB*, and at the point *A*, let the angle *BAD* be constructed equal to it, as is the case in the third figure;
let *AE* be drawn at right angles to *AD*, let *AB* be again bisected at *F*, let *FG* be drawn at right angles to *AB*, and let *GB* be joined.

Then, since *AF* is again equal to *FB*,
and *FG* is common,
the two sides *AF*, *FG* are equal to the two sides *BF*, *FG*;
and the angle *AFG* is equal to the angle *BFG*;
therefore the base *AG* is equal to the base *BG*. [I. 4]

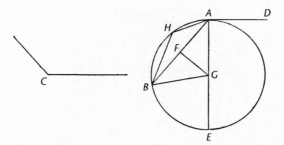

Therefore the circle described with centre *G* and distance *GA* will pass through *B* also; let it so pass, as *AEB*.

Now, since *AD* is drawn at right angles to the diameter *AE* from its extremity,
<div align="center">*AD* touches the circle *AEB*.</div> [III. 16, Por.]

And *AB* has been drawn across from the point of contact at *A*;
therefore the angle *BAD* is equal to the angle constructed in the alternate segment *AHB* of the circle. [III. 32]

But the angle *BAD* is equal to the angle at *C*.

Therefore the angle in the segment *AHB* is also equal to the angle at *C*.

Therefore on the given straight line *AB* the segment *AHB* of a circle has been described admitting an angle equal to the angle at *C*.

<div align="right">Q.E.F.</div>

Proposition 34

From a given circle to cut off a segment admitting an angle equal to a given rectilineal angle.

Let *ABC* be the given circle, and the angle at *D* the given rectilineal angle;
thus it is required to cut off from the circle *ABC* a segment admitting an angle equal to the given rectilineal angle, the angle at *D*.

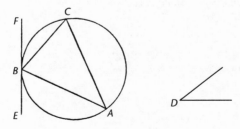

Let *EF* be drawn touching *ABC* at the point *B*, and on the straight line *FB,* and at the point *B* on it, let the angle *FBC* be constructed equal to the angle at *D*. [I. 23]

Then, since a straight line *EF* touches the circle *ABC,*
and *BC* has been drawn across from the point of contact at *B,*
the angle *FBC* is equal to the angle constructed in the alternate segment *BAC*. [III. 32]

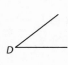

But the angle *FBC* is equal to the angle at *D*;
 therefore the angle in the segment *BAC* is equal to the angle at *D*.

Therefore from the given circle *ABC* the segment *BAC* has been cut off admitting an angle equal to the given rectilineal angle, the angle at *D*.

 Q.E.F.

Proposition 35

If in a circle two straight lines cut one another, the rectangle contained by the segments of the one is equal to the rectangle contained by the segments of the other.

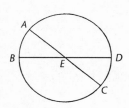

For in the circle *ABCD* let the two straight lines *AC, BD* cut one another at the point *E*;
I say that the rectangle contained by *AE, EC* is equal to the rectangle contained by *DE, EB*.

If now *AC, BD* are through the centre, so that *E* is the centre of the circle *ABCD*,
 it is manifest that, *AE, EC, DE, EB* being equal,
 the rectangle contained by *AE, EC* is also equal to the rectangle contained by *DE, EB*.

Next let *AC, DB* not be through the centre;
let the centre of *ABCD* be taken, and let it be *F*;
from *F* let *FG, FH* be drawn perpendicular to the straight lines *AC, DB*,
and let *FB, FC, FE* be joined.

Then, since a straight line *GF* through the centre cuts a straight line *AC* not through the centre at right angles,
 it also bisects it; **[III. 3]**
 therefore *AG* is equal to *GC*.

Since, then, the straight line *AC* has been cut into equal parts at *G* and into unequal parts at *E*,
the rectangle contained by *AE, EC* together with the square on *EG* is equal to the square on *GC*; **[II. 5]**
Let the square on *GF* be added;
therefore the rectangle *AE, EC* together with the squares on *GE, GF* is equal to the squares on *CG, GF*.

But the square on *FE* is equal to the squares on *EG, GF,* and the square on *FC* is equal to the squares on *CG, GF;* [I. 47]
therefore the rectangle *AE, EC* together with the square on *FE* is equal to the square on *FC.*

And *FC* is equal to *FB;*
therefore the rectangle *AE, EC* together with the square on *EF* is equal to the square on *FB.*

For the same reason, also,
the rectangle *DE, EB* together with the square on *FE* is equal to the square on *FB.*

But the rectangle *AE, EC* together with the square on *FE* was also proved equal to the square on *FB;*
therefore the rectangle *AE, EC* together with the square on *FE* is equal to the rectangle *DE, EB* together with the square on *FE.*

Let the square on *FE* be subtracted from each;
therefore the rectangle contained by *AE, EC* which remains is equal to the rectangle contained by *DE, EB.*

Therefore etc.

Q.E.D.

Proposition 36

If a point be taken outside a circle and from it there fall on the circle two straight lines, and if one of them cut the circle and the other touch it, the rectangle contained by the whole of the straight line which cuts the circle and the straight line intercepted on it outside between the point and the convex circumference will be equal to the square on the tangent.

For let a point *D* be taken outside the circle *ABC,* and from *D* let the two straight lines *DCA, DB* fall on the circle *ABC;* let *DCA* cut the circle *ABC* and let *BD* touch it;
I say that the rectangle contained by *AD, DC* is equal to the square on *DB.*

Then *DCA* is either through the centre or not through the centre.

First let it be through the centre, and let *F* be the centre of the circle *ABC;* let *FB* be joined;
therefore the angle *FBD* is right. [III. 18]

And, since AC has been bisected at F, and CD is added to it,
the rectangle AD, DC together with the square on FC is equal to the square on FD. [II. 6]

But FC is equal to FB;
therefore the rectangle AD, DC together with the square on FB is equal to the square on FD.

And the squares on FB, BD are equal to the square on FD; [I. 47]
therefore the rectangle AD, DC together with the square on FB is equal to the squares on FB, BD.

Let the square on FB be subtracted from each;
therefore the rectangle AD, DC which remains is equal to the square on the tangent DB.

Again, let DCA not be through the centre of the circle ABC;
let the centre E be taken, and from E let EF be drawn perpendicular to AC;
let EB, EC, ED be joined.

Then the angle EBD is right. [III. 18]
And, since a straight line EF through the centre cuts a straight line AC not through the centre at right angles,
it also bisects it ; [III. 3]
therefore AF is equal to FC.

Now, since the straight line AC has been bisected at the point F, and CD is added to it,
the rectangle contained by AD, DC together with the square on FC is equal to the square on FD. [II. 6]

Let the square on FE be added to each;
therefore the rectangle AD, DC together with the squares on CF, FE is equal to the squares on FD, FE.

But the square on EC is equal to the squares on CF, FE, for the angle EFC is right; [I. 47]
and the square on ED is equal to the squares on DF, FE;
therefore the rectangle AD, DC together with the square on EC is equal to the square on ED.

And EC is equal to EB;
therefore the rectangle AD, DC together with the square on EB is equal to the square on ED.

But the squares on EB, BD are equal to the square on ED, for the angle EBD is right; [I. 47]
therefore the rectangle AD, DC together with the square on EB is equal to the squares on EB, BD.

Let the square on EB be subtracted from each;
therefore the rectangle AD, DC which remains is equal to the square on DB.

Therefore etc.

Q.E.D.

Proposition 37

If a point be taken outside a circle and from the point there fall on the circle two straight lines, if one of them cut the circle, and the other fall on it, and if further the rectangle contained by the whole of the straight line which cuts the circle and the straight line intercepted on it outside between the point and the convex circumference be equal to the square on the straight line which falls on the circle, the straight line which falls on it will touch the circle.

For let a point D be taken outside the circle ABC; from D let the two straight lines DCA, DB fall on the circle ACB;

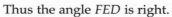

let DCA cut the circle and DB fall on it; and let the rectangle AD, DC be equal to the square on DB.

I say that DB touches the circle ABC.

For let DE be drawn touching ABC; let the centre of the circle ABC be taken, and let it be F; let FE, FB, FD be joined.

<div align="center">Thus the angle FED is right.</div> [III. 18]

Now, since DE touches the circle ABC, and DCA cuts it, the rectangle AD, DC is equal to the square on DE. [III. 36]

But the rectangle AD, DC was also equal to the square on DB;

<div align="center">therefore the square on DE is equal to the square on DB;</div>

<div align="center">therefore DE is equal to DB.</div>

And FE is equal to FB;

<div align="center">therefore the two sides DE, EF are equal to the two sides DB, BF;</div>

and FD is the common base of the triangles;

<div align="center">therefore the angle DEF is equal to the angle DBF.</div> [I. 8]

But the angle DEF is right;

<div align="center">therefore the angle DBF is also right.</div>

And FB produced is a diameter; and the straight line drawn at right angles to the diameter of a circle, from its extremity, touches the circle; [III. 16, Por.]

<div align="center">therefore DB touches the circle.</div>

Similarly this can be proved to be the case even if the centre be on AC.

Therefore etc.

<div align="right">Q.E.D.</div>

Book IV

Definitions

1. A rectilineal figure is said to be *inscribed in a rectilineal figure* when the respective angles of the inscribed figure lie on the respective sides of that in which it is inscribed.

2. Similarly a figure is said to be *circumscribed about a figure* when the respective sides of the circumscribed figure pass through the respective angles of that about which it is circumscribed.

3. A rectilineal figure is said to be *inscribed in a circle* when each angle of the inscribed figure lies on the circumference of the circle.

4. A rectilineal figure is said to be *circumscribed about a circle,* when each side of the circumscribed figure touches the circumference of the circle.

5. Similarly a circle is said to be *inscribed in a figure* when the circumference of the circle touches each side of the figure in which it is inscribed.

6. A circle is said to be *circumscribed about a figure* when the circumference of the circle passes through each angle of the figure about which it is circumscribed.

7. A straight line is said to be *fitted into a circle* when its extremities are on the circumference of the circle.

Proposition 1

Into a given circle to fit a straight line equal to a given straight line which is not greater than the diameter of the circle.

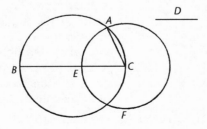

Let ABC be the given circle, and D the given straight line not greater than the diameter of the circle;

thus it is required to fit into the circle ABC a straight line equal to the straight line D.

Let a diameter BC of the circle ABC be drawn.

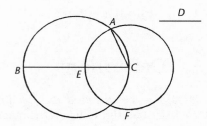

Then, if *BC* is equal to *D*, that which was enjoined will have been done; for *BC* has been fitted into the circle *ABC* equal to the straight line *D*.

But, if *BC* is greater than *D*, let *CE* be made equal to *D*,
and with centre *C* and distance *CE* let the circle *EAF* be described;
let *CA* be joined.

Then, since the point *C* is the centre of the circle *EAF*,
<div align="center">*CA* is equal to *CE*.</div>

But *CE* is equal to *D*;
<div align="center">therefore *D* is also equal to *CA*.</div>

Therefore into the given circle *ABC* there has been fitted *CA* equal to the given straight line *D*.

<div align="right">Q.E.F.</div>

Proposition 2

In a given circle to inscribe a triangle equiangular with a given triangle.

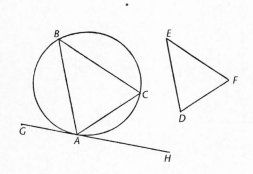

Let *ABC* be the given circle, and *DEF* the given triangle;
thus it is required to inscribe in the circle *ABC* a triangle equiangular with the triangle *DEF*.

Let *GH* be drawn touching the circle *ABC* at *A*; **[III. 16, Por.]**
on the straight line *AH*, and at the point *A* on it, let the angle *HAC* be constructed equal to the angle *DEF*,
and on the straight line *AG*, and at the point *A* on it, let the angle *GAB* be constructed equal to the angle *DFE*; **[I. 23]**
let *BC* be joined.

Then, since a straight line *AH* touches the circle *ABC*,
and from the point of contact at *A* the straight line *AC* is drawn across in the circle,
therefore the angle *HAC* is equal to the angle *ABC* in the alternate segment of the circle. **[III. 32]**

But the angle *HAC* is equal to the angle *DEF*;
<div align="center">therefore the angle *ABC* is also equal to the angle *DEF*.</div>

For the same reason

the angle *ACB* is also equal to the angle *DFE*;
therefore the remaining angle *BAC* is also equal to the remaining angle *EDF*. [I. 32]

Therefore in the given circle there has been inscribed a triangle equiangular with the given triangle.

Q.E.F.

Proposition 3

About a given circle to circumscribe a triangle equiangular with a given triangle.

Let *ABC* be the given circle, and *DEF* the given triangle;
thus it is required to circumscribe about the circle *ABC* a triangle equiangular with the triangle *DEF*.

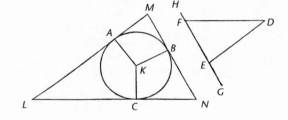

Let *EF* be produced in both directions to the points G, *H*,
let the centre *K* of the circle *ABC* be taken, [III. 1]
and let the straight line *KB* be drawn across at random;
on the straight line *KB*, and at the point *K* on it, let the angle *BKA* be constructed equal to the angle *DEG*,
and the angle *BKC* equal to the angle *DFH*; [I. 23]
and through the points *A, B, C* let *LAM, MBN, NCL* be drawn touching the circle *ABC*. [III. 16, Por.]

Now, since *LM, MN, NL* touch the circle *ABC* at the points *A, B, C*,
and *KA, KB, KC* have been joined from the centre *K* to the points *A, B, C*,

therefore the angles at the points *A, B, C* are right. [III. 18]

And, since the four angles of the quadrilateral *AMBK* are equal to four right angles, inasmuch as *AMBK* is in fact divisible into two triangles,
and the angles *KAM, KBM* are right,

therefore the remaining angles *AKB, AMB* are equal to two right angles.

But the angles *DEG, DEF* are also equal to two right angles; [I. 13]

therefore the angles *AKB, AMB* are equal to the angles *DEG, DEF*,
of which the angle *AKB* is equal to the angle *DEG*;
therefore the angle *AMB* which remains is equal to the angle *DEF* which remains.

Similarly it can be proved that the angle *LNB* is also equal to the angle *DFE*;

therefore the remaining angle *MLN* is equal to the angle *EDF*. [I. 32]

Therefore the triangle *LMN* is equiangular with the triangle *DEF*;
and it has been circumscribed about the circle *ABC*.

Therefore about a given circle there has been circumscribed a triangle equiangular with the given triangle.

Q.E.F.

Proposition 4

In a given triangle to inscribe a circle.

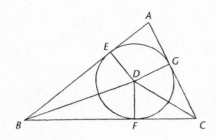

Let *ABC* be the given triangle;
thus it is required to inscribe a circle in the triangle *ABC*.

Let the angles *ABC*, *ACB* be bisected by the straight lines *BD*, *CD*, [I. 9]
and let these meet one another at the point *D*;
from *D* let *DE*, *DF*, *DG* be drawn perpendicular to the straight lines *AB*, *BC*, *CA*.

Now, since the angle *ABD* is equal to the angle *CBD*,

and the right angle *BED* is also equal to the right angle *BFD*,
EBD, *FBD* are two triangles having two angles equal to two angles and one side equal to one side, namely that subtending one of the equal angles, which is *BD* common to the triangles;
therefore they will also have the remaining sides equal to the remaining sides;
 [I. 26]

therefore *DE* is equal to *DF*.

For the same reason

DG is also equal to *DF*.

Therefore the three straight lines *DE*, *DF*, *DG* are equal to one another;
therefore the circle described with centre *D* and distance one of the straight lines *DE*, *DF*, *DG* will pass also through the remaining points, and will touch the straight lines *AB*, *BC*, *CA*, because the angles at the points *E*, *F*, *G* are right.

For, if it cuts them, the straight line drawn at right angles to the diameter of the circle from its extremity will be found to fall within the circle:
which was proved absurd; [III. 16]
therefore the circle described with centre *D* and distance one of the straight lines *DE*, *DF*, *DG* will not cut the straight lines *AB*, *BC*, *CA*;
therefore it will touch them, and will be the circle inscribed in the triangle *ABC*.
 [IV. Def. 5]

Let it be inscribed, as *FGE*.

Therefore in the given triangle *ABC* the circle *EFG* has been inscribed.

Q.E.F.

Proposition 5

About a given triangle to circumscribe a circle.

Let *ABC* be the given triangle;
thus it is required to circumscribe a circle about the given triangle *ABC*.

Let the straight lines *AB*, *AC* be bisected at the points *D*, *E*, [I. 10]
and from the points *D*, *E* let *DF*, *EF* be drawn at right angles to *AB*, *AC*;

they will then meet within the triangle *ABC*, or on the straight line *BC*, or outside *BC*.

First let them meet within at *F*, and let *FB*, *FC*, *FA* be joined.

Then, since *AD* is equal to *DB*,
and *DF* is common and at right angles,
therefore the base *AF* is equal to the base *FB*. [I. 4]

Similarly we can prove that *CF* is also equal to *AF*;
so that *FB* is also equal to *FC*;
therefore the three straight lines *FA*, *FB*, *FC* are equal to one another.

Therefore the circle described with centre *F* and distance one of the straight lines *FA*, *FB*, *FC* will pass also through the remaining points, and the circle will have been circumscribed about the triangle *ABC*.

Let it be circumscribed, as *ABC*.

Next, let *DF*, *EF* meet on the straight line *BC* at *F*, as is the case in the second figure; and let *AF* be joined.

Then, similarly, we shall prove that the point *F* is the centre of the circle circumscribed about the triangle *ABC*.

Again, let *DF*, *EF* meet outside the triangle *ABC* at *F*, as is the case in the third figure, and let *AF*, *BF*, *CF* be joined.

Then again, since *AD* is equal to *DB*,
and *DF* is common and at right angles,
therefore the base *AF* is equal to the base *BF*. [I. 4]

Similarly we can prove that *CF* is also equal to *AF*;
so that *BF* is also equal to *FC*;

therefore the circle described with centre *F* and distance one of the straight lines *FA*, *FB*, *FC* will pass also through the remaining points, and will have been circumscribed about the triangle *ABC*.

Therefore about the given triangle a circle has been circumscribed.

Q.E.F.

And it is manifest that, when the centre of the circle falls within the triangle, the angle *BAC*, being in a segment greater than the semicircle, is less than a right angle;
when the centre falls on the straight line *BC*, the angle *BAC*, being in a semicircle, is right;
and when the centre of the circle falls outside the triangle, the angle *BAC*, being in a segment less than the semicircle, is greater than a right angle. [III. 31]

Proposition 6

In a given circle to inscribe a square.

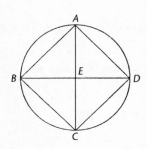

Let *ABCD* be the given circle;
thus it is required to inscribe a square in the circle *ABCD*.

Let two diameters *AC, BD* of the circle *ABCD* be drawn at right angles to one another, and let *AB, BC, CD, DA* be joined.

Then, since *BE* is equal to *ED*, for *E* is the centre, and *EA* is common and at right angles,
therefore the base *AB* is equal to the base *AD*. [I. 4]

For the same reason
each of the straight lines *BC, CD* is also equal to each of the straight lines *AB, AD*;
therefore the quadrilateral *ABCD* is equilateral.

I say next that it is also right-angled.

For, since the straight line *BD* is a diameter of the circle *ABCD*,
therefore *BAD* is a semicircle;
therefore the angle *BAD* is right. [III. 31]

For the same reason
each of the angles *ABC, BCD, CDA* is also right;
therefore the quadrilateral *ABCD* is right-angled.

But it was also proved equilateral;
therefore it is a square; [I. Def. 22]
and it has been inscribed in the circle *ABCD*.

Therefore in the given circle the square *ABCD* has been inscribed.

Q.E.F.

Proposition 7

About a given circle to circumscribe a square.

Let *ABCD* be the given circle;
thus it is required to circumscribe a square about the circle *ABCD*.

Let two diameters *AC, BD* of the circle *ABCD* be drawn at right angles to one another, and through the points *A, B, C, D* let *FG, GH, HK, KF* be drawn touching the circle *ABCD*.
[III. 16, Por.]

Then, since *FG* touches the circle *ABCD*,
and *EA* has been joined from the centre *E* to the point of contact at *A*,
therefore the angles at *A* are right. [III. 18]

For the same reason
the angles at the points *B, C, D* are also right.

Now, since the angle *AEB* is right,
and the angle *EBG* is also right,
therefore *GH* is parallel to *AC*. [I. 28]

For the same reason
AC is also parallel to *FK*,
so that *GH* is also parallel to *FK*. [I. 30]

Similarly we can prove that
each of the straight lines *GF*, *HK* is parallel to *BED*.

Therefore *GK*, *GC*, *AK*, *FB*, *BK* are parallelograms;
therefore *GF* is equal to *HK*, and *GH* to *FK*. [I. 34]

And, since *AC* is equal to *BD*,
and *AC* is also equal to each of the straight lines *GH*, *FK*,
while *BD* is equal to each of the straight lines *GF*, *HK*, [I. 34]
therefore the quadrilateral *FGHK* is equilateral.

I say next that it is also right-angled.

For, since *GBEA* is a parallelogram,
and the angle *AEB* is right,
therefore the angle *AGB* is also right. [I. 34]

Similarly we can prove that
the angles at *H*, *K*, *F* are also right.

Therefore *FGHK* is right-angled.

But it was also proved equilateral;
therefore it is a square;
and it has been circumscribed about the circle *ABCD*.

Therefore about the given circle a square has been circumscribed.

Q.E.F.

Proposition 8

In a given square to inscribe a circle.

Let *ABCD* be the given square;
thus it is required to inscribe a circle in the given
square *ABCD*.

Let the straight lines *AD*, *AB* be bisected at the points
E, *F* respectively, [I. 10]
through *E* let *EH* be drawn parallel to either *AB* or
CD, and through *F* let *FK* be drawn parallel to either
AD or *BC*; [I. 31]
therefore each of the figures *AK*, *KB*, *AH*, *HD*, *AG*, *GC*,
BG, *GD* is a parallelogram, and their opposite sides are
evidently equal. [I. 34]

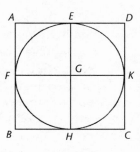

Now, since *AD* is equal to *AB*,
and *AE* is half of *AD*, and *AF* half of *AB*,
therefore *AE* is equal to *AF*,
so that the opposite sides are also equal;

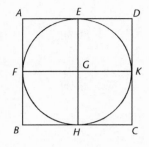

therefore *FG* is equal to *GE*.

Similarly we can prove that each of the straight lines *GH*, *GK* is equal to each of the straight lines *FG*, *GE*; therefore the four straight lines *GE*, *GF*, *GH*, *GK* are equal to one another.

Therefore the circle described with centre G and distance one of the straight lines *GE*, *GF*, *GH*, *GK* will pass also through the remaining points.

And it will touch the straight lines *AB*, *BC*, *CD*, *DA*, because the angles at *E*, *F*, *H*, *K* are right.

For, if the circle cuts *AB*, *BC*, *CD*, *DA*, the straight line drawn at right angles to the diameter of the circle from its extremity will fall within the circle: which was proved absurd; [III. 16]
therefore the circle described with centre G and distance one of the straight lines *GE*, *GF*, *GH*, *GK* will not cut the straight lines *AB*, *BC*, *CD*, *DA*.

Therefore it will touch them, and will have been inscribed in the square *ABCD*.

Therefore in the given square a circle has been inscribed.

Q.E.F.

Proposition 9

About a given square to circumscribe a circle.

Let *ABCD* be the given square;
thus it is required to circumscribe a circle about the square *ABCD*.

For let *AC*, *BD* be joined, and let them cut one another at *E*.

Then, since *DA* is equal to *AB*, and *AC* is common,
therefore the two sides *DA*, *AC* are equal to the two sides *BA*, *AC*;
and the base *DC* is equal to the base *BC*;
therefore the angle *DAC* is equal to the angle *BAC*. [I. 8]

Therefore the angle *DAB* is bisected by *AC*.

Similarly we can prove that each of the angles *ABC*, *BCD*, *CDA* is bisected by the straight lines *AC*, *DB*.

Now, since the angle *DAB* is equal to the angle *ABC*,
and the angle *EAB* is half the angle *DAB*,
and the angle *EBA* half the angle *ABC*,
therefore the angle *EAB* is also equal to the angle *EBA*;
so that the side *EA* is also equal to *EB*. [I. 6]

Similarly we can prove that each of the straight lines *EA*, *EB* is equal to each of the straight lines *EC*, *ED*.

Therefore the four straight lines *EA*, *EB*, *EC*, *ED* are equal to one another.

Therefore the circle described with centre *E* and distance one of the straight lines *EA*, *EB*, *EC*, *ED* will pass also through the remaining points;

and it will have been circumscribed about the square *ABCD*.

Let it be circumscribed, as *ABCD*.

Therefore about the given square a circle has been circumscribed.

<div align="right">Q.E.F.</div>

Proposition 10

To construct an isosceles triangle having each of the angles at the base double of the remaining one.

Let any straight line *AB* be set out, and let it be cut at the point *C* so that the rectangle contained by *AB*, *BC* is equal to the square on *CA*; [II. 11]
with centre *A* and distance *AB* let the circle *BDE* be described,
and let there be fitted in the circle *BDE* the straight line *BD* equal to the straight line *AC* which is not greater than the diameter of the circle *BDE*. [IV. 1]

Let *AD*, *DC* be joined, and let the circle *ACD* be circumscribed about the triangle *ACD*. [IV. 5]

Then, since the rectangle *AB*, *BC* is equal to the square on *AC*,
and *AC* is equal to *BD*,
 therefore the rectangle *AB*, *BC* is equal to the square on *BD*.

And, since a point *B* has been taken outside the circle *ACD*,
and from *B* the two straight lines *BA*, *BD* have fallen on the circle *ACD*, and one of them cuts it, while the other falls on it,
and the rectangle *AB*, *BC* is equal to the square on *BD*,
 therefore *BD* touches the circle *ACD*. [III. 37]

Since, then, *BD* touches it, and *DC* is drawn across from the point of contact at *D*,
therefore the angle *BDC* is equal to the angle *DAC* in the alternate segment of the circle. [III. 32]

Since, then, the angle *BDC* is equal to the angle *DAC*,
let the angle *CDA* be added to each;
 therefore the whole angle *BDA* is equal to the two angles *CDA*, *DAC*.

But the exterior angle *BCD* is equal to the angles *CDA*, *DAC*; [I. 32]
 therefore the angle *BDA* is also equal to the angle *BCD*.

But the angle *BDA* is equal to the angle *CBD*, since the side *AD* is also equal to *AB*; [I. 5]
 so that the angle *DBA* is also equal to the angle *BCD*.

 Therefore the three angles *BDA*, *DBA*, *BCD* are equal to one another.

And, since the angle *DBC* is equal to the angle *BCD*,
 the side *BD* is also equal to the side *DC*. [I. 6]
But *BD* is by hypothesis equal to *CA*;
 therefore *CA* is also equal to *CD*,

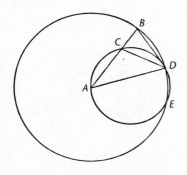

so that the angle CDA is also equal to the angle DAC;
[I. 5]

therefore the angles CDA, DAC are double of the angle DAC.

But the angle BCD is equal to the angles CDA, DAC; therefore the angle BCD is also double of the angle CAD.

But the angle BCD is equal to each of the angles BDA, DBA;
therefore each of the angles BDA, DBA is also double of the angle DAB.

Therefore the isosceles triangle ABD has been constructed having each of the angles at the base DB double of the remaining one.

Q.E.F.

Proposition 11

In a given circle to inscribe an equilateral and equiangular pentagon.

Let ABCDE be the given circle;
thus it is required to inscribe in the circle ABCDE an equilateral and equiangular pentagon.

Let the isosceles triangle FGH be set out having each of the angles at G, H double of the angle at F; [IV. 10]
let there be inscribed in the circle ABCDE the triangle ACD equiangular with the triangle FGH, so that the angle CAD is equal to the angle at F and the angles at G, H respectively equal to the angles ACD, CDA; [IV. 2]
therefore each of the angles ACD, CDA is also double of the angle CAD.

Now let the angles ACD, CDA be bisected respectively by the straight lines CE, DB,
[I. 9]
and let AB, BC, DE, EA be joined.

Then, since each of the angles ACD, CDA is double of the angle CAD,
and they have been bisected by the straight lines CE, DB,
therefore the five angles DAC, ACE, ECD, CDB, BDA are equal to one another.

But equal angles stand on equal circumferences; [III. 26]
therefore the five circumferences AB, BC, CD, DE, EA are equal to one another.

But equal circumferences are subtended by equal straight lines; [III. 29]
therefore the five straight lines AB, BC, CD, DE, EA are equal to one another;
therefore the pentagon ABCDE is equilateral.

I say next that it is also equiangular.

For, since the circumference AB is equal to the circumference DE, let BCD be added to each;
therefore the whole circumference ABCD is equal to the whole circumference EDCB.

And the angle *AED* stands on the circumference *ABCD,* and the angle *BAE* on the circumference *EDCB;*

therefore the angle *BAE* is also equal to the angle *AED.* [III. 27]

For the same reason

each of the angles *ABC, BCD, CDE* is also equal to each of the angles *BAE, AED;*

therefore the pentagon *ABCDE* is equiangular.

But it was also proved equilateral;

therefore in the given circle an equilateral and equiangular pentagon has been inscribed.

Q.E.F.

Proposition 12

About a given circle to circumscribe an equilateral and equiangular pentagon.

Let *ABCDE* be the given circle;
thus it is required to circumscribe an equilateral and equiangular pentagon about the circle *ABCDE.*

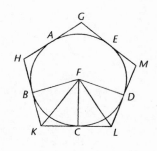

Let *A, B, C, D, E* be conceived to be the angular points of the inscribed pentagon, so that the circumferences *AB, BC, CD, DE, EA* are equal; [IV. 11]
through *A, B, C, D, E* let *GH, HK, KL, LM, MG* be drawn touching the circle; [III. 16, Por.]
let the centre *F* of the circle *ABCDE* be taken, [III. 1]
and let *FB, FK, FC, FL, FD* be joined.

Then, since the straight line *KL* touches the circle *ABCDE* at *C,*
and *FC* has been joined from the centre *F* to the point of contact at *C,*

therefore *FC* is perpendicular to *KL;* [III. 18]
therefore each of the angles at *C* is right.

For the same reason

the angles at the points *B, D* are also right.

And, since the angle *FCK* is right,

therefore the square on *FK* is equal to the squares on *FC, CK.* [I. 47]

For the same reason

the square on *FK* is also equal to the squares on *FB, BK;*
so that the squares on *FC, CK* are equal to the squares on *FB, BK,*
of which the square on *FC* is equal to the square on *FB;*
therefore the square on *CK* which remains is equal to the square on *BK.*

Therefore *BK* is equal to *CK.*

And, since *FB* is equal to *FC,*
and *FK* common,
the two sides *BF, FK* are equal to the two sides *CF, FK;* and the base *BK* equal to the base *CK;*

therefore the angle *BFK* is equal to the angle *KFC,* [I. 8]
and the angle *BKF* to the angle *FKC.*

Therefore the angle *BFC* is double of the angle *KFC,*

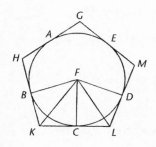

and the angle *BKC* of the angle *FKC*.

For the same reason
the angle *CFD* is also double of the angle *CFL*,
and the angle *DLC* of the angle *FLC*.

Now, since the circumference *BC* is equal to *CD*,
the angle *BFC* is also equal to the angle *CFD*. [III. 27]

And the angle *BFC* is double of the angle *KFC*, and the angle *DFC* of the angle *LFC*;
therefore the angle *KFC* is also equal to the angle *LFC*.

But the angle *FCK* is also equal to the angle *FCL*;
therefore *FKC*, *FLC* are two triangles having two angles equal to two angles and one side equal to one side, namely *FC* which is common to them;
therefore they will also have the remaining sides equal to the remaining sides, and the remaining angle to the remaining angle; [I. 26]
therefore the straight line *KC* is equal to *CL*,
and the angle *FKC* to the angle *FLC*.

And, since *KC* is equal to *CL*,
therefore *KL* is double of *KC*.

For the same reason it can be proved that
HK is also double of *BK*.

And *BK* is equal to *KC*;
therefore *HK* is also equal to *KL*.

Similarly each of the straight lines *HG*, *GM*, *ML* can also be proved equal to each of the straight lines *HK*, *KL*;
therefore the pentagon *GHKLM* is equilateral.

I say next that it is also equiangular.

For, since the angle *FKC* is equal to the angle *FLC*,
and the angle *HKL* was proved double of the angle *FKC*,
and the angle *KLM* double of the angle *FLC*,
therefore the angle *HKL* is also equal to the angle *KLM*.

Similarly each of the angles *KHG*, *HGM*, *GML* can also be proved equal to each of the angles *HKL*, *KLM*;
therefore the five angles *GHK*, *HKL*, *KLM*, *LMG*, *MGH* are equal to one another.

Therefore the pentagon *GHKLM* is equiangular.

And it was also proved equilateral; and it has been circumscribed about the circle *ABCDE*.

Q.E.F.

Proposition 13

In a given pentagon, which is equilateral and equiangular, to inscribe a circle.

Let *ABCDE* be the given equilateral and equiangular pentagon;
thus it is required to inscribe a circle in the pentagon *ABCDE*.

For let the angles *BCD*, *CDE* be bisected by the straight lines *CF*, *DF* respectively;

and from the point *F*, at which the straight lines *CF, DF* meet one another, let the straight lines *FB, FA, FE* be joined.

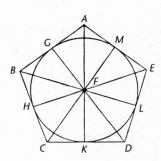

Then, since *BC* is equal to *CD*, and *CF* common,
 the two sides *BC, CF* are equal to the two sides *DC, CF*;
and the angle *BCF* is equal to the angle *DCF*;
 therefore the base *BF* is equal to the base *DF*,
 and the triangle *BCF* is equal to the triangle *DCF*,
and the remaining angles will be equal to the remaining angles, namely those which the equal sides subtend. [I. 4]

 Therefore the angle *CBF* is equal to the angle *CDF*.

And, since the angle *CDE* is double of the angle *CDF*,
and the angle *CDE* is equal to the angle *ABC*,
while the angle *CDF* is equal to the angle *CBF*;
 therefore the angle *CBA* is also double of the angle *CBF*;
 therefore the angle *ABF* is equal to the angle *FBC*;
 therefore the angle *ABC* has been bisected by the straight line *BF*.

Similarly it can be proved that the angles *BAE, AED* have also been bisected by the straight lines *FA, FE* respectively.

Now let *FG, FH, FK, FL, FM* be drawn from the point *F* perpendicular to the straight lines *AB, BC, CD, DE, EA*.

Then, since the angle *HCF* is equal to the angle *KCF*,
and the right angle *FHC* is also equal to the angle *FKC*,
FHC, FKC are two triangles having two angles equal to two angles and one side equal to one side, namely *FC* which is common to them and subtends one of the equal angles;
 therefore they will also have the remaining sides equal to the remaining sides;
 [I. 26]
 therefore the perpendicular *FH* is equal to the perpendicular *FK*.

Similarly it can be proved that each of the straight lines *FL, FM, FG* is also equal to each of the straight lines *FH, FK*;
 therefore the five straight lines *FG, FH, FK, FL, FM* are equal to one another.

Therefore the circle described with centre *F* and distance one of the straight lines *FG, FH, FK, FL, FM* will pass also through the remaining points;
and it will touch the straight lines *AB, BC, CD, DE, EA*, because the angles at the points *G, H, K, L, M* are right.

For, if it does not touch them, but cuts them, it will result that the straight line drawn at right angles to the diameter of the circle from its extremity falls within the circle: which was proved absurd. [III. 16]

Therefore the circle described with centre *F* and distance one of the straight lines *FG, FH, FK, FL, FM* will not cut the straight lines *AB, BC, CD, DE, EA*;
 therefore it will touch them.

Let it be described, as *GHKLM*.

Therefore in the given pentagon, which is equilateral and equiangular, a circle has been inscribed.

 Q.E.F.

Proposition 14

About a given pentagon, which is equilateral and equiangular, to circumscribe a circle.

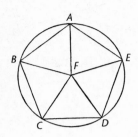

Let *ABCDE* be the given pentagon, which is equilateral and equiangular;
thus it is required to circumscribe a circle about the pentagon *ABCDE*.

Let the angles *BCD*, *CDE* be bisected by the straight lines *CF*, *DF* respectively, and from the point *F*, at which the straight lines meet, let the straight lines *FB*, *FA*, *FE* be joined to the points *B*, *A*, *E*.

Then in manner similar to the preceding it can be proved that the angles *CBA*, *BAE*, *AED* have also been bisected by the straight lines *FB*, *FA*, *FE* respectively.

Now, since the angle *BCD* is equal to the angle *CDE*,
and the angle *FCD* is half of the angle *BCD*,
and the angle *CDF* half of the angle *CDE*,
therefore the angle *FCD* is also equal to the angle *CDF*,
so that the side *FC* is also equal to the side *FD*. [I. 6]

Similarly it can be proved that each of the straight lines *FB*, *FA*, *FE* is also equal to each of the straight lines *FC*, *FD*;
therefore the five straight lines *FA*, *FB*, *FC*, *FD*, *FE* are equal to one another.

Therefore the circle described with centre *F* and distance one of the straight lines *FA*, *FB*, *FC*, *FD*, *FE* will pass also through the remaining points, and will have been circumscribed.

Let it be circumscribed, and let it be *ABCDE*.

Therefore about the given pentagon, which is equilateral and equiangular, a circle has been circumscribed.

Q.E.F.

Proposition 15

In a given circle to inscribe an equilateral and equiangular hexagon.

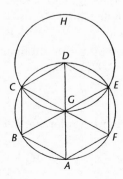

Let *ABCDEF* be the given circle;
thus it is required to inscribe an equilateral and equiangular hexagon in the circle *ABCDEF*.

Let the diameter *AD* of the circle *ABCDEF* be drawn;
let the centre *G* of the circle be taken, and with centre *D* and distance *DG* let the circle *EGCH* be described;
let *EG*, *CG* be joined and carried through to the points *B*, *F*,
and let *AB*, *BC*, *CD*, *DE*, *EF*, *FA* be joined.

I say that the hexagon *ABCDEF* is equilateral and equiangular.

For, since the point G is the centre of the circle *ABCDEF,*
<div align="center">GE is equal to GD.</div>

Again, since the point *D* is the centre of the circle *GCH,*
<div align="center">DE is equal to DG.</div>

But *GE* was proved equal to *GD;*
<div align="center">therefore *GE* is also equal to *ED;*</div>
<div align="center">therefore the triangle *EGD* is equilateral;</div>
and therefore its three angles *EGD, GDE, DEG* are equal to one another, inasmuch as, in isosceles triangles, the angles at the base are equal to one another. [I. 5]

And the three angles of the triangle are equal to two right angles; [I. 32]
<div align="center">therefore the angle *EGD* is one-third of two right angles.</div>

Similarly, the angle *DGC* can also be proved to be one-third of two right angles.

And, since the straight line *CG* standing on *EB* makes the adjacent angles *EGC, CGB* equal to two right angles,
<div align="center">therefore the remaining angle *CGB* is also one-third of two right angles.</div>

<div align="center">Therefore the angles *EGD, DGC, CGB* are equal to one another;</div>
<div align="center">so that the angles vertical to them, the angles *BGA, AGF, FGE* are equal. [I. 15]</div>

Therefore the six angles *EGD, DGC, CGB, BGA, AGF, FGE* are equal to one another.

But equal angles stand on equal circumferences; [III. 26]
therefore the six circumferences *AB, BC, CD, DE, EF, FA* are equal to one another.

And equal circumferences are subtended by equal straight lines; [III. 29]
<div align="center">therefore the six straight lines are equal to one another;</div>
<div align="center">therefore the hexagon *ABCDEF* is equilateral.</div>

I say next that it is also equiangular.

For, since the circumference *FA* is equal to the circumference *ED,*
let the circumference *ABCD* be added to each;
<div align="center">therefore the whole *FABCD* is equal to the whole *EDCBA;*</div>
and the angle *FED* stands on the circumference *FABCD,*
and the angle *AFE* on the circumference *EDCBA;*
<div align="center">therefore the angle *AFE* is equal to the angle *DEF.* [III. 27]</div>

Similarly it can be proved that the remaining angles of the hexagon *ABCDEF* are also severally equal to each of the angles *AFE, FED;*
<div align="center">therefore the hexagon *ABCDEF* is equiangular.</div>

But it was also proved equilateral;
and it has been inscribed in the circle *ABCDEF.*

Therefore in the given circle an equilateral and equiangular hexagon has been inscribed.

<div align="right">Q.E.F.</div>

PORISM. From this it is manifest that the side of the hexagon is equal to the radius of the circle.

And, in like manner as in the case of the pentagon, if through the points of division on the circle we draw tangents to the circle, there will be circumscribed

about the circle an equilateral and equiangular hexagon in conformity with what was explained in the case of the pentagon.

And further by means similar to those explained in the case of the pentagon we can both inscribe a circle in a given hexagon and circumscribe one about it.

<div align="right">Q.E.F.</div>

Proposition 16

In a given circle to inscribe a fifteen-angled figure which shall be both equilateral and equiangular.

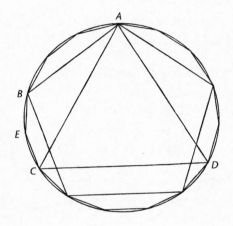

Let *ABCD* be the given circle;
thus it is required to inscribe in the circle *ABCD* a fifteen-angled figure which shall be both equilateral and equiangular.

In the circle *ABCD* let there be inscribed a side *AC* of the equilateral triangle inscribed in it, and a side *AB* of an equilateral pentagon;
therefore, of the equal segments of which there are fifteen in the circle *ABCD*, there will be five in the circumference *ABC* which is one-third of the circle, and there will be three in the circumference *AB* which is one-fifth of the circle;
therefore in the remainder *BC* there will be two of the equal segments.

Let *BC* be bisected at *E*; [III. 30]
 therefore each of the circumferences *BE*, *EC* is a fifteenth of the circle *ABCD*.

If therefore we join *BE*, *EC* and fit into the circle *ABCD* straight lines equal to them and in contiguity, a fifteen-angled figure which is both equilateral and equiangular will have been inscribed in it.

<div align="right">Q.E.F.</div>

COROLLARY. And, in like manner as in the case of the pentagon, if through the points of division on the circle we draw tangents to the circle, there will be circumscribed about the circle a fifteen-angled figure which is equilateral and equiangular.

And further, by proofs similar to those in the case of the pentagon, we can both inscribe a circle in the given fifteen-angled figure and circumscribe one about it.

<div align="right">Q.E.F.</div>

Book V

Definitions

1. A magnitude is a *part* of a magnitude, the less of the greater, when it measures the greater.

2. The greater is a *multiple* of the less when it is measured by the less.

3. A *ratio* is a sort of relation in respect of size between two magnitudes of the same kind.

4. Magnitudes are said to *have a ratio* to one another which are capable, when multiplied, of exceeding one another.

5. Magnitudes are said to *be in the same ratio*, the first to the second and the third to the fourth, when, if any equimultiples whatever be taken of the first and third, and any equimultiples whatever of the second and fourth, the former equimultiples alike exceed, are alike equal to, or alike fall short of, the latter equimultiples respectively taken in corresponding order.

6. Let magnitudes which have the same ratio be called *proportional*.

7. When, of the equimultiples, the multiple of the first magnitude exceeds the multiple of the second, but the multiple of the third does not exceed the multiple of the fourth, then the first is said to *have a greater ratio* to the second than the third has to the fourth.

8. A proportion in three terms is the least possible.

9. When three magnitudes are proportional, the first is said to have to the third the *duplicate ratio* of that which it has to the second.

10. When four magnitudes are [continuously][1] proportional, the first is said to have to the fourth the *triplicate ratio* of that which it has to the second, and so on continually, whatever be the proportion.

11. The term *corresponding magnitudes* is used of antecedents in relation to antecedents, and of consequents in relation to consequents.

12. *Alternate ratio* means taking the antecedent in relation to the antecedent and the consequent in relation to the consequent.[2]

13. *Inverse ratio* means taking the consequent as antecedent in relation to the antecedent as consequent.[3]

1. The adjective "continuously" was inserted by Heath; it has no correspondent in the Greek text. Magnitudes a, b, c, d are in "continuous" proportion when $a:b :: b:c :: c:d$. —Ed.

2. If $a:b :: c:d$, the proportion $a:c :: b:d$ is obtained "alternately" (*alternando*), as proved in Proposition V. 16. —Ed.

3. If $a:b :: c:d$, the proportion $b:a :: d:c$ is obtained "inversely" (*invertendo*), as proved in Proposition V. 7, Porism. —Ed.

14. *Composition of a ratio* means taking the antecedent together with the consequent as one in relation to the consequent by itself.[4]

15. *Separation of a ratio* means taking the excess by which the antecedent exceeds the consequent in relation to the consequent by itself.[5]

16. *Conversion of a ratio* means taking the antecedent in relation to the excess by which the antecedent exceeds the consequent.[6]

17. A ratio *ex aequali* arises when, there being several magnitudes and another set equal to them in multitude which taken two and two are in the same proportion, as the first is to the last among the first magnitudes, so is the first to the last among the second magnitudes;

 Or, in other words, it means taking the extreme terms by virtue of the removal of the intermediate terms.[7]

18. A *perturbed proportion* arises when, there being three magnitudes and another set equal to them in multitude, as antecedent is to consequent among the first magnitudes, so is antecedent to consequent among the second magnitudes, while, as the consequent is to a third among the first magnitudes, so is a third to the antecedent among the second magnitudes.[8]

Proposition 1

If there be any number of magnitudes whatever which are, respectively, equi-multiples of any magnitudes equal in multitude, then, whatever multiple one of the magnitudes is of one, that multiple also will all be of all.

4. If $a:b :: c:d$, the proportion $a+b:b :: c+d:d$ is obtained by composition (*componendo*). See Propositions V. 17 and V. 18. —Ed.

5. If $a:b :: c:d$, the proportion $a-b:b :: c-d:d$ is obtained by separation (*separando*). See Propositions V. 17 and V.18. —Ed.

6. If $a:b :: c:d$, the proportion $a:a-b :: c:c-d$ is obtained by conversion (*convertendo*). See Proposition V. 19. —Ed.

7. If $\qquad\qquad a:b :: d:e$
and also $\qquad\qquad b:c :: e:f$
then the proportion $\qquad\quad a:c :: d:f$
is obtained *ex aequali*. Note the similar disposition of middle terms b and e. The example can be extended to any number of pairs of magnitudes, as proved in Proposition V. 22. —Ed.

8. If $\qquad\qquad a:b :: e:f$
and also $\qquad\qquad b:c :: d:e$
then the proportion is said to be perturbed, since the middle terms b and e are oppositely disposed. The proportion $\qquad a:c :: d:f$
is obtained *ex aequali* in this case too, as proved in Proposition V. 23. —Ed.

Let any number of magnitudes whatever
AB, *CD* be respectively equimultiples of any
magnitudes *E*, *F* equal in multitude;
I say that, whatever multiple *AB* is of *E*,
that multiple will *AB*, *CD* also be of *E*, *F*.

For, since *AB* is the same multiple of *E* that *CD* is of *F*, as many magnitudes as
there are in *AB* equal to *E*, so many also are there in *CD* equal to *F*.

Let *AB* be divided into the magnitudes *AG*, *GB* equal to *E*,
and *CD* into *CH*, *HD* equal to *F*;
then the multitude of the magnitudes *AG*, *GB* will be equal to the multitude of
the magnitudes *CH*, *HD*.

Now, since *AG* is equal to *E*, and *CH* to *F*,
therefore *AG* is equal to *E*, and *AG*, *CH* to *E*, *F*.

For the same reason
GB is equal to *E*, and *GB*, *HD* to *E*, *F*;
therefore, as many magnitudes as there are in *AB* equal to *E*, so many also are
there in *AB*, *CD* equal to *E*, *F*;
therefore, whatever multiple *AB* is of *E*, that multiple will *AB*, *CD* also be of *E*, *F*.

Therefore etc.

Q.E.D.

Proposition 2

If a first magnitude be the same multiple of a second that a third is of a fourth,
and a fifth also be the same multiple of the second that a sixth is of the fourth,
the sum of the first and fifth will also be the same multiple of the second that the
sum of the third and sixth is of the fourth.

Let a first magnitude, *AB*, be the same multiple
of a second, *C*, that a third, *DE*, is of a fourth, *F*,
and let a fifth, *BG*, also be the same multiple of
the second, *C*, that a sixth, *EH*, is of the fourth *F*;
I say that the sum of the first and fifth, *AG*, will
be the same multiple of the second, *C*, that the
sum of the third and sixth, *DH*, is of the fourth, *F*.

For, since *AB* is the same multiple of *C* that *DE* is of *F*,
therefore, as many magnitudes as there are in *AB* equal to *C*, so many also are
there in *DE* equal to *F*.

For the same reason also,
as many as there are in *BG* equal to *C*, so many are there also in *EH* equal to *F*;
therefore, as many as there are in the whole *AG* equal to *C*, so many also are
there in the whole *DH* equal to *F*.

Therefore, whatever multiple *AG* is of *C*, that multiple also is *DH* of *F*.

Therefore the sum of the first and fifth, *AG*, is the same multiple of the second, *C*,
that the sum of the third and sixth, *DH*, is of the fourth, *F*.

Therefore etc.

Q.E.D.

Proposition 3

If a first magnitude be the same multiple of a second that a third is of a fourth, and if equimultiples be taken of the first and third, then also ex aequali *the magnitudes taken will be equimultiples respectively, the one of the second, and the other of the fourth.*

Let a first magnitude A be the same multiple of a second B that a third C is of a fourth D, and let equimultiples EF, GH be taken of A, C;
I say that EF is the same multiple of B that GH is of D.

For, since EF is the same multiple of A that GH is of C, therefore, as many magnitudes as there are in EF equal to A, so many also are there in GH equal to C.

Let EF be divided into the magnitudes EK, KF equal to A,
and GH into the magnitudes GL, LH equal to C;
then the multitude of the magnitudes EK, KF will be equal to the multitude of the magnitudes GL, LH.

And, since A is the same multiple of B that C is of D,
while EK is equal to A, and GL to C,
 therefore EK is the same multiple of B that GL is of D.

For the same reason
 KF is the same multiple of B that LH is of D.

Since, then, a first magnitude EK is the same multiple of a second B that a third GL is of a fourth D,
and a fifth KF is also the same multiple of the second B that a sixth LH is of the fourth D,
therefore the sum of the first and fifth, EF, is also the same multiple of the second B that the sum of the third and sixth, GH, is of the fourth D. [V. 2]

Therefore etc.

 Q.E.D.

Proposition 4

If a first magnitude have to a second the same ratio as a third to a fourth, any equimultiples whatever of the first and third will also have the same ratio to any equimultiples whatever of the second and fourth respectively, taken in corresponding order.

For let a first magnitude A have to a second B the same ratio as a third C to a fourth D;

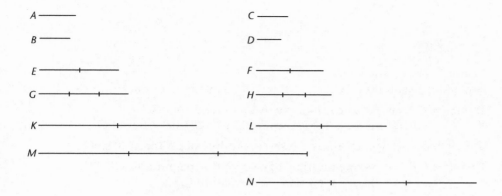

and let equimultiples E, F be taken of A, C,
and G, H other, chance, equimultiples of B, D;
I say that, as E is to G, so is F to H.

For let equimultiples K, L be taken of E, F,
and other, chance, equimultiples M, N of G, H.

Since E is the same multiple of A that F is of C,
and equimultiples K, L of E, F have been taken,
 therefore K is the same multiple of A that L is of C. [V. 3]

For the same reason M is the same multiple of B that N is of D.

And, since, as A is to B, so is C to D,
and of A, C equimultiples K, L have been taken,
and of B, D other, chance, equimultiples M, N,
 therefore, if K is in excess of M, L also is in excess of N,
 if it is equal, equal,
 and if less, less. [V. Def. 5]

And K, L are equimultiples of E, F,
and M, N other, chance, equimultiples of G, H;
 therefore, as E is to G, so is F to H. [V. Def. 5]

Therefore etc.

Q.E.D.

Proposition 5

*If a magnitude be the same multiple of a magnitude that a part subtracted is of
a part subtracted, the remainder will also be the same multiple of the remainder
that the whole is of the whole.*

For let the magnitude AB be the same
multiple of the magnitude CD that the part
AE subtracted is of the part CF subtracted;
I say that the remainder EB is also the same
multiple of the remainder FD that the
whole AB is of the whole CD.

For, whatever multiple AE is of CF, let EB be made that multiple of CG.

Then, since AE is the same multiple of CF that EB is of GC,
therefore AE is the same multiple of CF that AB is of GF. [v. 1]

But, by the assumption, AE is the same multiple of CF that AB is of CD.

Therefore AB is the same multiple of each of the magnitudes GF, CD;
therefore GF is equal to CD.

Let CF be subtracted from each;
therefore the remainder GC is equal to the remainder FD.

And, since AE is the same multiple of CF that EB is of GC,
and GC is equal to DF,
therefore AE is the same multiple of CF that EB is of FD.

But, by hypothesis,
AE is the same multiple of CF that AB is of CD;
therefore EB is the same multiple of FD that AB is of CD.

That is, the remainder EB will be the same multiple of the remainder FD that the whole AB is of the whole CD.

Therefore etc.

Q.E.D.

Proposition 6

If two magnitudes be equimultiples of two magnitudes, and any magnitudes subtracted from them be equimultiples of the same, the remainders also are either equal to the same or equimultiples of them.

For let two magnitudes AB, CD be equimultiples of two magnitudes E, F,
and let AG, CH subtracted from them be equimultiples of the same two E, F;
I say that the remainders also, GB, HD, are either equal to E, F or equimultiples of them.

For, first, let GB be equal to E;
I say that HD is also equal to F.

For let CK be made equal to F.

Since AG is the same multiple of E that CH is of F,
while GB is equal to E and KC to F,
therefore AB is the same multiple of E that KH is of F. [v. 2]

But, by hypothesis, AB is the same multiple of E that CD is of F;
therefore KH is the same multiple of F that CD is of F.

Since, then, each of the magnitudes KH, CD is the same multiple of F,

therefore *KH* is equal to *CD*.

Let *CH* be subtracted from each;
therefore the remainder *KC* is equal to the remainder *HD*.

But *F* is equal to *KC*;
therefore *HD* is also equal to *F*.

Hence, if *GB* is equal to *E*, *HD* is also equal to *F*.

Similarly we can prove that, even if *GB* be a multiple of *E*, *HD* is also the same multiple of *F*.

Therefore etc.

Q.E.D.

Proposition 7

Equal magnitudes have to the same the same ratio, as also has the same to equal magnitudes.

Let *A*, *B* be equal magnitudes and *C* any other, chance, magnitude;
I say that each of the magnitudes *A*, *B* has the same ratio to *C*, and *C* has the same ratio to each of the magnitudes *A*, *B*.

For let equimultiples *D*, *E* of *A*, *B* be taken,
and of *C* another, chance, multiple *F*.

Then, since *D* is the same multiple of *A* that *E* is of *B*, while *A* is equal to *B*,
therefore *D* is equal to *E*.

But *F* is another, chance, magnitude.

If therefore *D* is in excess of *F*, *E* is also in excess of *F*,
if equal to it, equal;
and, if less, less.

And *D*, *E* are equimultiples of *A*, *B*,
while *F* is another, chance, multiple of *C*;
therefore, as *A* is to *C*, so is *B* to *C*. [V. Def. 5]

I say next that *C* also has the same ratio to each of the magnitudes *A*, *B*.

For, with the same construction, we can prove similarly that *D* is equal to *E*;
and *F* is some other magnitude.

If therefore *F* is in excess of *D*, it is also in excess of *E*,
if equal, equal;
and, if less, less.

And *F* is a multiple of *C*,
while *D*, *E* are other, chance, equimultiples of *A*, *B*;
therefore, as *C* is to *A*, so is *C* to *B*. [V. Def. 5]

Therefore etc.

Q.E.D.

PORISM. From this it is manifest that, if any magnitudes are proportional, they will also be proportional inversely. [V. Def. 13]

Proposition 8

Of unequal magnitudes, the greater has to the same a greater ratio than the less has; and the same has to the less a greater ratio than it has to the greater.

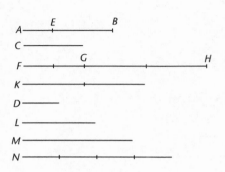

Let *AB, C* be unequal magnitudes, and let *AB* be greater;

let *D* be another, chance, magnitude;

I say that *AB* has to *D* a greater ratio than *C* has to *D*, and *D* has to *C* a greater ratio than it has to *AB*.

For, since *AB* is greater than *C*, let *BE* be made equal to *C*;

then the less of the magnitudes *AE, EB*, if multiplied, will sometime be greater than *D*. [V. Def. 4]

[*Case 1*]

First, let *AE* be less than *EB*;

let *AE* be multiplied, and let *FG* be a multiple of it which is greater than *D*;

then, whatever multiple *FG* is of *AE*, let *GH* be made the same multiple of *EB* and *K* of *C*;

and let *L* be taken double of *D*, *M* triple of it, and successive multiples increasing by one, until what is taken is a multiple of *D* and the first that is greater than *K*.

Let it be taken, and let it be *N* which is quadruple of *D* and the first multiple of it that is greater than *K*.

Then, since *K* is less than *N* first,

therefore *K* is not less than *M*.

And, since *FG* is the same multiple of *AE* that *GH* is of *EB*,

therefore *FG* is the same multiple of *AE* that *FH* is of *AB*. [V. 1]

But *FG* is the same multiple of *AE* that *K* is of *C*;

therefore *FH* is the same multiple of *AB* that *K* is of *C*;

therefore *FH, K* are equimultiples of *AB, C*.

Again, since *GH* is the same multiple of *EB* that *K* is of *C*,

and *EB* is equal to *C*,

therefore *GH* is equal to *K*.

But *K* is not less than *M*;

therefore neither is *GH* less than *M*.

And *FG* is greater than *D*;

therefore the whole *FH* is greater than *D, M* together.

But *D, M* together are equal to *N*,

inasmuch as *M* is triple of *D*, and *M, D* together are quadruple of *D*,

while *N* is also quadruple of *D*;

whence *M, D* together are equal to *N*.

But *FH* is greater than *M, D*;

therefore *FH* is in excess of *N*,

while *K* is not in excess of *N*.

And *FH, K* are equimultiples of *AB, C*, while *N* is another, chance, multiple of *D*;

therefore *AB* has to *D* a greater ratio than *C* has to *D*. **[V. Def. 7]**

I say next, that *D* also has to *C* a greater ratio than *D* has to *AB*.

For, with the same construction, we can prove similarly that *N* is in excess of *K*, while *N* is not in excess of *FH*.

And *N* is a multiple of *D*,

while *FH, K* are other, chance, equimultiples of *AB, C*;

therefore *D* has to *C* a greater ratio than *D* has to *AB*. **[V. Def. 7]**

[*Case 2*]

Again, let *AE* be greater than *EB*.

Then the less, *EB*, if multiplied, will sometime be greater than *D*. **[V. Def. 4]**

Let it be multiplied, and let *GH* be a multiple of *EB* and greater than *D*;

and, whatever multiple *GH* is of *EB*, let *FG* be made the same multiple of *AE*, and *K* of *C*.

Then we can prove similarly that *FH, K* are equimultiples of *AB, C*;

and, similarly, let *N* be taken as a multiple of *D* but the first that is greater than *FG*,

so that *FG* is again not less than *M*.

But *GH* is greater than *D*;

therefore the whole *FH* is in excess of *D, M*, that is, of *N*.

Now *K* is not in excess of *N*, inasmuch as *FG* also, which is greater than *GH*, that is, than *K*, is not in excess of *N*.

And in the same manner, by following the above argument, we complete the demonstration.

Therefore etc.

<div align="right">Q.E.D.</div>

Proposition 9

Magnitudes which have the same ratio to the same are equal to one another; and magnitudes to which the same has the same ratio are equal.

For let each of the magnitudes *A, B* have the same ratio to *C*;

I say that *A* is equal to *B*.

For, otherwise, each of the magnitudes *A, B* would not have had the same ratio to *C*; **[V. 8]**

A —————— B ——————
C ——————

but it has;

therefore *A* is equal to *B*.

Again, let *C* have the same ratio to each of the magnitudes *A*, *B*;
I say that *A* is equal to *B*.

For, otherwise, *C* would not have had the same ratio to each of the magnitudes *A*, *B*;

[V. 8]

but it has;

therefore *A* is equal to *B*.

Therefore etc.

Q.E.D.

Proposition 10

Of magnitudes which have a ratio to the same, that which has a greater ratio is greater; and that to which the same has a greater ratio is less.

A ———————————— B ——————
C ——————————

For let *A* have to *C* a greater ratio than *B* has to *C*;
I say that *A* is greater than *B*.

For, if not, *A* is either equal to *B* or less.

Now *A* is not equal to *B*;
for in that case each of the magnitudes *A*, *B* would have had the same ratio to *C*;

[V. 7]

but they have not;

therefore *A* is not equal to *B*.

Nor again is *A* less than *B*;
for in that case *A* would have had to *C* a less ratio than *B* has to *C*; [V. 8]
but it has not;

therefore *A* is not less than *B*.

But it was proved not to be equal either;

therefore *A* is greater than *B*.

Again, let *C* have to *B* a greater ratio than *C* has to *A*;
I say that *B* is less than *A*.

For, if not, it is either equal or greater.

Now *B* is not equal to *A*;
for in that case *C* would have had the same ratio to each of the magnitudes A, B;

[V. 7]

but it has not;

therefore *A* is not equal to *B*.

Nor again is *B* greater than *A*;
for in that case *C* would have had to *B* a less ratio than it has to A; **[V. 8]**
but it has not;

therefore *B* is not greater than *A*.

But it was proved that it is not equal either;

therefore *B* is less than *A*.

Therefore etc.

Q.E.D.

Proposition 11

Ratios which are the same with the same ratio are also the same with one another.

```
A ———————        C ————        E ———
B ————           D ————        F ——
G ———————————    H ———————————  K ——————
L ———————————    M———————————   N ————————
```

For, as *A* is to *B*, so let *C* be to *D*, and, as *C* is to *D*, so let *E* be to *F*;
I say that, as *A* is to *B*, so is *E* to *F*.

For of *A*, *C*, *E* let equimultiples *G*, *H*, *K* be taken,
and of *B*, *D*, *F* other, chance, equimultiples *L*, *M*, *N*.

Then since, as *A* is to *B*, so is *C* to *D*,
and of *A*, *C* equimultiples *G*, *H* have been taken,
and of *B*, *D* other, chance, equimultiples *L*, *M*,

therefore, if *G* is in excess of *L*, *H* is also in excess of *M*,
if equal, equal,
and if less, less.

Again, since, as *C* is to *D*, so is *E* to *F*,
and of *C*, *E* equimultiples *H*, *K* have been taken,
and of *D*, *F* other, chance, equimultiples *M*, *N*,

therefore, if *H* is in excess of *M*, *K* is also in excess of *N*,
if equal, equal,
and if less, less.

But we saw that, if *H* was in excess of *M*, *G* was also in excess of *L*;
if equal, equal; and if less, less;
so that, in addition,

if *G* is in excess of *L*, *K* is also in excess of *N*,
if equal, equal,
and if less, less.

And *G*, *K* are equimultiples of *A*, *E*,
while *L*, *N* are other, chance, equimultiples of *B*, *F*;

therefore, as *A* is to *B*, so is *E* to *F*.

Therefore etc.

Q.E.D.

Proposition 12

If any number of magnitudes be proportional, as one of the antecedents is to one of the consequents, so will all the antecedents be to all the consequents.

Let any number of magnitudes A, B, C, D, E, F be proportional, so that, as A is to B, so is C to D and E to F;
I say that, as A is to B, so are A, C, E to B, D, F.

For of A, C, E let equimultiples G, H, K be taken,
and of B, D, F other, chance, equimultiples L, M, N.

Then since, as A is to B, so is C to D, and E to F,
and of A, C, E equimultiples G, H, K have been taken,
and of B, D, F other, chance, equimultiples L, M, N,
therefore, if G is in excess of L, H is also in excess of M, and K of N,
if equal, equal,
and if less, less;
so that, in addition,
if G is in excess of L, then G, H, K are in excess of L, M, N,
if equal, equal,
and if less, less.

Now G and G, H, K are equimultiples of A and A, C, E, since, if any number of magnitudes whatever are respectively equimultiples of any magnitudes equal in multitude, whatever multiple one of the magnitudes is of one, that multiple also will all be of all. [V. 1]

For the same reason
L and L, M, N are also equimultiples of B and B, D, F;
therefore, as A is to B, so are A, C, E to B, D, F. [V. Def. 5]

Therefore etc.

Q.E.D.

Proposition 13

If a first magnitude have to a second the same ratio as a third to a fourth, and the third have to the fourth a greater ratio than a fifth has to a sixth, the first will also have to the second a greater ratio than the fifth to the sixth.

For let a first magnitude A have to a second B the same ratio as a third C has to a fourth D,
and let the third C have to the fourth D a greater ratio than a fifth E has to a sixth F;

A —————— C —————— M —————— G ——————————
B ——— D ——— N ————— K ——————

E —————— H ——————————
F ———— L ——————————

I say that the first A will also have to the second B a greater ratio than the fifth E to the sixth F.

For, since there are some equimultiples of C, E,
and of D, F other, chance, equimultiples, such that the multiple of C is in excess of the multiple of D,
while the multiple of E is not in excess of the multiple of F, [V. Def. 7]
let them be taken,
and let G, H be equimultiples of C, E,
and K, L other, chance, equimultiples of D, F, so that G is in excess of K, but H is not in excess of L;
and, whatever multiple G is of C, let M be also that multiple of A,
and, whatever multiple K is of D, let N be also that multiple of B.

Now, since, as A is to B, so is C to D,
and of A, C equimultiples M, G have been taken,
and of B, D other, chance, equimultiples N, K,
 therefore, if M is in excess of N, G is also in excess of K,
 if equal, equal,
 and if less, less. [V. Def. 5]

But G is in excess of K;
 therefore M is also in excess of N.

But H is not in excess of L;
and M, H are equimultiples of A, E,
and N, L other, chance, equimultiples of B, F;
 therefore A has to B a greater ratio than E has to F. [V. Def. 7]

Therefore etc.

 Q.E.D.

Proposition 14

If a first magnitude have to a second the same ratio as a third has to a fourth, and the first be greater than the third, the second will also be greater than the fourth; if equal, equal; and if less, less.

For let a first magnitude A have the same
ratio to a second B as a third C has to a A —————— C ——————
fourth D; and let A be greater than C;
I say that B is also greater than D. B —————— D ——————

For, since A is greater than C,
and B is another, chance, magnitude,
 therefore A has to B a greater ratio than C has to B. [V. 8]

But, as A is to B, so is C to D;
therefore C has also to D a greater ratio than C has to B. [V. 13]

But that to which the same has a greater ratio is less; [V. 10]
therefore D is less than B;
so that B is greater than D.

Similarly we can prove that, if A be equal to C, B will also be equal to D;
and, if A be less than C, B will also be less than D.

Therefore etc.

Q.E.D.

Proposition 15

Parts have the same ratio as the same multiples of them taken in corresponding order.

For let AB be the same multiple of C that DE is of F;
I say that, as C is to F, so is AB to DE.

For, since AB is the same multiple of C that DE is of F, as many magnitudes as there are in AB equal to C, so many are there also in DE equal to F.

Let AB be divided into the magnitudes AG, GH, HB equal to C,
and DE into the magnitudes DK, KL, LE equal to F;
then the multitude of the magnitudes AG, GH, HB will be equal to the multitude of the magnitudes DK, KL, LE.

And, since AG, GH, HB are equal to one another,
and DK, KL, LE are also equal to one another,
therefore, as AG is to DK, so is GH to KL, and HB to LE. [V. 7]

Therefore, as one of the antecedents is to one of the consequents, so will all the antecedents be to all the consequents; [V. 12]
therefore, as AG is to DK, so is AB to DE.

But AG is equal to C and DK to F;
therefore, as C is to F, so is AB to DE.

Therefore etc.

Q.E.D.

Proposition 16

If four magnitudes be proportional, they will also be proportional alternately.

Let *A, B, C, D* be four proportional magnitudes, so that, as *A* is to *B*, so is *C* to *D*;
I say that they will also be so alternately [V. Def. 12], that is, as *A* is to *C*, so is *B* to *D*.

For of *A, B* let equimultiples *E, F* be taken,
and of *C, D* other, chance, equimultiples *G, H*.

Then, since *E* is the same multiple of *A* that *F* is of *B*,
and parts have the same ratio as the same multiples of them, [V. 15]
 therefore, as *A* is to *B*, so is *E* to *F*.

But as *A* is to *B*, so is *C* to *D*;
 therefore also, as *C* is to *D*, so is *E* to *F*. [V. 11]

Again, since *G, H* are equimultiples of *C, D*,
 therefore, as *C* is to *D*, so is *G* to *H*. [V. 15]

But, as *C* is to *D*, so is *E* to *F*;
 therefore also, as *E* is to *F*, so is *G* to *H*. [V. 11]

But, if four magnitudes be proportional, and the first be greater than the third,
the second will also be greater than the fourth;
if equal, equal;
and if less, less. [V. 14]

 Therefore, if *E* is in excess of *G*, *F* is also in excess of *H*,
 if equal, equal,
 and if less, less.

Now *E, F* are equimultiples of *A, B*,
and *G, H* other, chance, equimultiples of *C, D*;
 therefore, as *A* is to *C*, so is *B* to *D*. [V. Def. 5]

Therefore etc.

 Q.E.D.

Proposition 17

If magnitudes be proportional componendo, *they will also be proportional*
separando.

Let *AB, BE, CD, DF* be magnitudes proportional *componendo* [V. Def. 14], so that, as
AB is to *BE*, so is *CD* to *DF*;
I say that they will also be proportional *separando* [V. Def. 15], that is, as *AE* is to *EB*,
so is *CF* to *DF*.

For of *AE, EB, CF, FD* let equimultiples *GH, HK, LM, MN* be taken, and of *EB, FD* other, chance, equimultiples, *KO, NP.*

Then, since *GH* is the same multiple of *AE* that *HK* is of *EB*,
 therefore *GH* is the same multiple of *AE* that *GK* is of *AB*. [V. 1]

But *GH* is the same multiple of *AE* that *LM* is of *CF*;
 therefore *GK* is the same multiple of *AB* that *LM* is of *CF*.

Again, since *LM* is the same multiple of *CF* that *MN* is of *FD*,
 therefore *LM* is the same multiple of *CF* that *LN* is of *CD*. [V. 1]

But *LM* was the same multiple of *CF* that *GK* is of *AB*;
 therefore *GK* is the same multiple of *AB* that *LN* is of *CD*.

 Therefore *GK, LN* are equimultiples of *AB, CD*.

Again, since *HK* is the same multiple of *EB* that *MN* is of *FD*,
and *KO* is also the same multiple of *EB* that *NP* is of *FD*,
 therefore the sum *HO* is also the same multiple of *EB* that *MP* is of *FD*. [V. 2]

And, since, as *AB* is to *BE*, so is *CD* to *DF*,
and of *AB, CD* equimultiples *GK, LN* have been taken,
and of *EB, FD* equimultiples *HO, MP*,
 therefore, if *GK* is in excess of *HO*, *LN* is also in excess of *MP*,
 if equal, equal,
 and if less, less.

Let *GK* be in excess of *HO*;
then, if *HK* be subtracted from each,
 GH is also in excess of *KO*.[1]

But we saw that, if *GK* was in excess of *HO*, *LN* was also in excess of *MP*;
 therefore *LN* is also in excess of *MP*,
and, if *MN* be subtracted from each,
 LM is also in excess of *NP*;
 so that, if *GH* is in excess of *KO*, *LM* is also in excess of *NP*.

Similarly we can prove that,
 if *GH* be equal to *KO*, *LM* will also be equal to *NP*,
 and if less, less.

And *GH, LM* are equimultiples of *AE, CF*,
while *KO, NP* are other, chance, equimultiples of *EB, FD*;
 therefore, as *AE* is to *EB*, so is *CF* to *FD*.

Therefore etc.

 Q.E.D.

1. The wording in this paragraph reproduces Heath's 1908 translation and agrees with the Greek. The very different wording given in the Encyclopaedia Brittanica edition thus has no textual basis. —Ed.

Proposition 18

If magnitudes be proportional separando, *they will also be proportional* componendo.

Let *AE, EB, CF, FD* be magnitudes proportional *separando* [V. **Def. 15**], so that, as *AE* is to *EB*, so is *CF* to *FD*;
I say that they will also be proportional *componendo* [V. **Def. 14**], that is, as *AB* is to *BE*, so is *CD* to *FD*.

For, if *CD* be not to *DF* as *AB* to *BE*,
then, as *AB* is to *BE*, so will *CD* be either to some magnitude less than *DF* or to a greater.

First, let it be in that ratio to a less magnitude *DG*.

Then, since, as *AB* is to *BE*, so is *CD* to *DG*,
> they are magnitudes proportional *componendo*;
> so that they will also be proportional *separando*. [V. 17]

Therefore, as *AE* is to *EB*, so is *CG* to *GD*.

But also, by hypothesis,
> as *AE* is to *EB*, so is *CF* to *FD*.

> Therefore also, as *CG* is to *GD*, so is *CF* to *FD*. [V. 11]

But the first *CG* is greater than the third *CF*;
> therefore the second *GD* is also greater than the fourth *FD*. [V. 14]

But it is also less: which is impossible.

> Therefore, as *AB* is to *BE*, so is not *CD* to a less magnitude than *FD*.

Similarly we can prove that neither is it in that ratio to a greater;
> it is therefore in that ratio to *FD* itself.

Therefore etc.

Q.E.D.

Proposition 19

If, as a whole is to a whole, so is a part subtracted to a part subtracted, the remainder will also be to the remainder as whole to whole.

For, as the whole *AB* is to the whole *CD*, so let the part *AE* subtracted be to the part *CF* subtracted;
I say that the remainder *EB* will also be to the remainder *FD* as the whole *AB* to the whole *CD*.

For since, as *AB* is to *CD*, so is *AE* to *CF*,

alternately also, as BA is to AE, so is DC to CF. [V. 16]

And, since the magnitudes are proportional *componendo*, they will also be proportional *separando*, [V. 17]

that is, as BE is to EA, so is DF to CF,

and, alternately,

as BE is to DF, so is EA to FC. [V. 16]

But, as AE is to CF, so by hypothesis is the whole AB to the whole CD.

Therefore also the remainder EB will be to the remainder FD as the whole AB is to the whole CD. [V. 11]

Therefore etc.

Q.E.D.

⟨**PORISM.** From this it is manifest that, if magnitudes be proportional *componendo* [V. Def. 14], they will also be proportional *convertendo* [V. Def. 16].⟩

Proposition 20

If there be three magnitudes, and others equal to them in multitude, which taken two and two are in the same ratio, and if ex aequali *the first be greater than the third, the fourth will also be greater than the sixth; if equal, equal; and, if less, less.*

Let there be three magnitudes A, B, C, and others D, E, F equal to them in multitude, which taken two and two are in the same ratio, so that,

as A is to B, so is D to E,

and,

as B is to C, so is E to F;

and let A be greater than C ex aequali; [V. Def. 17]

I say that D will also be greater than F;

if A is equal to C, equal; and, if less, less.

For, since A is greater than C,

and B is some other magnitude,

and the greater has to the same a greater ratio than the less has, [V. 8]

therefore A has to B a greater ratio than C has to B.

But, as A is to B, so is D to E,

and, as C is to B, inversely, so is F to E;

therefore D has also to E a greater ratio than F has to E. [V. 13]

But, of magnitudes which have a ratio to the same, that which has a greater ratio is greater; [V. 10]

therefore D is greater than F.

Similarly we can prove that, if A be equal to C, D will also be equal to F; and if less, less.

Therefore etc.

Q.E.D.

Proposition 21

If there be three magnitudes, and others equal to them in multitude, which taken two and two together are in the same ratio, and the proportion of them be perturbed, then, if ex aequali *the first magnitude is greater than the third, the fourth will also be greater than the sixth; if equal, equal; and if less, less.*

$$A \text{———————} \qquad D \text{———————}$$
$$B \text{—————} \qquad E \text{———————}$$
$$C \text{—————} \qquad F \text{—————}$$

Let there be three magnitudes A, B, C, and others D, E, F equal to them in multitude, which taken two and two are in the same ratio, and let the proportion of them be perturbed **[V. Def. 18]**, so that,

as A is to B, so is E to F,

and,

as B is to C, so is D to E,

and let A be greater than C *ex aequali*; **[V. Def. 17]**
I say that D will also be greater than F;

if A is equal to C, equal; and if less, less.

For, since A is greater than C, and B is some other magnitude,

therefore A has to B a greater ratio than C has to B. **[V. 8]**

But, as A is to B, so is E to F,
and, as C is to B, inversely, so is E to D.

Therefore also E has to F a greater ratio than E has to D. **[V. 13]**

But that to which the same has a greater ratio is less; **[V. 10]**

therefore F is less than D;

therefore D is greater than F.

Similarly we can prove that, if A be equal to C, D will also be equal to F; and if less, less.

Therefore etc.

Q.E.D.

Proposition 22

If there be any number of magnitudes whatever, and others equal to them in multitude, which taken two and two together are in the same ratio, they will also be in the same ratio ex aequali.

Let there be any number of magnitudes *A, B, C*, and others *D, E, F* equal to them in multitude, which taken two and two together are in the same ratio, so that,

as *A* is to *B*, so is *D* to *E*,

and,

as *B* is to *C*, so is *E* to *F*;

I say that they will also be in the same ratio *ex aequali* [**V. Def. 17**],
[that is, as *A* is to *C*, so is *D* to *F*].

For of *A, D* let equimultiples *G, H* be taken,
and of *B, E* other, chance, equimultiples *K, L*;
and, further, of *C, F* other, chance, equimultiples *M, N*.

Then, since, as *A* is to *B*, so is *D* to *E*,
and of *A, D* equimultiples *G, H* have been taken,
and of *B, E* other, chance, equimultiples *K, L*,

therefore, as *G* is to *K*, so is *H* to *L*. [**V. 4**]

For the same reason also,

as *K* is to *M*, so is *L* to *N*.

Since, then, there are three magnitudes *G, K, M*, and others *H, L, N* equal to them in multitude, which taken two and two together are in the same ratio,

therefore, *ex aequali*, if *G* is in excess of *M, H* is also in excess of *N*;

if equal, equal;

and if less, less. [**V. 20**]

And *G, H* are equimultiples of *A, D*,
and *M, N* other, chance, equimultiples of *C, F*.

Therefore, as *A* is to *C*, so is *D* to *F*. [**V. Def. 5**]

Therefore etc.

Q.E.D.

Proposition 23

If there be three magnitudes, and others equal to them in multitude, which taken two and two together are in the same ratio, and the proportion of them be perturbed, they will also be in the same ratio ex aequali.

Let there be three magnitudes *A, B, C*, and others equal to them in multitude,

which, taken two and two together, are in the same proportion, namely D, E, F; and let the proportion of them be perturbed [V. Def. 18], so that,

as A is to B, so is E to F,

and,

as B is to C, so is D to E;

I say that, as A is to C, so is D to F.

Of A, B, D let equimultiples G, H, K be taken, and of C, E, F other, chance, equimultiples L, M, N.

Then, since G, H are equimultiples of A, B, and parts have the same ratio as the same multiples of them, [V. 15]

therefore, as A is to B, so is G to H.

For the same reason also,

as E is to F, so is M to N.

And, as A is to B, so is E to F;

therefore also, as G is to H, so is M to N. [V. 11]

Next, since, as B is to C, so is D to E,

alternately, also, as B is to D, so is C to E. [V. 16]

And, since H, K are equimultiples of B, D, and parts have the same ratio as their equimultiples,

therefore, as B is to D, so is H to K. [V. 15]

But, as B is to D, so is C to E;

therefore also, as H is to K, so is C to E. [V. 11]

Again, since L, M are equimultiples of C, E,

therefore, as C is to E, so is L to M. [V. 15]

But, as C is to E, so is H to K;

therefore also, as H is to K, so is L to M, [V. 11]

and, alternately, as H is to L, so is K to M. [V. 16]

But it was also proved that,

as G is to H, so is M to N.

Since, then, there are three magnitudes G, H, L, and others equal to them in multitude K, M, N, which taken two and two together are in the same ratio, and the proportion of them is perturbed,

therefore, *ex aequali*, if G is in excess of L, K is also in excess of N;

if equal, equal;

and if less, less. [V. 21]

And G, K are equimultiples of A, D, and L, N of C, F.

Therefore, as A is to C, so is D to F.

Therefore etc.

Q.E.D.

Proposition 24

If a first magnitude have to a second the same ratio as a third has to a fourth,

and also a fifth have to the second the same ratio as a sixth to the fourth, the first and fifth added together will have to the second the same ratio as the third and sixth have to the fourth.

Let a first magnitude *AB* have to a second *C* the same ratio as a third *DE* has to a fourth *F*;
and let also a fifth *BG* have to the second *C* the same ratio as a sixth *EH* has to the fourth *F*;
I say that the first and fifth added together, *AG*, will have to the second *C* the same ratio as the third and sixth, *DH*, has to the fourth *F*.

For since, as *BG* is to *C*, so is *EH* to *F*,
 inversely, as *C* is to *BG*, so is *F* to *EH*.

Since, then, as *AB* is to *C*, so is *DE* to *F*,
and, as *C* is to *BG*, so is *F* to *EH*,
 therefore, *ex aequali*, as *AB* is to *BG*, so is *DE* to *EH*. [V. 22]

And, since the magnitudes are proportional *separando*, they will also be proportional *componendo*; [V. 18]
 therefore, as *AG* is to *GB*, so is *DH* to *HE*.

But also, as *BG* is to *C*, so is *EH* to *F*;
 therefore, *ex aequali*, as *AG* is to *C*, so is *DH* to *F*

 [V. 22]

Therefore etc.

 Q.E.D.

Proposition 25

If four magnitudes be proportional, the greatest and the least are greater than the remaining two.

Let the four magnitudes *AB*, *CD*, *E*, *F* be proportional so that, as *AB* is to *CD*, so is *E* to *F*, and let *AB* be the greatest of them and *F* the least;
I say that *AB*, *F* are greater than *CD*, *E*.

For let *AG* be made equal to *E*, and *CH* equal to *F*.

Since, as *AB* is to *CD*, so is *E* to *F*,
and *E* is equal to *AG*, and *F* to *CH*,
 therefore, as *AB* is to *CD*, so is *AG* to *CH*.

And since, as the whole *AB* is to the whole *CD*, so is the part *AG* subtracted to the part *CH* subtracted,

the remainder *GB* will also be to the remainder *HD* as the whole *AB* is to the whole *CD*. [V. 19]

But *AB* is greater than *CD*;

therefore *GB* is also greater than *HD*.

And, since *AG* is equal to *E*, and *CH* to *F*,

therefore *AG*, *F* are equal to *CH*, *E*.

And if, *GB*, *HD* being unequal, and *GB* greater,
AG, *F* be added to *GB*
and *CH*, *E* be added to *HD*,

it follows that *AB*, *F* are greater than *CD*, *E*.

Therefore etc.

Q.E.D.

Book VI

Definitions

1. *Similar rectilineal figures* are such as have their angles severally equal and the sides about the equal angles proportional.

⟨2. *Reciprocally related* figures.⟩[1]

3. A straight line is said to have been *cut in extreme and mean ratio* when, as the whole line is to the greater segment, so is the greater to the less.

4. The *height* of any figure is the perpendicular drawn from the vertex to the base.

Proposition 1

Triangles and parallelograms which are under the same height are to one another as their bases.

Let *ABC, ACD* be triangles and *EC, CF* parallelograms under the same height; I say that, as the base *BC* is to the base *CD*, so is the triangle *ABC* to the triangle *ACD*, and the parallelogram *EC* to the parallelogram *CF*.

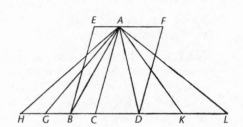

For let *BD* be produced in both directions to the points *H, L* and let any number of straight lines *BG, GH* be made equal to the base *BC*, and any number of straight lines *DK, KL* equal to the base *CD*;
let *AG, AH, AK, AL* be joined.

Then, since *CB, BG, GH* are equal to one another,
the triangles *ABC, AGB, AHG* are also equal to one another. [I. 38]

Therefore, whatever multiple the base *HC* is of the base *BC*, that multiple also is the triangle *AHC* of the triangle *ABC*.

For the same reason,
whatever multiple the base *LC* is of the base *CD*, that multiple also is the triangle *ALC* of the triangle *ACD*;
and, if the base *HC* is equal to the base *CL*, the triangle *AHC* is also equal to the triangle *ACL*, [I. 38]

1. Definition 2 is regarded as spurious. It is never used by Euclid. According to Heath, its Greek text conveys no intelligible meaning, and Heath does not translate it. —Ed.

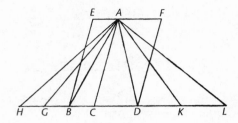

if the base *HC* is in excess of the base *CL*, the triangle *AHC* is also in excess of the triangle *ACL*,
and, if less, less.

Thus, there being four magnitudes, two bases *BC, CD* and two triangles *ABC, ACD,*
equimultiples have been taken of the base *BC* and the triangle *ABC*, namely the base *HC* and the triangle *AHC*,
and of the base *CD* and the triangle *ADC* other, chance, equimultiples, namely the base *LC* and the triangle *ALC*;
and it has been proved that,
if the base *HC* is in excess of the base *CL*, the triangle *AHC* is also in excess of the triangle *ALC*;
if equal, equal;
and, if less, less.

Therefore, as the base *BC* is to the base *CD*, so is the triangle *ABC* to the triangle *ACD*. [V. Def. 5]

Next, since the parallelogram *EC* is double of the triangle *ABC*, [I. 41]
and the parallelogram *FC* is double of the triangle *ACD*,
while parts have the same ratio as the same multiples of them, [V. 15]
therefore, as the triangle *ABC* is to the triangle *ACD*, so is the parallelogram *EC* to the parallelogram *FC*.

Since, then, it was proved that, as the base *BC* is to *CD*, so is the triangle *ABC* to the triangle *ACD*,
and, as the triangle *ABC* is to the triangle *ACD*, so is the parallelogram *EC* to the parallelogram *CF*,
therefore also, as the base *BC* is to the base *CD*, so is the parallelogram *EC* to the parallelogram *FC*. [V. 11]

Therefore etc.

Q.E.D.

Proposition 2

If a straight line be drawn parallel to one of the sides of a triangle, it will cut the sides of the triangle proportionally; and, if the sides of the triangle be cut proportionally, the line joining the points of section will be parallel to the remaining side of the triangle.

For let *DE* be drawn parallel to *BC*, one of the sides of the triangle *ABC*;

I say that, as *BD* is to *DA*, so is *CE* to *EA*.

For let *BE*, *CD* be joined.

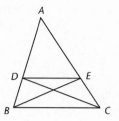

Therefore the triangle *BDE* is equal to the triangle *CDE*; for they are on the same base *DE* and in the same parallels *DE*, *BC*. [I. 38]

And the triangle *ADE* is another area.

But equals have the same ratio to the same; [V. 7] therefore, as the triangle *BDE* is to the triangle *ADE*, so is the triangle *CDE* to the triangle *ADE*.

But, as the triangle *BDE* is to *ADE*, so is *BD* to *DA*; for, being under the same height, the perpendicular drawn from *E* to *AB*, they are to one another as their bases. [VI. 1]

For the same reason also,

 as the triangle *CDE* is to *ADE*, so is *CE* to *EA*.

 Therefore also, as *BD* is to *DA*, so is *CE* to *EA*. [V. 11]

Again, let the sides *AB*, *AC* of the triangle *ABC* be cut proportionally, so that, as *BD* is to *DA*, so is *CE* to *EA*; and let *DE* be joined.

I say that *DE* is parallel to *BC*.

For, with the same construction, since, as *BD* is to *DA*, so is *CE* to *EA*, but, as *BD* is to *DA*, so is the triangle *BDE* to the triangle *ADE*, and, as *CE* is to *EA*, so is the triangle *CDE* to the triangle *ADE*, [VI. 1] therefore also, as the triangle *BDE* is to the triangle *ADE*, so is the triangle *CDE* to the triangle *ADE*. [V. 11]

 Therefore each of the triangles *BDE, CDE* has the same ratio to *ADE*.

 Therefore the triangle *BDE* is equal to the triangle *CDE*; [V. 9] and they are on the same base *DE*.

But equal triangles which are on the same base are also in the same parallels. [I. 39]

 Therefore *DE* is parallel to *BC*.

Therefore etc.

 Q.E.D.

Proposition 3

If an angle of a triangle be bisected and the straight line cutting the angle cut the base also, the segments of the base will have the same ratio as the remaining sides of the triangle; and, if the segments of the base have the same ratio as the remaining sides of the triangle, the straight line joined from the vertex to the point of section will bisect the angle of the triangle.

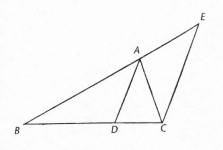

Let *ABC* be a triangle, and let the angle *BAC* be bisected by the straight line *AD*;
I say that, as *BD* is to *CD*, so is *BA* to *AC*.

For let *CE* be drawn through *C* parallel to *DA*, and let *BA* be carried through and meet it at *E*.

Then, since the straight line *AC* falls upon the parallels *AD*, *EC*,
the angle *ACE* is equal to the angle *CAD*. [I. 29]

But the angle *CAD* is by hypothesis equal to the angle *BAD*;
therefore the angle *BAD* is also equal to the angle *ACE*.

Again, since the straight line *BAE* falls upon the parallels *AD*, *EC*,
the exterior angle *BAD* is equal to the interior angle *AEC*. [I. 29]

But the angle *ACE* was also proved equal to the angle *BAD*;
therefore the angle *ACE* is also equal to the angle *AEC*,
so that the side *AE* is also equal to the side *AC*. [I. 6]

And, since *AD* has been drawn parallel to *EC*, one of the sides of the triangle *BCE*,
therefore, proportionally, as *BD* is to *DC*, so is *BA* to *AE*.

But *AE* is equal to *AC*; [VI. 2]
therefore, as *BD* is to *DC*, so is *BA* to *AC*.

Again, let *BA* be to *AC* as *BD* to *DC*, and let *AD* be joined;
I say that the angle *BAC* has been bisected by the straight line *AD*.

For, with the same construction,
since, as *BD* is to *DC*, so is *BA* to *AC*,
and also, as *BD* is to *DC*, so is *BA* to *AE*: for *AD* has been drawn parallel to *EC*,
one of the sides of the triangle *BCE*: [VI. 2]
therefore also, as *BA* is to *AC*, so is *BA* to *AE*. [V. 11]

Therefore *AC* is equal to *AE*, [V. 9]
so that the angle *AEC* is also equal to the angle *ACE*. [I. 5]

But the angle *AEC* is equal to the exterior angle *BAD*, [I. 29]
and the angle *ACE* is equal to the alternate angle *CAD*; [I. 29]
therefore the angle *BAD* is also equal to the angle *CAD*.

Therefore the angle *BAC* has been bisected by the straight line *AD*.

Therefore etc.

Q.E.D.

Proposition 4

In equiangular triangles the sides about the equal angles are proportional, and those are corresponding sides which subtend the equal angles.

Let *ABC*, *DCE* be equiangular triangles having the angle *ABC* equal to the angle *DCE*, the angle *BAC* to the angle *CDE*, and further the angle *ACB* to the angle *CED*;
I say that in the triangles *ABC*, *DCE* the sides about the equal angles are proportional, and those are corresponding sides which subtend the equal angles.

For let *BC* be placed in a straight line with *CE*.

Then, since the angles *ABC*, *ACB* are less than
two right angles, [I. 17]
and the angle *ACB* is equal to the angle *DEC*,
therefore the angles *ABC*, *DEC* are less than two
right angles;
 therefore *BA*, *ED*, when produced, will meet.
 [I. Post. 5]

Let them be produced and meet at *F*.

Now, since the angle *DCE* is equal to the angle
ABC,
 BF is parallel to *CD*. [I. 28]

Again, since the angle *ACB* is equal to the angle *DEC*,
 AC is parallel to *FE*. [I. 28]

Therefore *FACD* is a parallelogram;
 therefore *FA* is equal to *DC*, and *AC* to *FD*. [I. 34]

And, since *AC* has been drawn parallel to *FE*, one side of the triangle *FBE*,
 therefore, as *BA* is to *AF*, so is *BC* to *CE*. [VI. 2]

But *AF* is equal to *CD*;
 therefore, as *BA* is to *CD*, so is *BC* to *CE*,
 and alternately, as *AB* is to *BC*, so is *DC* to *CE*. [V. 16]

Again, since *CD* is parallel to *BF*,
 therefore, as *BC* is to *CE*, so is *FD* to *DE*. [VI. 2]

But *FD* is equal to *AC*;
 therefore, as *BC* is to *CE*, so is *AC* to *DE*,
 and alternately, as *BC* is to *CA*, so is *CE* to *ED*. [V. 16]

Since, then, it was proved that,
as *AB* is to *BC*, so is *DC* to *CE*,
and, as *BC* is to *CA*, so is *CE* to *ED*;
 therefore, *ex aequali*, as *BA* is to *AC*, so is *CD* to *DE*. [V. 22]

Therefore etc.

Q.E.D.

Proposition 5

*If two triangles have their sides proportional, the triangles will be equiangular
and will have those angles equal which the corresponding sides subtend.*

Let *ABC*, *DEF* be two triangles having
their sides proportional, so that,
as *AB* is to *BC*, so is *DE* to *EF*,
as *BC* is to *CA*, so is *EF* to *FD*,
and further, as *BA* is to *AC*, so is *ED* to *DF*;
I say that the triangle *ABC* is equi-
angular with the triangle *DEF*, and

they will have those angles equal which the corresponding sides subtend, namely the angle *ABC* to the angle *DEF*, the angle *BCA* to the angle *EFD*, and further the angle *BAC* to the angle *EDF*.

For on the straight line *EF*, and at the points *E*, *F* on it, let there be constructed the angle *FEG* equal to the angle *ABC*, and the angle *EFG* equal to the angle *ACB*;

[I. 23]

therefore the remaining angle at *A* is equal to the remaining angle at *G*. [I. 32]

Therefore the triangle *ABC* is equiangular with the triangle *GEF*.

Therefore in the triangles *ABC*, *GEF* the sides about the equal angles are proportional, and those are corresponding sides which subtend the equal angles;

[VI. 4]

therefore, as *AB* is to *BC*, so is *GE* to *EF*.

But, as *AB* is to *BC*, so by hypothesis is *DE* to *EF*;

therefore, as *DE* is to *EF*, so is *GE* to *EF*. [V. 11]

Therefore each of the straight lines *DE*, *GE* has the same ratio to *EF*;

therefore *DE* is equal to *GE*. [V. 9]

For the same reason

DF is also equal to *GF*.

Since then *DE* is equal to *EG*,
and *EF* is common,

the two sides *DE*, *EF* are equal to the two sides *GE*, *EF*;
and the base *DF* is equal to the base *FG*;

therefore the angle *DEF* is equal to the angle *GEF*, [I. 8]

and the triangle *DEF* is equal to the triangle *GEF*,

and the remaining angles are equal to the remaining angles, namely those which the equal sides subtend. [I. 4]

Therefore the angle *DFE* is also equal to the angle *GFE*,
and the angle *EDF* to the angle *EGF*.

And, since the angle *FED* is equal to the angle *GEF*,
while the angle *GEF* is equal to the angle *ABC*,

therefore the angle *ABC* is also equal to the angle *DEF*.

For the same reason

the angle *ACB* is also equal to the angle *DFE*,
and further, the angle at *A* to the angle at *D*;

therefore the triangle *ABC* is equiangular with the triangle *DEF*.

Therefore etc.

Q.E.D.

Proposition 6

If two triangles have one angle equal to one angle and the sides about the equal angles proportional, the triangles will be equiangular and will have those angles equal which the corresponding sides subtend.

Let *ABC, DEF* be two triangles having one angle *BAC* equal to one angle *EDF* and the sides about the equal angles proportional, so that,
as *BA* is to *AC*, so is *ED* to *DF*;
I say that the triangle *ABC* is equiangular with the triangle *DEF*, and will have the angle *ABC* equal to the angle *DEF*, and the angle *ACB* to the angle *DFE*.

For on the straight line *DF*, and at the points *D, F* on it, let there be constructed the angle *FDG* equal to either of the angles *BAC, EDF*, and the angle *DFG* equal to the angle *ACB*; [I. 23]
 therefore the remaining angle at *B* is equal to the remaining angle at *G*. [I. 32]

Therefore the triangle *ABC* is equiangular with the triangle *DGF*.

Therefore, proportionally, as *BA* is to *AC*, so is *GD* to *DF*. [VI. 4]

But, by hypothesis, as *BA* is to *AC*, so also is *ED* to *DF*;
 therefore also, as *ED* is to *DF*, so is *GD* to *DF*. [V. 11]

Therefore *ED* is equal to *DG*; [V. 9]
and *DF* is common;
 therefore the two sides *ED, DF* are equal to the two sides *GD, DF*;
and the angle *EDF* is equal to the angle *GDF*;
 therefore the base *EF* is equal to the base *GF*,
 and the triangle *DEF* is equal to the triangle *DGF*,
and the remaining angles will be equal to the remaining angles, namely those which the equal sides subtend. [I. 4]

Therefore the angle *DFG* is equal to the angle *DFE*,
 and the angle *DGF* to the angle *DEF*.

But the angle *DFG* is equal to the angle *ACB*;
 therefore the angle *ACB* is also equal to the angle *DFE*.

And, by hypothesis, the angle *BAC* is also equal to the angle *EDF*;
therefore the remaining angle at *B* is also equal to the remaining angle at *E*;
[1. 32]
 therefore the triangle *ABC* is equiangular with the triangle *DEF*.

Therefore etc.

Q.E.D.

Proposition 7

If two triangles have one angle equal to one angle, the sides about other angles proportional, and the remaining angles either both less or both not less than a right angle, the triangles will be equiangular and will have those angles equal, the sides about which are proportional.

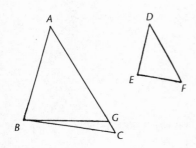

Let *ABC, DEF* be two triangles having one angle equal to one angle, the angle *BAC* to the angle *EDF,* the sides about other angles *ABC, DEF* proportional, so that,
as *AB* is to *BC,* so is *DE* to *EF,*
and, first, each of the remaining angles at *C, F* less than a right angle;
I say that the triangle *ABC* is equiangular with the triangle *DEF,* the angle *ABC* will be equal to the angle *DEF,* and the remaining angle, namely the angle at *C,* equal to the remaining angle, the angle at *F.*

For, if the angle *ABC* is unequal to the angle *DEF,* one of them is greater.

Let the angle *ABC* be greater;
and on the straight line *AB,* and at the point *B* on it, let the angle *ABG* be constructed equal to the angle *DEF.* [I. 23]

Then, since the angle *A* is equal to *D,*
and the angle *ABG* to the angle *DEF,*
 therefore the remaining angle *AGB* is equal to the remaining angle *DFE.* [I. 32]

 Therefore the triangle *ABG* is equiangular with the triangle *DEF.*

 Therefore, as *AB* is to *BG,* so is *DE* to *EF.* [VI. 4]

But, as *DE* is to *EF,* so by hypothesis is *AB* to *BC;*
 therefore *AB* has the same ratio to each of the straight lines *BC, BG;* [V. 11]
 therefore *BC* is equal to *BG,* [V. 9]
 so that the angle at *C* is also equal to the angle *BGC.* [I. 5]

But, by hypothesis, the angle at *C* is less than a right angle;
 therefore the angle *BGC* is also less than a right angle;
 so that the angle *AGB* adjacent to it is greater than a right angle. [I. 13]

And it was proved equal to the angle at *F;*
 therefore the angle at *F* is also greater than a right angle.

But it is by hypothesis less than a right angle: which is absurd.

 Therefore the angle *ABC* is not unequal to the angle *DEF;*
 therefore it is equal to it.

But the angle at *A* is also equal to the angle at *D;*
 therefore the remaining angle at *C* is equal to the remaining angle at *F.* [I. 32]

 Therefore the triangle *ABC* is equiangular with the triangle *DEF.*

But, again, let each of the angles at *C, F* be supposed not less than a right angle;
I say again that, in this case too, the triangle *ABC* is equiangular with the triangle *DEF.*

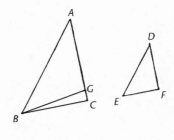

For, with the same construction, we can prove similarly that

BC is equal to BG;

so that the angle at C is also equal to the angle BGC.

[I. 5]

But the angle at C is not less than a right angle; therefore neither is the angle BGC less than a right angle.

Thus in the triangle BGC the two angles are not less than two right angles: which is impossible. [I. 17]

Therefore, once more, the angle ABC is not unequal to the angle DEF; therefore it is equal to it.

But the angle at A is also equal to the angle at D; therefore the remaining angle at C is equal to the remaining angle at F. [I. 32]

Therefore the triangle ABC is equiangular with the triangle DEF.

Therefore etc.

Q.E.D.

Proposition 8

If in a right-angled triangle a perpendicular be drawn from the right angle to the base, the triangles adjoining the perpendicular are similar both to the whole and to one another.

Let ABC be a right-angled triangle having the angle BAC right, and let AD be drawn from A perpendicular to BC;
I say that each of the triangles ABD, ADC is similar to the whole ABC and, further, they are similar to one another.

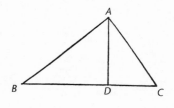

For, since the angle BAC is equal to the angle ADB, for each is right,
and the angle at B is common to the two triangles ABC and ABD,
therefore the remaining angle ACB is equal to the remaining angle BAD; [I. 32]
therefore the triangle ABC is equiangular with the triangle ABD.

Therefore, as BC which subtends the right angle in the triangle ABC is to BA which subtends the right angle in the triangle ABD, so is AB itself which subtends the angle at C in the triangle ABC to BD which subtends the equal angle BAD in the triangle ABD, and so also is AC to AD which subtends the angle at B common to the two triangles. [VI. 4]

Therefore the triangle ABC is both equiangular to the triangle ABD and has the sides about the equal angles proportional.

Therefore the triangle ABC is similar to the triangle ABD. [VI. Def. 1]

Similarly we can prove that
the triangle ABC is also similar to the triangle ADC;
therefore each of the triangles ABD, ADC is similar to the whole ABC.

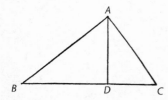

I say next that the triangles *ABD*, *ADC* are also similar to one another.

For, since the right angle *BDA* is equal to the right angle *ADC*,
and moreover the angle *BAD* was also proved equal to the angle at *C*,
therefore the remaining angle at *B* is also equal to the remaining angle *DAC*; [I. 32]
therefore the triangle *ABD* is equiangular with the triangle *ADC*.

Therefore, as *BD* which subtends the angle *BAD* in the triangle *ABD* is to *DA* which subtends the angle at *C* in the triangle *ADC* equal to the angle *BAD*, so is *AD* itself which subtends the angle at *B* in the triangle *ABD* to *DC* which subtends the angle *DAC* in the triangle *ADC* equal to the angle at *B*,
and so also is *BA* to *AC*, these sides subtending the right angles; [VI. 4]
therefore the triangle *ABD* is similar to the triangle *ADC*. [VI. Def. 1]

Therefore etc.

Q.E.D.

PORISM. From this it is clear that, if in a right-angled triangle a perpendicular be drawn from the right angle to the base, the straight line so drawn is a mean proportional between the segments of the base.

Proposition 9

From a given straight line to cut off a prescribed part.

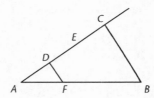

Let *AB* be the given straight line;
thus it is required to cut off from *AB* a prescribed part.

Let the third part be that prescribed.

Let a straight line *AC* be drawn through from *A* containing with *AB* any angle;
let a point *D* be taken at random on *AC*, and let *DE*, *EC* be made equal to *AD*. [I. 3]

Let *BC* be joined, and through *D* let *DF* be drawn parallel to it. [I. 31]

Then, since *FD* has been drawn parallel to *BC*, one of the sides of the triangle *ABC*,
therefore, proportionally, as *CD* is to *DA*, so is *BF* to *FA*. [VI. 2]

But *CD* is double of *DA*;
therefore *BF* is also double of *FA*;
therefore *BA* is triple of *AF*.

Therefore from the given straight line *AB* the prescribed third part *AF* has been cut off.

Q.E.F.

Proposition 10

To cut a given uncut straight line similarly to a given cut straight line.

Let *AB* be the given uncut straight line, and *AC* the
straight line cut at the points *D, E*; and let them be
so placed as to contain any angle;
let *CB* be joined, and through *D, E* let *DF, EG* be
drawn parallel to *BC*, and through *D* let *DHK* be
drawn parallel to *AB*. [I. 31]

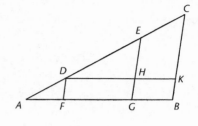

Therefore each of the figures *FH, HB* is a
parallelogram;
 therefore *DH* is equal to *FG* and *HK* to *GB*. [I. 34]

Now, since the straight line *HE* has been drawn
parallel to *KC*, one of the sides of the triangle *DKC*,
 therefore, proportionally, as *CE* is to *ED*, so is *KH* to *HD*. [VI. 2]

But *KH* is equal to *BG*, and *HD* to *GF*;
 therefore, as *CE* is to *ED*, so is *BG* to *GF*.

Again, since *FD* has been drawn parallel to *GE*, one of the sides of the triangle *AGE*,
 therefore, proportionally, as *ED* is to *DA*, so is *GF* to *FA*. [VI. 2]

But it was also proved that, as *CE* is to *ED*, so is *BG* to *GF*;
 therefore, as *CE* is to *ED*, so is *BG* to *GF*,
 and, as *ED* is to *DA*, so is *GF* to *FA*.

Therefore the given uncut straight line *AB* has been cut similarly to the given cut
straight line *AC*.

 Q.E.F.

Proposition 11

To two given straight lines to find a third proportional.

Let *BA, AC* be the two given straight lines, and let them
be placed so as to contain any angle;
thus it is required to find a third proportional to *BA, AC*.

For let them be produced to the points *D, E*,
and let *BD* be made equal to *AC*; [I. 3]
let BC be joined,
and through D let DE be drawn parallel to it. [I. 31]

Since, then, *BC* has been drawn parallel to *DE*, one of the
sides of the triangle *ADE*,
 proportionally, as *AB* is to *BD*, so is *AC* to *CE*. [VI. 2]

But *BD* is equal to *AC*;
 therefore, as *AB* is to *AC*, so is *AC* to *CE*.

Therefore to two given straight lines *AB, AC* a third proportional to them, *CE*,
has been found.

 Q.E.F.

Proposition 12

To three given straight lines to find a fourth proportional.

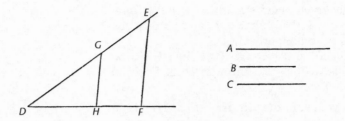

Let *A, B, C* be the three given straight lines;
thus it is required to find a fourth proportional to *A, B, C*.

Let two straight lines *DE, DF* be set out containing any angle *EDF*;
let *DG* be made equal to *A, GE* equal to *B*, and further *DH* equal to *C*;
let *GH* be joined, and let *EF* be drawn through *E* parallel to it. [I. 31]

Since, then, *GH* has been drawn parallel to *EF*, one of the sides of the triangle *DEF*,

therefore, as *DG* is to *GE*, so is *DH* to *HF*. [VI. 2]

But *DG* is equal to *A, GE* to *B*, and *DH* to *C*;

therefore, as *A* is to *B*, so is *C* to *HF*.

Therefore to the three given straight lines *A, B, C* a fourth proportional *HF* has been found.

Q.E.F.

Proposition 13

To two given straight lines to find a mean proportional.

Let *AB, BC* be the two given straight lines;
thus it is required to find a mean proportional to *AB, BC*.

Let them be placed in a straight line, and let the semi-circle *ADC* be described on *AC*;
let *BD* be drawn from the point *B* at right angles to the straight line *AC*,
and let *AD, DC* be joined.

Since the angle *ADC* is an angle in a semicircle, it is right. [III. 31]

And, since, in the right-angled triangle *ADC, DB* has been drawn from the right angle perpendicular to the base,
therefore *DB* is a mean proportional between the segments of the base, *AB, BC*.
[VI. 8, Por.]

Therefore to the two given straight lines *AB, BC* a mean proportional *DB* has been found.

Q.E.F.

Proposition 14

In equal and equiangular parallelograms the sides about the equal angles are reciprocally proportional; and equiangular parallelograms in which the sides about the equal angles are reciprocally proportional are equal.

Let *AB, BC* be equal and equiangular parallelograms having the angles at *B* equal,
and let *DB, BE* be placed in a straight line;
 therefore *FB, BG* are also in a straight line. [I. 14]

I say that, in *AB, BC*, the sides about the equal angles are reciprocally proportional, that is to say, that, as *DB* is to *BE*, so is *GB* to *BF*.

For let the parallelogram *FE* be completed.

Since, then, the parallelogram *AB* is equal to the parallelogram *BC*,
and *FE* is another area,
 therefore, as *AB* is to *FE*, so is *BC* to *FE*. [V. 7]

But, as *AB* is to *FE*, so is *DB* to *BE*, [VI. 1]
and, as *BC* is to *FE*, so is *GB* to *BF*, [VI. 1]
 therefore also, as *DB* is to *BE*, so is *GB* to *BF*. [V. 11]

Therefore in the parallelograms *AB, BC* the sides about the equal angles are reciprocally proportional.

Next, let *GB* be to *BF* as *DB* to *BE*;
I say that the parallelogram *AB* is equal to the parallelogram *BC*.

For since, as *DB* is to *BE*, so is *GB* to *BF*,
while, as *DB* is to *BE*, so is the parallelogram *AB* to the parallelogram *FE*, [VI. 1]
and, as *GB* is to *BF*, so is the parallelogram *BC* to the parallelogram *FE*, [VI. 1]
 therefore also, as *AB* is to *FE*, so is *BC* to *FE*; [V. 11]
 therefore the parallelogram *AB* is equal to the parallelogram *BC*. [V. 9]

Therefore etc.

Q.E.D.

Proposition 15

In equal triangles which have one angle equal to one angle the sides about the equal angles are reciprocally proportional; and those triangles which have one angle equal to one angle, and in which the sides about the equal angles are reciprocally proportional, are equal.

Let *ABC, ADE* be equal triangles having one angle equal to one angle, namely the angle *BAC* to the angle *DAE*;
I say that in the triangles *ABC, ADE* the sides about the equal angles are reciprocally proportional,
that is to say, that, as *CA* is to *AD*, so is *EA* to *AB*.

For let them be placed so that *CA* is in a straight line with *AD*;

therefore *EA* is also in a straight line with *AB*. [I. 14]

Let *BD* be joined.

Since, then, the triangle *ABC* is equal to the triangle *ADE*, and *BAD* is another area,
therefore, as the triangle *CAB* is to the triangle *BAD*, so is the triangle *EAD* to the triangle *BAD*. [V. 7]

But, as *CAB* is to *BAD*, so is *CA* to *AD*, [VI. 1]
and, as *EAD* is to *BAD*, so is *EA* to *AB*. [VI. 1]

Therefore also, as *CA* is to *AD*, so is *EA* to *AB*. [V. 11]

Therefore in the triangles *ABC*, *ADE* the sides about the equal angles are reciprocally proportional.

Next, let the sides of the triangles *ABC*, *ADE* be reciprocally proportional, that is to say, let *EA* be to *AB* as *CA* to *AD*;
I say that the triangle *ABC* is equal to the triangle *ADE*.

For, if *BD* be again joined,
since, as *CA* is to *AD*, so is *EA* to *AB*,
while, as *CA* is to *AD*, so is the triangle *ABC* to the triangle *BAD*,
and, as *EA* is to *AB*, so is the triangle *EAD* to the triangle *BAD*, [VI. 1]
therefore, as the triangle *ABC* is to the triangle *BAD*, so is the triangle *EAD* to the triangle *BAD*. [V. 11]

Therefore each of the triangles *ABC*, *EAD* has the same ratio to *BAD*.

Therefore the triangle *ABC* is equal to the triangle *EAD*. [V. 9]
Therefore etc.

Q.E.D.

Proposition 16

If four straight lines be proportional, the rectangle contained by the extremes is equal to the rectangle contained by the means; and, if the rectangle contained by the extremes be equal to the rectangle contained by the means, the four straight lines will be proportional.

Let the four straight lines *AB*, *CD*, *E*, *F* be proportional, so that, as *AB* is to *CD*, so is *E* to *F*;
I say that the rectangle contained by *AB*, *F* is equal to the rectangle contained by *CD*, *E*.

Let *AG*, *CH* be drawn from the points *A*, *C* at right angles to the straight lines *AB*, *CD*, and let *AG* be made equal to *F*, and *CH* equal to *E*.

Let the parallelograms *BG*, *DH* be completed.

Then since, as *AB* is to *CD*, so is *E* to *F*,
while *E* is equal to *CH*, and *F* to *AG*,
 therefore, as *AB* is to *CD*, so is *CH* to *AG*.

Therefore in the parallelograms *BG*, *DH* the sides about the equal angles are reciprocally proportional.

But those equiangular parallelograms in which the sides about the equal angles are reciprocally proportional are equal; [VI. 14]
 therefore the parallelogram *BG* is equal to the parallelogram *DH*.
And *BG* is the rectangle *AB*, *F*, for *AG* is equal to *F*;
and *DH* is the rectangle *CD*, *E*, for *E* is equal to *CH*;
therefore the rectangle contained by *AB*, *F* is equal to the rectangle contained by *CD*, *E*.

Next, let the rectangle contained by *AB*, *F* be equal to the rectangle contained by *CD*, *E*;
I say that the four straight lines will be proportional, so that, as *AB* is to *CD*, so is *E* to *F*.

For, with the same construction,
since the rectangle *AB*, *F* is equal to the rectangle *CD*, *E*,
and the rectangle *AB*, *F* is *BG*, for *AG* is equal to *F*,
and the rectangle *CD*, *E* is *DH*, for *CH* is equal to *E*,
 therefore *BG* is equal to *DH*.

And they are equiangular.

But in equal and equiangular parallelograms the sides about the equal angles are reciprocally proportional. [VI. 14]
 Therefore, as *AB* is to *CD*, so is *CH* to *AG*.

But *CH* is equal to *E*, and *AG* to *F*;
 therefore, as *AB* is to *CD*, so is *E* to *F*.

Therefore etc.

 Q.E.D.

Proposition 17

If three straight lines be proportional, the rectangle contained by the extremes is equal to the square on the mean; and, if the rectangle contained by the extremes be equal to the square on the mean, the three straight lines will be proportional.

Let the three straight lines *A*, *B*, *C* be proportional, so that, as *A* is to *B*, so is *B* to *C*;
I say that the rectangle contained by *A*, *C* is equal to the square on *B*.

Let *D* be made equal to *B*.

Then, since, as *A* is to *B*, so is *B* to *C*,
and *B* is equal to *D*,

therefore, as *A* is to *B*, so is *D* to *C*.

But, if four straight lines be proportional, the rectangle contained by the extremes
is equal to the rectangle contained by the means. **[VI. 16]**

Therefore the rectangle *A*, *C* is equal to the rectangle *B*, *D*.

But the rectangle *B*, *D* is the square on *B*, for *B* is equal to *D*;
therefore the rectangle contained by *A*, *C* is equal to the square on *B*.

Next, let the rectangle *A*, *C* be equal to the square on *B*;
I say that, as *A* is to *B*, so is *B* to *C*.

For, with the same construction,
since the rectangle *A*, *C* is equal to the square on *B*,
while the square on *B* is the rectangle *B*, *D*, for *B* is equal to *D*,

therefore the rectangle *A*, *C* is equal to the rectangle *B*, *D*.

But, if the rectangle contained by the extremes be equal to that contained by the
means, the four straight lines are proportional. **[VI. 16]**

Therefore, as *A* is to *B*, so is *D* to *C*.

But *B* is equal to *D*;

therefore, as *A* is to *B*, so is *B* to *C*.

Therefore etc.

Q.E.D.

Proposition 18

*On a given straight line to describe a rectilineal figure similar and similarly
situated to a given rectilineal figure.*

Let *AB* be the given straight line and *CE* the given rectilineal figure;
thus it is required to describe on the straight line *AB* a rectilineal figure similar
and similarly situated to the rectilineal figure *CE*.

Let *DF* be joined, and on the straight line *AB*, and at the points *A*, *B* on it, let the
angle *GAB* be constructed equal to the angle at *C*, and the angle *ABG* equal to
the angle *CDF*. **[I. 23]**

Therefore the remaining angle *CFD* is equal to the angle *AGB*; [I. 32]
therefore the triangle *FCD* is equiangular with the triangle *GAB*.

Therefore, proportionally, as *FD* is to *GB*, so is *FC* to *GA*, and *CD* to *AB*.

Again, on the straight line *BG*, and at the points *B*, *G* on it, let the angle *BGH* be constructed equal to the angle *DFE*, and the angle *GBH* equal to the angle *FDE*.
 [I. 23]

Therefore the remaining angle at *E* is equal to the remaining angle at *H*; [I. 32]
therefore the triangle *FDE* is equiangular with the triangle *GBH*;
therefore, proportionally, as *FD* is to *GB*, so is *FE* to *GH*, and *ED* to *HB*. [VI. 4]

But it was also proved that, as *FD* is to *GB*, so is *FC* to *GA*, and *CD* to *AB*; therefore also, as *FC* is to *AG*, so is *CD* to *AB*, and *FE* to *GH*, and further *ED* to *HB*.

And, since the angle *CFD* is equal to the angle *AGB*,
and the angle *DFE* to the angle *BGH*,
therefore the whole angle *CFE* is equal to the whole angle *AGH*.

For the same reason
the angle *CDE* is also equal to the angle *ABH*.

And the angle at *C* is also equal to the angle at *A*,
and the angle at *E* to the angle at *H*.

Therefore *AH* is equiangular with *CE*;
and they have the sides about their equal angles proportional;
therefore the rectilineal figure *AH* is similar to the rectilineal figure *CE*. [VI. Def. 1]

Therefore on the given straight line *AB* the rectilineal figure *AH* has been described similar and similarly situated to the given rectilineal figure *CE*.

Q.E.F.

Proposition 19

Similar triangles are to one another in the duplicate ratio of the corresponding sides.

Let *ABC*, *DEF* be similar triangles having the angle at *B* equal to the angle at *E*, and such that, as *AB* is to *BC*, so is *DE* to *EF*, so that *BC* corresponds to *EF*;
 [V. Def. 11]
I say that the triangle *ABC* has to the triangle *DEF* a ratio duplicate of that which *BC* has to *EF*.

For let a third proportional *BG* be taken to *BC*, *EF*, so that, as *BC* is to *EF*, so is *EF* to *BG*; [VI. 11]
and let *AG* be joined.

Since then, as *AB* is to *BC*, so is *DE* to *EF*,
therefore, alternately, as *AB* is to *DE*, so is *BC* to *EF*. [V. 16]

But, as *BC* is to *EF*, so is *EF* to *BG*;
therefore also, as *AB* is to *DE*, so is *EF* to *BG*. [V. 11]

Therefore in the triangles *ABG*, *DEF* the sides about the equal angles are reciprocally proportional.

But those triangles which have one angle equal to one angle, and in which the sides about the equal angles are reciprocally proportional, are equal; [VI. 15]
therefore the triangle *ABG* is equal to the triangle *DEF*.

Now since, as *BC* is to *EF*, so is *EF* to *BG*,
and, if three straight lines be proportional, the first has to the third a ratio duplicate of that which it has to the second, [V. Def. 9]
therefore *BC* has to *BG* a ratio duplicate of that which *CB* has to *EF*.

But, as *CB* is to *BG*, so is the triangle *ABC* to the triangle *ABG*; [VI. 1]
therefore the triangle *ABC* also has to the triangle *ABG* a ratio duplicate of that which *BC* has to *EF*.

But the triangle *ABG* is equal to the triangle *DEF*;
therefore the triangle *ABC* also has to the triangle *DEF* a ratio duplicate of that which *BC* has to *EF*.

Therefore etc.

Q.E.D.

PORISM. From this it is manifest that, if three straight lines be proportional, then, as the first is to the third, so is the figure described on the first to that which is similar and similarly described on the second.

Proposition 20

Similar polygons are divided into similar triangles, and into triangles equal in multitude and in the same ratio as the wholes, and the polygon has to the polygon a ratio duplicate of that which the corresponding side has to the corresponding side.

Let *ABCDE*, *FGHKL* be similar polygons, and let *AB* correspond to *FG*;
I say that the polygons *ABCDE*, *FGHKL* are divided into similar triangles, and into triangles equal in multitude and in the same ratio as the wholes, and the polygon *ABCDE* has to the polygon *FGHKL* a ratio duplicate of that which *AB* has to *FG*.

Let *BE*, *EC*, *GL*, *LH* be joined.

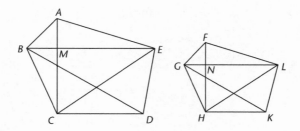

Now, since the polygon *ABCDE* is similar to the polygon *FGHKL*,
the angle *BAE* is equal to the angle *GFL*;
and, as *BA* is to *AE*, so is *GF* to *FL*. [VI. Def. 1]

Since then *ABE*, *FGL* are two triangles having one angle equal to one angle and the sides about the equal angles proportional,
therefore the triangle *ABE* is equiangular with the triangle *FGL*; [VI. 6]
so that it is also similar; [VI. 4 and Def. 1]
therefore the angle *ABE* is equal to the angle *FGL*.

But the whole angle *ABC* is also equal to the whole angle *FGH* because of the similarity of the polygons;
therefore the remaining angle *EBC* is equal to the angle *LGH*.

And, since, because of the similarity of the triangles *ABE*, *FGL*,
as *EB* is to *BA*, so is *LG* to *GF*,
and moreover also, because of the similarity of the polygons,
as *AB* is to *BC*, so is *FG* to *GH*,
therefore, *ex aequali*, as *EB* is to *BC*, so is *LG* to *GH*; [V. 22]
that is, the sides about the equal angles *EBC*, *LGH* are proportional;
therefore the triangle *EBC* is equiangular with the triangle *LGH*, [VI. 6]
so that the triangle *EBC* is also similar to the triangle *LGH*. [VI. 4 and Def. 1]

For the same reason
the triangle *ECD* is also similar to the triangle *LHK*.

Therefore the similar polygons *ABCDE*, *FGHKL* have been divided into similar triangles, and into triangles equal in multitude.

I say that they are also in the same ratio as the wholes, that is, in such manner that the triangles are proportional, and *ABE*, *EBC*, *ECD* are antecedents, while *FGL*, *LGH*, *LHK* are their consequents, and that the polygon *ABCDE* has to the polygon *FGHKL* a ratio duplicate of that which the corresponding side has to the corresponding side, that is *AB* to *FG*.

For let *AC*, *FH* be joined.

Then since, because of the similarity of the polygons,
the angle *ABC* is equal to the angle *FGH*,
and, as *AB* is to *BC*, so is *FG* to *GH*,
the triangle *ABC* is equiangular with the triangle *FGH*; [VI. 6]
therefore the angle *BAC* is equal to the angle *GFH*,
and the angle *BCA* to the angle *GHF*.

And, since the angle *BAM* is equal to the angle *GFN*,
and the angle *ABM* is also equal to the angle *FGN*,
therefore the remaining angle *AMB* is also equal to the remaining angle *FNG*; [I. 32]

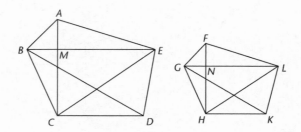

therefore the triangle *ABM* is equiangular with the triangle *FGN*.

Similarly we can prove that
the triangle *BMC* is also equiangular with the triangle *GNH*.

Therefore, proportionally,
as *AM* is to *MB*, so is *FN* to *NG*,
and, as *BM* is to *MC*, so is *GN* to *NH*;
so that, in addition, *ex aequali*,
as *AM* is to *MC*, so is *FN* to *NH*.

But, as *AM* is to *MC*, so is the triangle *ABM* to *MBC*, and *AME* to *EMC*; for they are to one another as their bases. [VI. 1]

Therefore also, as one of the antecedents is to one of the consequents, so are all the antecedents to all the consequents; [V. 12]
therefore, as the triangle *AMB* is to *BMC*, so is *ABE* to *CBE*.

But, as *AMB* is to *BMC*, so is *AM* to *MC*;
therefore also, as *AM* is to *MC*, so is the triangle *ABE* to the triangle *EBC*.

For the same reason also,
as *FN* is to *NH*, so is the triangle *FGL* to the triangle *GLH*.

And, as *AM* is to *MC*, so is *FN* to *NH*;
therefore also, as the triangle *ABE* is to the triangle *BEC*, so is the triangle *FGL* to the triangle *GLH*;
and, alternately, as the triangle *ABE* is to the triangle *FGL*, so is the triangle *BEC* to the triangle *GLH*.

Similarly we can prove, if *BD*, *GK* be joined, that, as the triangle *BEC* is to the triangle *LGH*, so also is the triangle *ECD* to the triangle *LHK*.

And since, as the triangle *ABE* is to the triangle *FGL*, so is *EBC* to *LGH*, and further *ECD* to *LHK*,
therefore also, as one of the antecedents is to one of the consequents, so are all the antecedents to all the consequents; [V. 12]
therefore, as the triangle *ABE* is to the triangle *FGL*, so is the polygon *ABCDE* to the polygon *FGHKL*.

But the triangle *ABE* has to the triangle *FGL* a ratio duplicate of that which the corresponding side *AB* has to the corresponding side *FG*; for similar triangles are in the duplicate ratio of the corresponding sides. [VI. 19]

Therefore the polygon *ABCDE* also has to the polygon *FGHKL* a ratio duplicate of that which the corresponding side *AB* has to the corresponding side *FG*.

Therefore etc.

Q.E.D.

PORISM. Similarly also it can be proved in the case of quadrilaterals that they are in the duplicate ratio of the corresponding sides. And it was also proved in the case of triangles; therefore also, generally, similar rectilineal figures are to one another in the duplicate ratio of the corresponding sides.

Proposition 21

Figures which are similar to the same rectilineal figure are also similar to one another.

For let each of the rectilineal figures *A*, *B* be similar to *C*;
I say that *A* is also similar to *B*.

For, since *A* is similar to *C*,
it is equiangular with it and has the sides about the equal angles proportional.

[VI. Def. 1]

Again, since *B* is similar to *C*,
it is equiangular with it and has the sides about the equal angles proportional.

Therefore each of the figures *A*, *B* is equiangular with *C* and with *C* has the sides about the equal angles proportional;

therefore *A* is similar to *B*.

Q.E.D.

Proposition 22

If four straight lines be proportional, the rectilineal figures similar and similarly described upon them will also be proportional; and if the rectilineal figures similar and similarly described upon them be proportional, the straight lines will themselves also be proportional.

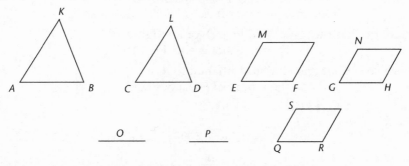

Let the four straight lines *AB, CD, EF, GH* be proportional, so that,
as *AB* is to *CD*, so is *EF* to *GH*,

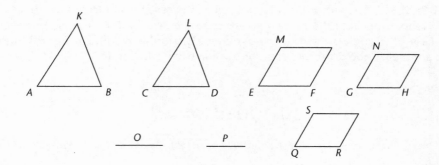

and let there be described on *AB*, *CD* the similar and similarly situated rectilineal figures *KAB*, *LCD*,
and on *EF*, *GH* the similar and similarly situated rectilineal figures *MF*, *NH*;
I say that, as *KAB* is to *LCD*, so is *MF* to *NH*.

For let there be taken a third proportional *O* to *AB*, *CD*,
and a third proportional *P* to *EF*, *GH*. [VI. 11]

Then since, as *AB* is to *CD*, so is *EF* to *GH*,
and, as *CD* is to *O*, so is *GH* to *P*,
 therefore, *ex aequali*, as *AB* is to *O*, so is *EF* to *P*. [V. 22]

But, as *AB* is to *O*, so is *KAB* to *LCD*, [VI. 19, Por.]
and, as *EF* is to *P*, so is *MF* to *NH*;
 therefore also, as *KAB* is to *LCD*, so is *MF* to *NH*. [V. 11]

Next, let *MF* be to *NH* as *KAB* is to *LCD*;
I say also that, as *AB* is to *CD*, so is *EF* to *GH*.

For, if *EF* is not to *GH* as *AB* to *CD*,
let *EF* be to *QR* as *AB* to *CD*, [VI. 12]
and on *QR* let the rectilineal figure *SR* be described similar and similarly situated
to either of the two *MF*, *NH*. [VI. 18]

Since then, as *AB* is to *CD*, so is *EF* to *QR*,
and there have been described on *AB*, *CD* the similar and similarly situated
figures *KAB*, *LCD*,
and on *EF*, *QR* the similar and similarly situated figures *MF*, *SR*,
 therefore, as *KAB* is to *LCD*, so is *MF* to *SR*.

But also, by hypothesis,
 as *KAB* is to *LCD*, so is *MF* to *NH*;
 therefore also, as *MF* is to *SR*, so is *MF* to *NH*. [V. 11]

Therefore *MF* has the same ratio to each of the figures *NH*, *SR*;
 therefore *NH* is equal to *SR*. [V. 9]

But it is also similar and similarly situated to it;
 therefore *GH* is equal to *QR*.

And, since, as *AB* is to *CD*, so is *EF* to *QR*,
while *QR* is equal to *GH*,
 therefore, as *AB* is to *CD*, so is *EF* to *GH*.

Therefore etc.

 Q.E.D.

Proposition 23

Equiangular parallelograms have to one another the ratio compounded of the ratios of their sides.

Let *AC, CF* be equiangular parallelograms having the angle *BCD* equal to the angle *ECG*;
I say that the parallelogram *AC* has to the parallelogram *CF* the ratio compounded of the ratios of the sides.

For let them be placed so that *BC* is in a straight line with *CG*;
therefore *DC* is also in a straight line with *CE*.

Let the parallelogram *DG* be completed;
let a straight line *K* be set out, and let it be contrived that,
as *BC* is to *CG*, so is *K* to *L*,
and, as *DC* is to *CE*, so is *L* to *M*. [VI. 12]

Then the ratios of *K* to *L* and of *L* to *M* are the same as the ratios of the sides, namely of *BC* to *CG* and of *DC* to *CE*.

But the ratio of *K* to *M* is compounded of the ratio of *K* to *L* and of that of *L* to *M*;
 so that *K* has also to *M* the ratio compounded of the ratios of the sides.

Now since, as *BC* is to *CG*, so is the parallelogram *AC* to the parallelogram *CH*,
 [VI. 1]
while, as *BC* is to *CG*, so is *K* to *L*,
 therefore also, as *K* is to *L*, so is *AC* to *CH*. [V. 11]

Again, since, as *DC* is to *CE*, so is the parallelogram *CH* to *CF*, [VI. 1]
while, as *DC* is to *CE*, so is *L* to *M*,
therefore also, as *L* is to *M*, so is the parallelogram *CH* to the parallelogram *CF*.
 [V. 11]

Since, then, it was proved that, as *K* is to *L*, so is the parallelogram *AC* to the parallelogram *CH*,
and, as *L* is to *M*, so is the parallelogram *CH* to the parallelogram *CF*,
 therefore, *ex aequali*, as *K* is to *M*, so is *AC* to the parallelogram *CF*.

But *K* has to *M* the ratio compounded of the ratios of the sides;
 therefore *AC* also has to *CF* the ratio compounded of the ratios of the sides.

Therefore etc.

 Q.E.D.

Proposition 24

In any parallelogram the parallelograms about the diameter are similar both to the whole and to one another.

Let *ABCD* be a parallelogram, and *AC* its diameter, and let *EG*, *HK* be parallelograms about *AC*;
I say that each of the parallelograms *EG*, *HK* is similar both to the whole *ABCD* and to the other.

For, since *EF* has been drawn parallel to *BC*, one of the sides of the triangle *ABC*,
proportionally, as *BE* is to *EA*, so is *CF* to *FA*.

[VI. 2]

Again, since *FG* has been drawn parallel to *CD*, one of the sides of the triangle *ACD*,
proportionally, as *CF* is to *FA*, so is *DG* to *GA*.

[VI. 2]

But it was proved that, as *CF* is to *FA*, so also is *BE* to *EA*;
therefore also, as *BE* is to *EA*, so is *DG* to *GA*,
and therefore, *componendo*,
as *BA* is to *AE*, so is *DA* to *AG*, [V. 18]
and, alternately,
as *BA* is to *AD*, so is *EA* to *AG*. [V. 16]

Therefore in the parallelograms *ABCD*, *EG*, the sides about the common angle *BAD* are proportional.

And, since *GF* is parallel to *DC*,
the angle *AFG* is equal to the angle *DCA*;
and the angle *DAC* is common to the two triangles *ADC*, *AGF*;
therefore the triangle *ADC* is equiangular with the triangle *AGF*.

For the same reason
the triangle *ACB* is also equiangular with the triangle *AFE*,
and the whole parallelogram *ABCD* is equiangular with the parallelogram *EG*.

Therefore, proportionally,
as *AD* is to *DC*, so is *AG* to *GF*,
as *DC* is to *CA*, so is *GF* to *FA*,
as *AC* is to *CB*, so is *AF* to *FE*,
and further, as *CB* is to *BA*, so is *FE* to *EA*.

And, since it was proved that, as *DC* is to *CA*, so is *GF* to *FA*,
and, as *AC* is to *CB*, so is *AF* to *FE*,
therefore, *ex aequali*, as *DC* is to *CB*, so is *GF* to *FE*. [V. 22]

Therefore in the parallelograms *ABCD*, *EG* the sides about the equal angles are proportional;
therefore the parallelogram *ABCD* is similar to the parallelogram *EG*. [VI. Def. 1]

For the same reason
the parallelogram *ABCD* is also similar to the parallelogram *KH*;
therefore each of the parallelograms *EG*, *HK* is similar to *ABCD*.

But figures similar to the same rectilineal figure are also similar to one another;

[VI. 21]

therefore the parallelogram *EG* is also similar to the parallelogram *HK*.

Therefore etc.

Q.E.D.

Proposition 25

To construct one and the same figure similar to a given rectilineal figure and equal to another given rectilineal figure.

Let ABC be the given rectilineal figure to which the figure to be constructed must be similar, and D that to which it must be equal;

thus it is required to construct one and the same figure similar to ABC and equal to D.

Let there be applied to BC the parallelogram BE equal to the triangle ABC,

[I. 44]

and to CE the parallelogram CM equal to D in the angle FCE which is equal to the angle CBL.

[I. 45]

> Therefore BC is in a straight line with CF, and LE with EM.

Now let GH be taken a mean proportional to BC, CF,

[VI. 13]

and on GH let KGH be described similar and similarly situated to ABC.

[VI. 18]

Then, since, as BC is to GH, so is GH to CF,

and, if three straight lines be proportional, as the first is to the third, so is the figure on the first to the similar and similarly situated figure described on the second,

[VI. 19, Por.]

> therefore, as BC is to CF, so is the triangle ABC to the triangle KGH.

But, as BC is to CF, so also is the parallelogram BE to the parallelogram EF.

[VI. 1]

Therefore also, as the triangle ABC is to the triangle KGH, so is the parallelogram BE to the parallelogram EF;

therefore, alternately, as the triangle ABC is to the parallelogram BE, so is the triangle KGB to the parallelogram EF.

[V. 16]

But the triangle ABC is equal to the parallelogram BE;

> therefore the triangle KGH is also equal to the parallelogram EF.

But the parallelogram EF is equal to D;

> therefore KGH is also equal to D.

And KGH is also similar to ABC.

Therefore one and the same figure KGH has been constructed similar to the given rectilineal figure ABC and equal to the other given figure D.

Q.E.D.

Proposition 26

If from a parallelogram there be taken away a parallelogram similar and similarly situated to the whole and having a common angle with it, it is about the same diameter with the whole.

For from the parallelogram *ABCD* let there be taken away the parallelogram *AF* similar and similarly situated to *ABCD*, and having the angle *DAB* common with it;
I say that *ABCD* is about the same diameter with *AF*.

For suppose it is not, but, if possible, let *AHC* be the diameter [of *ABCD*], let *GF* be produced and carried through to *H*, and let *HK* be drawn through *H* parallel to either of the straight lines *AD*, *BC*. [I. 31]

Since, then, *ABCD* is about the same diameter with *KG*,
therefore, as *DA* is to *AB*, so is *GA* to *AK*. [VI. 24]

But also, because of the similarity of *ABCD*, *EG*,
as *DA* is to *AB*, so is *GA* to *AE*;
therefore also,
as *GA* is to *AK*, so is *GA* to *AE*. [V. 11]

Therefore *GA* has the same ratio to each of the straight lines *AK*, *AE*.

Therefore *AE* is equal to *AK*, [V. 9]
the less to the greater: which is impossible.

Therefore *ABCD* cannot but be about the same diameter with *AF*;
therefore the parallelogram *ABCD* is about the same diameter with the parallelogram *AF*.

Therefore etc.

Q.E.D.

Proposition 27

Of all the parallelograms applied to the same straight line and deficient by parallelogrammic figures similar and similarly situated to that described on the half of the straight line, that parallelogram is greatest which is applied to the half of the straight line and is similar to the defect.

Let *AB* be a straight line and let it be bisected at *C*; let there be applied to the straight line *AB* the parallelogram *AD* deficient by the parallelogrammic figure *DB* described on the half of *AB*, that is, *CB*;
I say that, of the parallelograms applied to *AB* and deficient by parallelogrammic figures similar and similarly situated to *DB*, *AD* is greatest.

For let there be applied to the straight line *AB* the parallelogram *AF* deficient by the parallelogrammic

figure *FB* similar and similarly situated to *DB*;
I say that *AD* is greater than *AF*.

For, since the parallelogram *DB* is similar to the parallelogram *FB*,
they are about the same diameter. [VI. 26]

Let their diameter *DB* be drawn, and let the figure be described.

Then, since *CF* is equal to *FE*, [I. 43]
and *FB* is common,
therefore the whole *CH* is equal to the whole *KE*.

But *CH* is equal to *CG*, since *AC* is also equal to *CB*. [I. 36]

Therefore *GC* is also equal to *EK*.

Let *CF* be added to each;
therefore the whole *AF* is equal to the gnomon *LMN*;
so that the parallelogram *DB*, that is, *AD*, is greater than the parallelogram *AF*.

Therefore etc.

Q.E.D.

Proposition 28

To a given straight line to apply a parallelogram equal to a given rectilineal figure and deficient by a parallelogrammic figure similar to a given one: thus the given rectilineal figure must not be greater than the parallelogram described on the half of the straight line and similar to the defect.

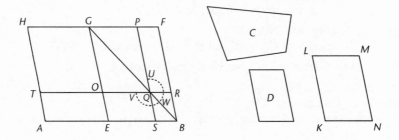

Let *AB* be the given straight line, *C* the given rectilineal figure to which the figure to be applied to *AB* is required to be equal, not being greater than the parallelogram described on the half of *AB* and similar to the defect, and *D* the parallelogram to which the defect is required to be similar;
thus it is required to apply to the given straight line *AB* a parallelogram equal to the given rectilineal figure *C* and deficient by a parallelogrammic figure which is similar to *D*.

Let *AB* be bisected at the point *E*, and on *EB* let *EBFG* be described similar and similarly situated to *D*; [VI. 18]
let the parallelogram *AG* be completed.

If then *AG* is equal to *C*, that which was enjoined will have been done;
for there has been applied to the given straight line *AB* the parallelogram *AG* equal

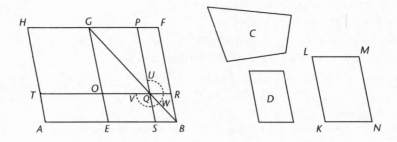

to the given rectilineal figure *C* and deficient by a parallelogrammic figure *GB* which is similar to *D*.

But, if not, let *HE* be greater than *C*.

Now *HE* is equal to *GB*;
> therefore *GB* is also greater than *C*.

Let *KLMN* be constructed at once equal to the excess by which *GB* is greater than *C* and similar and similarly situated to *D*. [VI. 25]

But *D* is similar to *GB*;
> therefore *KM* is also similar to *GB*. [VI. 21]

Let, then, *KL* correspond to *GE*, and *LM* to *GF*.

Now, since *GB* is equal to *C*, *KM*,
> therefore *GB* is greater. than *KM*;
> therefore also *GE* is greater than *KL*, and *GF* than *LM*.

Let *GO* be made equal to *KL*, and *GP* equal to *LM*;
and let the parallelogram *OGPQ* be completed;
> therefore it is equal and similar to *KM*.

Therefore *GQ* is also similar to *GB*; [VI. 21]
> therefore *GQ* is about the same diameter with *GB*. [VI. 26]

Let *GQB* be their diameter, and let the figure be described.

Then, since *BG* is equal to *C*, *KM*,
and in them *GQ* is equal to *KM*,
> therefore the remainder, the gnomon *UWV*, is equal to the remainder *C*.

And, since *PR* is equal to *OS*,
let *QB* be added to each;
> therefore the whole *PB* is equal to the whole *OB*.

But *OB* is equal to *TE*, since the side *AE* is also equal to the side *EB*; [I. 36]
> therefore *TE* is also equal to *PB*.

Let *OS* be added to each;
> therefore the whole *TS* is equal to the whole, the gnomon *VWU*.

But the gnomon *VWU* was proved equal to *C*;
> therefore *TS* is also equal to *C*.

Therefore to the given straight line *AB* there has been applied the parallelogram *ST* equal to the given rectilineal figure *C* and deficient by a parallelogrammic figure *QB* which is similar to *D*.

> Q.E.F.

Proposition 29

To a given straight line to apply a parallelogram equal to a given rectilineal figure and exceeding by a parallelogrammic figure similar to a given one.

Let *AB* be the given straight line, *C* the given rectilineal figure to which the figure to be applied to *AB* is required to be equal, and *D* that to which the excess is required to be similar;

thus it is required to apply to the straight line *AB* a parallelogram equal to the rectilineal figure *C* and exceeding by a parallelogrammic figure similar to *D*.

Let *AB* be bisected at *E*;
let there be described on *EB* the parallelogram *BF* similar and similarly situated to *D*;
and let *GH* be constructed at once equal to the sum of *BF*, *C* and similar and similarly situated to *D*. [VI. 25]

Let *KH* correspond to *FL* and *KG* to *FE*.

Now, since *GH* is greater than *FB*,
 therefore *KH* is also greater than *FL*, and *KG* than *FE*.

Let *FL*, *FE* be produced,
let *FLM* be equal to *KH*, and *FEN* to *KG*,
and let *MN* be completed;
 therefore *MN* is both equal and similar to *GH*.

But *GH* is similar to *EL*;
 therefore *MN* is also similar to *EL*; [VI. 21]
 therefore *EL* is about the same diameter with *MN*. [VI. 26]

Let their diameter *FO* be drawn, and let the figure be described.

Since *GH* is equal to *EL*, *C*,
while *GH* is equal to *MN*,
 therefore *MN* is also equal to *EL*, *C*.

Let *EL* be subtracted from each;
 therefore the remainder, the gnomon *XWV*, is equal to *C*.

Now, since *AE* is equal to *EB*,
 AN is also equal to *NB*, [I. 36]
 that is, to *LP*. [I. 43]

Let *EO* be added to each;
 therefore the whole *AO* is equal to the gnomon *VWX*.

But the gnomon *VWX* is equal to *C*;

therefore *AO* is also equal to *C*.

Therefore to the given straight line *AB* there has been applied the parallelogram *AO* equal to the given rectilineal figure *C* and exceeding by a parallelogrammic figure *QP* which is similar to *D*, since *PQ* is also similar to *EL*. [VI. 24]

Q.E.F.

Proposition 30

To cut a given finite straight line in extreme and mean ratio.

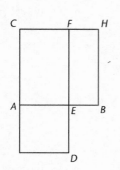

Let *AB* be the given finite straight line; thus it is required to cut *AB* in extreme and mean ratio.

On *AB* let the square *BC* be described; and let there be applied to *AC* the parallelogram *CD* equal to *BC* and exceeding by the figure *AD* similar to *BC*. [VI. 29]

Now *BC* is a square;

therefore *AD* is also a square.

And, since *BC* is equal to *CD*, let *CE* be subtracted from each; therefore the remainder *BF* is equal to the remainder *AD*.

But it is also equiangular with it; therefore in *BF*, *AD* the sides about the equal angles are reciprocally proportional; [VI. 14]

therefore, as *FE* is to *ED*, so is *AE* to *EB*.

But *FE* is equal to *AB*, and *ED* to *AE*.

Therefore, as *BA* is to *AE*, so is *AE* to *EB*.

And *AB* is greater than *AE*;

therefore *AE* is also greater than *EB*.

Therefore the straight line *AB* has been cut in extreme and mean ratio at *E*, and the greater segment of it is *AE*.

Q.E.F.

Proposition 31

In right-angled triangles the figure on the side subtending the right angle is equal to the similar and similarly described figures on the sides containing the right angle.

Let *ABC* be a right-angled triangle having the angle *BAC* right;
I say that the figure on *BC* is equal to the similar and similarly described figures on *BA*, *AC*.

Let *AD* be drawn perpendicular.

Then since, in the right-angled triangle *ABC*, *AD* has been drawn from the right angle at *A* perpendicular to the base *BC*,
the triangles *ABD*, *ADC* adjoining the perpendicular are similar both to the whole *ABC* and to one another. **[VI. 8]**

And, since *ABC* is similar to *ABD*,
therefore, as *CB* is to *BA*, so is *AB* to *BD*. **[VI. Def. 1]**

And, since three straight lines are proportional,
as the first is to the third, so is the figure on the first to the similar and similarly described figure on the second. **[VI. 19, Por.]**

Therefore, as *CB* is to *BD*, so is the figure on *CB* to the similar and similarly described figure on *BA*.

For the same reason also,
as *BC* is to *CD*, so is the figure on *BC* to that on *CA*;
so that, in addition,
as *BC* is to *BD*, *DC*, so is the figure on *BC* to the similar and similarly described figures on *BA*, *AC*. **[V. 24]**

But *BC* is equal to *BD*, *DC*;
therefore the figure on *BC* is also equal to the similar and similarly described figures on *BA*, *AC*.

Therefore etc.

Q.E.D.

Proposition 32

If two triangles having two sides proportional to two sides be placed together at one angle so that their corresponding sides are also parallel, the remaining sides

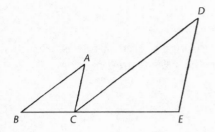

of the triangles will be in a straight line.

Let *ABC, DCE* be two triangles having the two sides *BA, AC* proportional to the two sides *DC, DE*, so that, as *AB* is to *AC*, so is *DC* to *DE*, and *AB* parallel to *DC*, and *AC* to *DE*; I say that *BC* is in a straight line with *CE*.

For, since *AB* is parallel to *DC*, and the straight line *AC* has fallen upon them, the alternate angles *BAC, ACD* are equal to one another. [I. 29]

For the same reason
 the angle *CDE* is also equal to the angle *ACD*;
 so that the angle *BAC* is equal to the angle *CDE*.

And, since *ABC, DCE* are two triangles having one angle, the angle at *A*, equal to one angle, the angle at *D*, and the sides about the equal angles proportional, so that, as *BA* is to *AC*, so is *CD* to *DE*, therefore the triangle *ABC* is equiangular with the triangle *DCE*; [VI. 6] therefore the angle *ABC* is equal to the angle *DCE*.

But the angle *ACD* was also proved equal to the angle *BAC*; therefore the whole angle *ACE* is equal to the two angles *ABC, BAC*.

Let the angle *ACB* be added to each; therefore the angles *ACE, ACB* are equal to the angles *BAC, ACB, CBA*.

But the angles *BAC, ABC, ACB* are equal to two right angles; [I. 32] therefore the angles *ACE, ACB* are also equal to two right angles.

Therefore with a straight line *AC*, and at the point *C* on it, the two straight lines *BC, CE* not lying on the same side make the adjacent angles *ACE, ACB* equal to two right angles;
 therefore *BC* is in a straight line with *CE*. [I. 14]

Therefore etc.

Q.E.D.

Proposition 33

In equal circles angles have the same ratio as the circumferences on which they

stand, whether they stand at the centres or at the circumferences.

Let *ABC, DEF* be equal circles, and let the angles *BGC, EHF* be angles at their centres G, H, and the angles *BAC, EDF* angles at the circumferences;
I say that, as the circumference *BC* is to the circumference *EF,* so is the angle *BGC* to the angle *EHF,* and the angle *BAC* to the angle *EDF.*

For let any number of consecutive circumferences *CK, KL* be made equal to the circumference *BC,*
and any number of consecutive circumferences *FM, MN* equal to the circumference *EF;*
and let *GK, GL, HM, HN* be joined.

Then, since the circumferences *BC, CK, KL* are equal to one another,
the angles *BGC, CGK, KGL* are also equal to one another; [III. 27]
therefore, whatever multiple the circumference *BL* is of *BC,* that multiple also is the angle *BGL* of the angle *BGC.*

For the same reason also,
whatever multiple the circumference *NE* is of *EF,* that multiple also is the angle *NHE* of the angle *EHF.*

If then the circumference *BL* is equal to the circumference *EN,* the angle *BGL* is also equal to the angle *EHN;* [III. 27]
if the circumference *BL* is greater than the circumference *EN,* the angle *BGL* is also greater than the angle *EHN;*
and, if less, less.

There being then four magnitudes, two circumferences *BC, EF,* and two angles *BGC, EHF,*
there have been taken, of the circumference *BC* and the angle *BGC* equimultiples, namely the circumference *BL* and the angle *BGL,*
and of the circumference *EF* and the angle *EHF* equimultiples, namely the circumference *EN* and the angle *EHN.*

And it has been proved that,
if the circumference *BL* is in excess of the circumference *EN,* the angle *BGL* is also in excess of the angle *EHN;*
if equal, equal;
and if less, less.

Therefore, as the circumference *BC* is to *EF,* so is the angle *BGC* to the angle *EHF.*
[V. Def. 5]

But, as the angle *BGC* is to the angle *EHF,* so is the angle *BAC* to the angle *EDF;* for they are doubles respectively.

Therefore also, as the circumference *BC* is to the circumference *EF,* so is the angle *BGC* to the angle *EHF,* and the angle *BAC* to the angle *EDF.*

Therefore etc.

Q.E.D.

Book VII

Definitions

1. An *unit* is that by virtue of which each of the things that exist is called one.

2. A *number* is a multitude composed of units.

3. A number is *a part* of a number, the less of the greater, when it measures the greater;

4. but *parts* when it does not measure it.

5. The greater number is a *multiple* of the less when it is measured by the less.

6. An *even number* is that which is divisible into two equal parts.

7. An *odd number* is that which is not divisible into two equal parts, or that which differs by an unit from an even number.

8. An *even-times even number* is that which is measured by an even number according to an even number.

9. An *even-times odd number* is that which is measured by an even number according to an odd number.

10. An *odd-times odd number* is that which is measured by an odd number according to an odd number.

11. A *prime number* is that which is measured by an unit alone.

12. Numbers *prime to one another* are those which are measured by an unit alone as a common measure.

13. A *composite number* is that which is measured by some number.

14. Numbers *composite to one another* are those which are measured by some number as a common measure.

15. A number is said to *multiply* a number when that which is multiplied is added to itself as many times as there are units in the other, and thus some number is produced.

16. And, when two numbers having multiplied one another make some number, the number so produced is called *plane,* and its *sides* are the numbers which have multiplied one another.

17. And, when three numbers having multiplied one another make some number, the number so produced is *solid,* and its sides are the numbers which have multiplied one another.

18. A *square number* is equal multiplied by equal, or a number which is contained by two equal numbers.

19. And a *cube* is equal multiplied by equal and again by equal, or a number which is contained by three equal numbers.

20. Numbers are *proportional* when the first is the same multiple, or the same part, or the same parts, of the second that the third is of the fourth.

21. *Similar plane* and *solid* numbers are those which have their sides proportional.

22. A *perfect number* is that which is equal to its own parts.

Proposition 1

Two unequal numbers being set out, and the less being continually subtracted in turn from the greater, if the number which is left never measures the one before it until an unit is left, the original numbers will be prime to one another.

For, the less of two unequal numbers *AB, CD* being continually subtracted from the greater, let the number which is left never measure the one before it until an unit is left;

I say that *AB, CD* are prime to one another, that is, that an unit alone measures *AB, CD.*

For, if *AB, CD* are not prime to one another, some number will measure them.

Let a number measure them, and let it be *E*;
let *CD*, measuring *BF*, leave *FA* less than itself,
let *AF*, measuring *DG*, leave *GC* less than itself,
and let *GC*, measuring *FH*, leave an unit *HA.*

Since, then, *E* measures *CD*, and *CD* measures *BF*,
therefore *E* also measures *BF*.

But it also measures the whole *BA*;
therefore it will also measure the remainder *AF*.

But *AF* measures *DG*;
therefore *E* also measures *DG*.

But it also measures the whole *DC*;
therefore it will also measure the remainder *CG*.

But *CG* measures *FH*;
therefore *E* also measures *FH*.

But it also measures the whole *FA*;
therefore it will also measure the remainder, the unit *AH*, though it is a number:
which is impossible.

Therefore no number will measure the numbers *AB, CD*;

therefore *AB, CD* are prime to one another. [VII. Def. 12]

Q.E.D.

Proposition 2

Given two numbers not prime to one another, to find their greatest common measure.

Let *AB, CD* be the two given numbers not prime to one another.

Thus it is required to find the greatest common measure of *AB, CD.*

If now *CD* measures *AB*—and it also measures itself—*CD* is a common measure of *CD, AB.*

And it is manifest that it is also the greatest; for no greater number than CD will measure CD.

But, if CD does not measure AB, then, the less of the numbers AB, CD being continually subtracted from the greater, some number will be left which will measure the one before it.

For an unit will not be left; otherwise AB, CD will be prime to one another, [VII. 1]
which is contrary to the hypothesis.

Therefore some number will be left which will measure the one before it.

Now let CD, measuring BE, leave EA less than itself,
let EA, measuring DF, leave FC less than itself,
and let CF measure AE.

Since, then, CF measures AE, and AE measures DF,
therefore CF will also measure DF.

But it also measures itself;
therefore it will also measure the whole CD.

But CD measures BE;
therefore CF also measures BE.

But it also measures EA;
therefore it will also measure the whole BA.

But it also measures CD;
therefore CF measures AB, CD.

Therefore CF is a common measure of AB, CD.

I say next that it is also the greatest.

For, if CF is not the greatest common measure of AB, CD, some number which is greater than CF will measure the numbers AB, CD.

Let such a number measure them, and let it be G.

Now, since G measures CD, while CD measures BE,
G also measures BE.

But it also measures the whole BA;
therefore it will also measure the remainder AE.

But AE measures DF;
therefore G will also measure DF.

But it also measures the whole DC;
therefore it will also measure the remainder CF, that is, the greater will measure the less: which is impossible.

Therefore no number which is greater than CF will measure the numbers AB, CD;
therefore CF is the greatest common measure of AB, CD.

Q.E.D.

PORISM. From this it is manifest that, if a number measure two numbers, it will also measure their greatest common measure.

Proposition 3

Given three numbers not prime to one another, to find their greatest common measure.

Let *A*, *B*, *C* be the three given numbers not prime to one another; thus it is required to find the greatest common measure of *A*, *B*, *C*.

For let the greatest common measure, *D*, of the two numbers *A*, *B* be taken;

[VII. 2]

then *D* either measures, or does not measure, *C*.

First, let it measure it.

But it measures *A*, *B* also;

therefore *D* measures *A*, *B*, *C*;

therefore *D* is a common measure of *A*, *B*, *C*.

I say that it is also the greatest.

For, if *D* is not the greatest common measure of *A*, *B*, *C*, some number which is greater than *D* will measure the numbers *A*, *B*, *C*.

Let such a number measure them, and let it be *E*.

Since, then, *E* measures *A*, *B*, *C*,

it will also measure *A*, *B*;

therefore it will also measure the greatest common measure of *A*, *B*. [VII. 2, Por.]

But the greatest common measure of *A*, *B* is *D*;

therefore *E* measures *D*, the greater the less:

which is impossible.

Therefore no number which is greater than *D* will measure the numbers *A*, *B*, *C*;

therefore *D* is the greatest common measure of *A*, *B*, *C*.

Next, let *D* not measure *C*;

I say first that *C*, *D* are not prime to one another.

For, since *A*, *B*, *C* are not prime to one another, some number will measure them.

Now that which measures *A*, *B*, *C* will also measure *A*, *B*, and will measure *D*, the greatest common measure of *A*, *B*. [VII. 2, Por.]

But it measures *C* also;

therefore some number will measure the numbers *D*, *C*;

therefore *D*, *C* are not prime to one another.

Let then their greatest common measure *E* be taken. [VII. 2]

Then, since *E* measures *D*,

and *D* measures *A*, *B*,

therefore *E* also measures *A*, *B*.

But it measures *C* also;

therefore *E* measures *A*, *B*, *C*;
therefore *E* is a common measure of *A*, *B*, *C*.

I say next that it is also the greatest.

For, if *E* is not the greatest common measure of *A*, *B*, *C*, some number which is greater than *E* will measure the numbers *A*, *B*, *C*.

Let such a number measure them, and let it be *F*.

Now, since *F* measures *A*, *B*, *C*,
it also measures *A*, *B*;
therefore it will also measure the greatest common measure of *A*, *B*. [VII. 2, Por.]

But the greatest common measure of *A*, *B* is *D*;
therefore *F* measures *D*.

And it measures *C* also;
therefore *F* measures *D*, *C*;
therefore it will also measure the greatest common measure of *D*, *C*. [VII. 2, Por.]

But the greatest common measure of *D*, *C* is *E*;
therefore *F* measures *E*, the greater the less:
which is impossible.

Therefore no number which is greater than *E* will measure the numbers *A*, *B*, *C*;
therefore *E* is the greatest common measure of *A*, *B*, *C*.

Q.E.D.

Proposition 4

Any number is either a part or parts of any number, the less of the greater.

Let *A*, *BC* be two numbers, and let *BC* be the less;
I say that *BC* is either a part, or parts, of *A*.

For *A*, *BC* are either prime to one another or not.

First, let *A*, *BC* be prime to one another.

Then, if *BC* be divided into the units in it, each unit of those in *BC* will be some part of *A*;
so that *BC* is parts of *A*.

Next let *A*, *BC* not be prime to one another;
then *BC* either measures, or does not measure, *A*.

If now *BC* measures *A*, *BC* is a part of *A*.

But, if not, let the greatest common measure *D* of *A*, *BC* be taken; [VII. 2]
and let *BC* be divided into the numbers equal to *D*, namely *BE*, *EF*, *FC*.

Now, since *D* measures *A*,
D is a part of *A*.

But *D* is equal to each of the numbers *BE*, *EF*, *FC*;
therefore each of the numbers *BE*, *EF*, *FC* is also a part of *A*;
so that *BC* is parts of *A*.

Therefore etc.

Q.E.D.

Proposition 5

If a number be a part of a number, and another be the same part of another, the sum will also be the same part of the sum that the one is of the one.

For let the number A be a part of BC,
and another, D, the same part of another EF that A is of BC;
I say that the sum of A, D is also the same part of the sum of BC, EF that A is of BC.

For since, whatever part A is of BC, D is also the same part of EF, therefore, as many numbers as there are in BC equal to A, so many numbers are there also in EF equal to D.

Let BC be divided into the numbers equal to A, namely BG, GC,
and EF into the numbers equal to D, namely EH, HF;
then the multitude of BG, GC will be equal to the multitude of EH, HF.

And, since BG is equal to A, and EH to D,
therefore BG, EH are also equal to A, D.

For the same reason
GC, HF are also equal to A, D.

Therefore, as many numbers as there are in BC equal to A, so many are there also in BC, EF equal to A, D.

Therefore, whatever multiple BC is of A, the same multiple also is the sum of BC, EF of the sum of A, D.

Therefore, whatever part A is to BC, the same part also is the sum of A, D of the sum of BC, EF.

Q.E.D.

Proposition 6

If a number be parts of a number, and another be the same parts of another, the sum will also be the same parts of the sum that the one is of the one.

For let the number AB be parts of the number C, and another, DE, the same parts of another, F, that AB is of C;
I say that the sum of AB, DE is also the same parts of the sum of C, F that AB is of C.

For since, whatever parts AB is of C, DE is also the same parts of F,
therefore, as many parts of C as there are in AB, so many parts of F are there also in DE.

Let AB be divided into the parts of C, namely AG, GB,
and DE into the parts of F, namely DH, HE;
thus the multitude of AG, GB will be equal to the multitude of DH, HE.

And since, whatever part AG is of C, the same part is DH of F also,

therefore, whatever part *AG* is of *C*, the same part also is the sum of *AG*, *DH* of the sum of *C*, *F*. [VII. 5]

For the same reason,
whatever part *GB* is of *C*, the same part also is the sum of *GB*, *HE* of the sum of *C*, *F*.

Therefore, whatever parts *AB* is of *C*, the same parts also is the sum of *AB*, *DE* of the sum of *C*, *F*.

Q.E.D.

Proposition 7

If a number be that part of a number, which a number subtracted is of a number subtracted, the remainder will also be the same part of the remainder that the whole is of the whole.

For let the number *AB* be that part of the number *CD* which *AE* subtracted is of *CF* subtracted;
I say that the remainder *EB* is also the same part of the remainder *FD* that the whole *AB* is of the whole *CD*.

For, whatever part *AE* is of *CF*, the same part also let *EB* be of *CG*.

Now since, whatever part *AE* is of *CF*, the same part also is *EB* of *CG*,
 therefore, whatever part *AE* is of *CF*, the same part also is *AB* of *GF*. [VII. 5]

But, whatever part *AE* is of *CF*, the same part also, by hypothesis, is *AB* of *CD*;
 therefore, whatever part *AB* is of *GF*, the same part is it of *CD* also;
 therefore *GF* is equal to *CD*.

Let *CF* be subtracted from each;
 therefore the remainder *GC* is equal to the remainder *FD*.

Now since, whatever part *AE* is of *CF*, the same part also is *EB* of *GC*,
while *GC* is equal to *FD*,
 therefore, whatever part *AE* is of *CF*, the same part also is *EB* of *FD*.

But, whatever part *AE* is of *CF*, the same part also is *AB* of *CD*;
therefore also the remainder *EB* is the same part of the remainder *FD* that the whole *AB* is of the whole *CD*.

Q.E.D.

Proposition 8

If a number be the same parts of a number that a number subtracted is of a number subtracted, the remainder will also be the same parts of the remainder that the whole is of the whole.

For let the number AB be the same parts of the number CD that AE subtracted is of CF subtracted;

I say that the remainder EB is also the same parts of the remainder FD that the whole AB is of the whole CD.

For let GH be made equal to AB.

Therefore, whatever parts GH is of CD, the same parts also is AE of CF.

Let GH be divided into the parts of CD, namely GK, KH, and AE into the parts of CF, namely AL, LE;
thus the multitude of GK, KH will be equal to the multitude of AL, LE.

Now since, whatever part GK is of CD, the same part also is AL of CF, while CD is greater than CF,
therefore GK is also greater than AL.

Let GM be made equal to AL.

Therefore, whatever part GK is of CD, the same part also is GM of CF; therefore also the remainder MK is the same part of the remainder FD that the whole GK is of the whole CD. [VII. 7]

Again, since, whatever part KH is of CD, the same part also is EL of CF, while CD is greater than CF,
therefore HK is also greater than EL.

Let KN be made equal to EL.

Therefore, whatever part KH is of CD, the same part also is KN of CF; therefore also the remainder NH is the same part of the remainder FD that the whole KH is of the whole CD. [VII. 7]

But the remainder MK was also proved to be the same part of the remainder FD that the whole GK is of the whole CD;
therefore also the sum of MK, NH is the same parts of DF that the whole HG is of the whole CD.

But the sum of MK, NH is equal to EB, and HG is equal to BA;
therefore the remainder EB is the same parts of the remainder FD that the whole AB is of the whole CD.

Q.E.D.

Proposition 9

If a number be a part of a number, and another be the same part of another, alternately also, whatever part or parts the first is of the third, the same part, or the same parts, will the second also be of the fourth.

For let the number A be a part of the number BC, and another, D, the same part of another, EF, that A is of BC; I say that, alternately also, whatever part or parts A is of D, the same part or parts is BC of EF also.

For since, whatever part A is of BC, the same part also is D of EF, therefore, as many numbers as there are in BC equal to A, so many also are there in EF equal to D.

Let BC be divided into the numbers equal to A, namely BG, GC, and EF into those equal to D, namely EH, HF;
 thus the multitude of BG, GC will be equal to the multitude of EH, HF.

Now, since the numbers BG, GC are equal to one another, and the numbers EH, HF are also equal to one another, while the multitude of BG, GC is equal to the multitude of EH, HF, therefore, whatever part or parts BG is of EH, the same part or the same parts is GC of HF also; so that, in addition, whatever part or parts BG is of EH, the same part also, or the same parts, is the sum BC of the sum EF. [VII. 5, 6]

But BG is equal to A, and EH to D; therefore, whatever part or parts A is of D, the same part or the same parts is BC of EF also.

Q.E.D.

Proposition 10

If a number be parts of a number, and another be the same parts of another, alternately also, whatever parts or part the first is of the third, the same parts or the same part will the second also be of the fourth.

For let the number AB be parts of the number C, and another, DE, the same parts of another, F; I say that, alternately also, whatever parts or part AB is of DE, the same parts or the same part is C of F also.

For since, whatever parts AB is of C, the same parts also is DE of F, therefore, as many parts of C as there are in AB, so many parts also of F are there in DE.

Let AB be divided into the parts of C, namely AG, GB, and DE into the parts of F, namely DH, HE;
 thus the multitude of AG, GB will be equal to the multitude of DH, HE.

Now since, whatever part AG is of C, the same part also is DH of F, alternately also, whatever part or parts AG is of DH, the same part or the same parts is C of F also. [VII. 9]

For the same reason also, whatever part or parts GB is of HE, the same part or the same parts is C of F also; so that, in addition, whatever parts or part AB is of DE, the same parts also, or the same part, is C of F. [VII. 5, 6]

Q.E.D.

Proposition 11

If, as whole is to whole, so is a number subtracted to a number subtracted, the remainder will also be to the remainder as whole to whole.

As the whole AB is to the whole CD, so let AE subtracted be to CF subtracted;
I say that the remainder EB is also to the remainder FD as the whole AB to the whole CD.

Since, as AB is to CD, so is AE to CF,
whatever part or parts AB is of CD, the same part or the same parts is AE of CF also; [VII. Def. 20]
Therefore also the remainder EB is the same part or parts of FD that AB is of CD. [VII. 7, 8]

Therefore, as EB is to FD, so is AB to CD. [VII. Def. 20]

Q.E.D.

Proposition 12

If there be as many numbers as we please in proportion, then, as one of the antecedents is to one of the consequents, so are all the antecedents to all the consequents.

Let A, B, C, D be as many numbers as we please in proportion, so that, as A is to B, so is C to D;
I say that, as A is to B, so are A, C to B, D.

For since, as A is to B, so is C to D,
whatever part or parts A is of B, the same part or parts is C of D also. [VII. Def. 20]

Therefore also the sum of A, C is the same part or the same parts of the sum of B, D that A is of B. [VII. 5, 6]

Therefore, as A is to B, so are A, C to B, D. [VII. Def. 20]

Q.E.D.

Proposition 13

If four numbers be proportional, they will also be proportional alternately.

Let the four numbers A, B, C, D be proportional, so that, as A is to B, so is C to D;
I say that they will also be proportional alternately, so that, as A is to C, so will B be to D.

For since, as A is to B, so is C to D,
therefore, whatever part or parts A is of B, the same part or the same parts is C of D also; [VII. Def. 20]

Therefore, alternately, whatever part or parts A is of C, the same part or the same parts is B of D also. [VII. 10]

Therefore, as A is to C, so is B to D. [VII. Def. 20]

Q.E.D.

Proposition 14

If there be as many numbers as we please, and others equal to them in multitude, which taken two and two are in the same ratio, they will also be in the same ratio ex aequali.

Let there be as many numbers as we please A, B, C, and others equal to them in multitude D, E, F, which taken two and two are in the same ratio, so that,
as A is to B, so is D to E,
and, as B is to C, so is E to F;
I say that, *ex aequali,*
as A is to C, so also is D to F.

For, since, as A is to B, so is D to E,
therefore, alternately,
as A is to D, so is B to E. [VII. 13]

Again, since, as B is to C, so is E to F,
therefore, alternately,
as B is to E, so is C to F. [VII. 13]

But, as B is to E, so is A to D;
therefore also, as A is to D, so is C to F.

Therefore, alternately,
as A is to C, so is D to F. [VII. 13]

Q.E.D.

Proposition 15

If an unit measure any number, and another number measure any other number the same number of times, alternately also, the unit will measure the third number the same number of times that the second measures the fourth.

For let the unit A measure any number BC,
and let another number D measure any other number EF the same number of times;
I say that, alternately also, the unit A measures the number D the same number of times that BC measures EF.

For, since the unit *A* measures the number *BC* the same number of times that *D* measures *EF*,
therefore, as many units as there are in *BC*, so many numbers equal to *D* are there in *EF* also.

Let *BC* be divided into the units in it, *BG, GH, HC*,
and *EF* into the numbers *EK, KL, LF* equal to *D*.

Thus the multitude of *BG, GH, HC* will be equal to the multitude of *EK, KL, LF*.

And, since the units *BG, GH, HC* are equal to one another,
and the numbers *EK, KL, LF* are also equal to one another,
while the multitude of the units *BG, GH, HC* is equal to the multitude of the numbers *EK, KL, LF*,
therefore, as the unit *BG* is to the number *EK*, so will the unit *GH* be to the number *KL*, and the unit *HC* to the number *LF*.

Therefore also, as one of the antecedents is to one of the consequents, so will all the antecedents be to all the consequents; [VII. 12]
therefore, as the unit *BG* is to the number *EK*, so is *BC* to *EF*.

But the unit *BG* is equal to the unit *A*,
and the number *EK* to the number *D*.

Therefore, as the unit *A* is to the number *D*, so is *BC* to *EF*.

Therefore the unit *A* measures the number *D* the same number of times that *BC* measures *EF*.

Q.E.D.

Proposition 16

If two numbers by multiplying one another make certain numbers, the numbers so produced will be equal to one another.

Let *A, B* be two numbers, and let *A* by multiplying *B* make *C*, and *B* by multiplying *A* make *D*;
I say that *C* is equal to *D*.

For, since *A* by multiplying *B* has made *C*, therefore *B* measures *C* according to the units in *A*.

But the unit *E* also measures the number *A* according to the units in it;
therefore the unit *E* measures *A* the same number of times that *B* measures *C*.

Therefore, alternately, the unit *E* measures the number *B* the same number of times that *A* measures *C*. [VII. 15]

Again, since *B* by multiplying *A* has made *D*,
therefore *A* measures *D* according to the units in *B*.

But the unit *E* also measures *B* according to the units in it;
therefore the unit *E* measures the number *B* the same number of times that *A* measures *D*.

But the unit *E* measured the number *B* the same number of times that *A* measures *C*;
therefore *A* measures each of the numbers *C*, *D* the same number of times.

Therefore *C* is equal to *D*.

Q.E.D.

Proposition 17

If a number by multiplying two numbers make certain numbers, the numbers so produced will have the same ratio as the numbers multiplied.

For let the number *A* be multiplying the two numbers *B*, *C* make *D*, *E*;
I say that, as *B* is to *C*, so is *D* to *E*.

For, since *A* by multiplying *B* has made *D*,
therefore *B* measures *D* according to the units in *A*.

But the unit *F* also measures the number *A* according to the units in it;
therefore the unit *F* measures the number *A* the same number of times that *B* measures *D*.

Therefore, as the unit *F* is to the number *A*, so is *B* to *D*. [VII. Def. 20]

For the same reason,
as the unit *F* is to the number *A*, so also is *C* to *E*;
therefore also, as *B* is to *D*, so is *C* to *E*.

Therefore, alternately, as *B* is to *C*, so is *D* to *E*. [VII. 13]

Q.E.D.

Proposition 18

If two numbers by multiplying any number make certain numbers, the numbers so produced will have the same ratio as the multipliers.

For let two numbers *A*, *B* by multiplying any number *C* make *D*, *E*;
I say that, as *A* is to *B*, so is *D* to *E*.

For, since *A* by multiplying *C* has made *D*,

therefore also C by multiplying A has made D.

[VII. 16]

For the same reason also,
C by multiplying B has made E.

Therefore, the number C by multiplying the two numbers A, B has made D, E.

Therefore, as A is to B, so is D to E.

[VII. 17]

Q.E.D.

Proposition 19

If four numbers be proportional, the number produced from the first and fourth will be equal to the number produced from the second and third; and, if the number produced from the first and fourth be equal to that produced from the second and third, the four numbers will be proportional.

Let A, B, C, D be four numbers in proportion, so that,
as A is to B, so is C to D;
and let A by multiplying D make E,
and let B by multiplying C make F;
I say that E is equal to F.

For let A by multiplying C make G.

Since, then, A by multiplying C has made G, and by multiplying D has made E,
the number A by multiplying the two numbers C, D has made G, E.

Therefore, as C is to D, so is G to E.　　[VII. 17]

But, as C is to D, so is A to B;
therefore also, as A is to B, so is G to E.

Again, since A by multiplying C has made G,
but, further, B has also by multiplying C made F,
the two numbers A, B by multiplying a certain number C have made G, F.

Therefore, as A is to B, so is G to F.

[VII. 18]

But further, as A is to B, so is G to E also;
therefore also, as G is to E, so is G to F.
Therefore G has to each of the numbers E, F the same ratio;
therefore E is equal to F.

[cf. V. 9]

Again, let E be equal to F;
I say that, as A is to B, so is C to D.

For, with the same construction, since E is equal to F,
therefore, as G is to E, so is G to F.

[cf. V. 7]

But, as G is to E, so is C to D,

[VII. 17]

and, as G is to F, so is A to B.

[VII. 18]

Therefore also, as A is to B, so is C to D.

Q.E.D.

Proposition 20

The least numbers of those which have the same ratio with them measure those which have the same ratio the same number of times, the greater the greater and the less the less.

For let *CD*, *EF* be the least numbers of those which have the same ratio with *A*, *B*;
I say that *CD* measures *A* the same number of times that *EF* measures *B*.

Now *CD* is not parts of *A*.

For, if possible, let it be so;
 therefore *EF* is also the same parts of *B* that *CD* is of *A*.
 [VII. 13 and Def. 20]

Therefore, as many parts of *A* as there are in *CD*, so many parts of *B* are there also in *EF*.

Let *CD* be divided into the parts of *A*, namely *CG*, *GD*,
and *EF* into the parts of *B*, namely *EH*, *HF*;
 thus the multitude of *CG*, *GD* will be equal to the multitude of *EH*, *HF*.

Now, since the numbers *CG*, *GD* are equal to one another, and the numbers *EH*, *HF* are also equal to one another, while the multitude of *CG*, *GD* is equal to the multitude of *EH*, *HF*,
 therefore, as *CG* is to *EH*, so is *GD* to *HF*.

Therefore also, as one of the antecedents is to one of the consequents, so will all the antecedents be to all the consequents. **[VII. 12]**

 Therefore, as *CG* is to *EH*, so is *CD* to *EF*.

 Therefore *CG*, *EH* are in the same ratio with *CD*, *EF*, being less than they: which is impossible, for by hypothesis *CD*, *EF* are the least numbers of those which have the same ratio with them.

 Therefore *CD* is not parts of *A*;
 therefore it is a part of it. **[VII. 4]**

And *EF* is the same part of *B* that *CD* is of *A*; **[VII. 13 and Def. 20]**
 therefore *CD* measures *A* the same number of times that *EF* measures *B*.

 Q.E.D.

Proposition 21

Numbers prime to one another are the least of those which have the same ratio with them.

Let *A*, *B* be numbers prime to one another;
I say that *A*, *B* are the least of those which have the same ratio with them.

For, if not, there will be some numbers less than *A*, *B* which are in the same ratio with *A*, *B*.

Let them be C, D.

Since, then, the least numbers of those which have the same ratio measure those which have the same ratio the same number of times, the greater the greater and the less the less, that is, the antecedent the antecedent and the consequent the consequent, [VII. 20]
therefore C measures A the same number of times that D measures B.

Now, as many times as C measures A, so many units let there be in E.

Therefore D also measures B according to the units in E.

And, since C measures A according to the units in E,
 therefore E also measures A according to the units in C. [VII. 16]

For the same reason
 E also measures B according to the units in D. [VII. 16]

Therefore E measures A, B which are prime to one another:
which is impossible. [VII. Def. 12]

Therefore there will be no numbers less than A, B which are in the same ratio with A, B.

Therefore A, B are the least of those which have the same ratio with them.

Q.E.D.

Proposition 22

The least numbers of those which have the same ratio with them are prime to one another.

Let A, B be the least numbers of those which have the same ratio with them;
I say that A, B are prime to one another.

For, if they are not prime to one another, some number will measure them.

Let some number measure them, and let it be C.

And, as many times as C measures A, so many units let there be in D,
and, as many times as C measures B, so many units let there be in E.

Since C measures A according to the units in D,
 therefore C by multiplying D has made A. [VII. Def. 15]

For the same reason also
 C by multiplying E has made B.

Thus the number C by multiplying the two numbers D, E has made A, B;
 therefore, as D is to E, so is A to B; [VII. 17]
therefore D, E are in the same ratio with A, B, being less than they:
which is impossible.

Therefore no number will measure the numbers A, B.

Therefore A, B are prime to one another.

Q.E.D.

Proposition 23

If two numbers be prime to one another, the number which measures the one of them will be prime to the remaining number.

Let *A*, *B* be two numbers prime to one another, and let any number *C* measure *A*;
I say that *C*, *B* are also prime to one another.

For, if *C*, *B* are not prime to one another,
 some number will measure *C*, *B*.

Let a number measure them, and let it be *D*.

Since *D* measures *C*, and *C* measures *A*,
 therefore *D* also measures *A*.

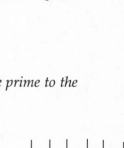

But it also measures *B*;
 therefore *D* measures *A*, *B* which are prime to one another:
which is impossible [VII. Def. 12]

 Therefore no number will measure the numbers *C*, *B*.

Therefore *C*, *B* are prime to one another.

 Q.E.D.

Proposition 24

If two numbers be prime to any number, their product also will be prime to the same.

For let the two numbers *A*, *B* be prime to any number *C*, and let *A* by multiplying *B* make *D*;
I say that *C*, *D* are prime to one another.

For, if *C*, *D* are not prime to one another, some number will measure *C*, *D*.

Let a number measure them, and let it be *E*.

Now, since *C*, *A* are prime to one another,
and a certain number *E* measures *C*,
 therefore *A*, *E* are prime to one another. [VII. 23]

As many times, then, as *E* measures *D*, so many units let there be in *F*;
 therefore *F* also measures *D* according to the units in *E*. [VII. 16]

 Therefore *E* by multiplying *F* has made *D*. [VII. Def. 15]

But, further, *A* by multiplying *B* has also made *D*;
 therefore the product of *E*, *F* is equal to the product of *A*, *B*.

But, if the product of the extremes be equal to that of the means, the four numbers are proportional; [VII. 19]
 therefore, as *E* is to *A*, so is *B* to *F*.

But *A*, *E* are prime to one another,

numbers which are prime to one another are also the least of those which have the same ratio, [VII. 21]
and the least numbers of those which have the same ratio with them measure those which have the same ratio the same number of times, the greater the greater and the less the less, that is, the antecedent the antecedent and the consequent the consequent; [VII. 20]

therefore E measures B.

But it also measures C;
therefore E measures B, C which are prime to one another: which is impossible. [VII. Def. 12]

Therefore no number will measure the numbers C, D.

Therefore C, D are prime to one another.

Q.E.D.

Proposition 25

If two numbers be prime to one another, the product of one of them into itself will be prime to the remaining one.

Let A, B be two numbers prime to one another,
and let A by multiplying itself make C;
I say that B, C are prime to one another.

For let D be made equal to A.

Since A, B are prime to one another, and A is equal to D,
therefore D, B are also prime to one another.

Therefore each of the two numbers D, A is prime to B;
therefore the product of D, A will also be prime to B. [VII. 24]

But the number which is the product of D, A is C.

Therefore C, B are prime to one another.

Q.E.D.

Proposition 26

If two numbers be prime to two numbers, both to each, their products also will be prime to one another.

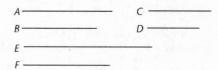

For let the two numbers A, B be prime to the two numbers C, D; both to each, and let A by multiplying B make E, and let C by multiplying D make F;
I say that E, F are prime to one another.

For, since each of the numbers A, B is prime to C,
> therefore the product of A, B will also be prime to C. **[VII. 24]**

But the product of A, B is E;
> therefore E, C are prime to one another.

For the same reason
> E, D are also prime to one another.

Therefore each of the numbers C, D is prime to E.
> Therefore the product of C, D will also be prime to E. **[VII. 24]**

But the product of C, D is F.

Therefore E, F are prime to one another.

<div align="right">Q.E.D.</div>

Proposition 27

If two numbers be prime to one another, and each by multiplying itself make a certain number, the products will be prime to one another; and, if the original numbers by multiplying the products make certain numbers, the latter will also be prime to one another ⟨and this is always the case with the extremes⟩.

Let A, B be two numbers prime to one another,
let A by multiplying itself make C, and by multiplying C make D,
and let B by multiplying itself make E, and by multiplying E make F;
I say that both C, E and D, F are prime to one another.

For, since A, B are prime to one another, and A by multiplying itself has made C,
> therefore C, B are prime to one another. **[VII. 25]**

Since, then, C, B are prime to one another,
and B by multiplying itself has made E,
> therefore C, E are prime to one another. **[VII. 25]**

Again, since A, B are prime to one another,
and B by multiplying itself has made E,
> therefore A, E are prime to one another. **[VII. 25]**

Since, then, the two numbers A, C are prime to the two numbers B, E, both to each,
> therefore also the product of A, C is prime to the product of B, E. **[VII. 26]**

And the product of A, C is D, and the product of B, E is F.

Therefore D, F are prime to one another.

<div align="right">Q.E.D.</div>

Proposition 28

If two numbers be prime to one another, the sum will also be prime to each of

them; and, if the sum of two numbers be prime to any one of them, the original numbers will also be prime to one another.

For let two numbers AB, BC prime to one another be added;

I say that the sum AC is also prime to each of the numbers AB, BC.

For, if CA, AB are not prime to one another, some number will measure CA, AB.

Let a number measure them, and let it be D.

Since, then, D measures CA, AB,

therefore it will also measure the remainder BC.

But it also measures BA;

therefore D measures AB, BC which are prime to one another:

which is impossible. [VII. Def. 12]

Therefore no number will measure the numbers CA, AB;

therefore CA, AB are prime to one another.

For the same reason

AC, CB are also prime to one another.

Therefore CA is prime to each of the numbers AB, BC.

Again, let CA, AB be prime to one another;

I say that AB, BC are also prime to one another.

For, if AB, BC are not prime to one another,

some number will measure AB, BC.

Let a number measure them, and let it be D.

Now, since D measures each of the numbers AB, BC, it will also measure the whole CA.

But it also measures AB;

therefore D measures CA, AB which are prime to one another:

which is impossible. [VII. Def. 12]

Therefore no number will measure the numbers AB, BC.

Therefore AB, BC are prime to one another.

Q.E.D.

Proposition 29

Any prime number is prime to any number which it does not measure.

Let A be a prime number, and let it not measure B;

I say that B, A are prime to one another.

For, if B, A are not prime to one another,

some number will measure them.

Let C measure them.

Since *C* measures *B*, and *A* does not measure *B*,
> therefore *C* is not the same with *A*.

Now, since *C* measures *B*, *A*,
> therefore it also measures *A* which is prime, though it is not the same with it:
which is impossible.

> Therefore no number will measure *B*, *A*.

Therefore *A*, *B* are prime to one another.

Q.E.D.

Proposition 30

If two numbers by multiplying one another make some number, and any prime number measure the product, it will also measure one of the original numbers.

For let the two numbers *A*, *B* by multiplying one another make *C*, and let any prime number *D* measure *C*;
I say that *D* measures one of the numbers *A*, *B*.

For let it not measure *A*.

Now *D* is prime;
> therefore *A*, *D* are prime to one another.
> > **[VII. 29]**

And, as many times as *D* measures *C*, so many units let there be in *E*.

Since then *D* measures *C* according to the units in *E*,
> therefore *D* by multiplying *E* has made *C*. **[VII. Def. 15]**

Further, *A* by multiplying *B* has also made *C*;
> therefore the product of *D*, *E* is equal to the product of *A*, *B*.

> Therefore, as *D* is to *A*, so is *B* to *E*. **[VII. 19]**

But *D*, *A* are prime to one another,
primes are also least, **[VII. 21]**
and the least measure the numbers which have the same ratio the same number
of times, the greater the greater and the less the less, that is, the antecedent the
antecedent and the consequent the consequent; **[VII. 20]**
> therefore *D* measures *B*.

Similarly we can also show that, if *D* do not measure *B*, it will measure *A*.

Therefore *D* measures one of the numbers *A*, *B*.

Q.E.D.

Proposition 31

Any composite number is measured by some prime number.

A ————————————

B ————————

C ——————

Let A be a composite number;
I say that A is measured by some prime number.

For, since A is composite,
some number will measure it.

Let a number measure it, and let it be B.

Now, if B is prime, what was enjoined will have been done.

But if it is composite, some number will measure it.

Let a number measure it, and let it be C.

Then, since C measures B,
and B measures A,
therefore C also measures A.

And, if C is prime, what was enjoined will have been done.

But if it is composite, some number will measure it.

Thus, if the investigation be continued in this way, some prime number will be found which will measure the number before it, which will also measure A.

For, if it is not found, an infinite series of numbers will measure the number A, each of which is less than the other:
which is impossible in numbers.

Therefore some prime number will be found which will measure the one before it, which will also measure A.

Therefore any composite number is measured by some prime number.

Q.E.D.

Proposition 32

Any number either is prime or is measured by some prime number.

A ————————————

Let A be a number;
I say that A either is prime or is measured by some prime number.

If now A is prime, that which was enjoined will have been done.

But if it is composite, some prime number will measure it. [VII. 31]

Therefore any number either is prime or is measured by some prime number.

Q.E.D.

Proposition 33

Given as many numbers as we please, to find the least of those which have the same ratio with them.

Let A, B, C be the given numbers, as many as we please;

thus it is required to find the least of those which have the same ratio with *A*, *B*, *C*.

A, *B*, *C* are either prime to one another or not.

Now, if *A*, *B*, *C* are prime to one another, they are the least of those which have the same ratio with them.

[VII. 21]

But, if not, let *D* the greatest common measure of *A*, *B*, *C* be taken, [VII. 3]
and, as many times as *D* measures the numbers *A*, *B*, *C* respectively, so many units let there be in the numbers *E*, *F*, *G* respectively.

Therefore the numbers *E*, *F*, *G* measure the numbers *A*, *B*, *C* respectively according to the units in *D*. [VII. 16]

Therefore *E*, *F*, *G* measure *A*, *B*, *C* the same number of times;
therefore *E*, *F*, *G* are in the same ratio with *A*, *B*, *C*. [VII. Def. 20]

I say next that they are the least that are in that ratio.

For, if *E*, *F*, *G* are not the least of those which have the same ratio with *A*, *B*, *C*, there will be numbers less than *E*, *F*, *G* which are in the same ratio with *A*, *B*, *C*.

Let them be *H*, *K*, *L*;
therefore *H* measures *A* the same number of times that the numbers *K*, *L* measure the numbers *B*, *C* respectively.

Now, as many times as *H* measures *A*, so many units let there be in *M*;
therefore the numbers *K*, *L* also measure the numbers *B*, *C* respectively according to the units in *M*.

And, since *H* measures *A* according to the units in *M*,
therefore *M* also measures *A* according to the units in *H*. [VII. 16]

For the same reason,
M also measures the numbers *B*, *C* according to the units in the numbers *K*, *L* respectively;
Therefore *M* measures *A*, *B*, *C*.

Now, since *H* measures *A* according to the units in *M*,
therefore *H* by multiplying *M* has made *A*. [VII. Def. 15]

For the same reason also,
E by multiplying *D* has made *A*.

Therefore the product of *E*, *D* is equal to the product of *H*, *M*.

Therefore, as *E* is to *H*, so is *M* to *D*. [VII. 19]

But *E* is greater than *H*;
therefore *M* is also greater than *D*.

And it measures *A*, *B*, *C*: which is impossible,
for by hypothesis, *D* is the greatest common measure of *A*, *B*, *C*.

Therefore there cannot be any numbers less than *E*, *F*, *G* which are in the same ratio with *A*, *B*, *C*.

Therefore *E*, *F*, *G* are the least of those which have the same ratio with *A*, *B*, *C*.

Q.E.D.

Proposition 34

Given two numbers, to find the least number which they measure.

Let *A*, *B* be the two given numbers;
thus it is required to find the least number which they
measure.

Now *A*, *B* are either prime to one another or not.

First, let *A*, *B* be prime to one another,
and let *A* by multiplying *B* make *C*;
therefore also *B* by multiplying *A* has made *C*.

Therefore *A*, *B* measure *C*.

I say next that it is also the least number they measure.

For, if not, *A*, *B* will measure some number which is less than *C*.

Let them measure *D*.

Then, as many times as *A* measures *D*, so many units let there be in *E*,
and, as many times as *B* measures *D*, so many units let there be in *F*;
therefore *A* by multiplying *E* has made *D*,
and *B* by multiplying *F* has made *D*; [VII. Def. 15]
therefore the product of *A*, *E* is equal to the product of *B*, *F*.

Therefore, as *A* is to *B*, so is *F* to *E*. [VII. 19]

But *A*, *B* are prime,
primes are also least, [VII. 21]
and the least measure the numbers which have the same ratio the same number
of times, the greater the greater and the less the less; [VII. 20]
therefore *B* measures *E*, as consequent consequent.

And, since *A* by multiplying *B*, *E* has made *C*, *D*,
therefore, as *B* is to *E*, so is *C* to *D*. [VII. 17]

But *B* measures *E*;
therefore *C* also measures *D*, the greater the less:
which is impossible.

Therefore *A*, *B* do not measure any number less than *C*;
therefore *C* is the least that is measured by *A*, *B*.

Next, let *A*, *B* not be prime to one another,
and let *F*, *E*, the least numbers of those which have the
same ratio with *A*, *B*, be taken; [VII. 33]
therefore the product of *A*, *E* is equal to the product
of *B*, *F*. [VII. 19]

And let *A* by multiplying *E* make *C*;
therefore also *B* by multiplying *F* has made *C*;
therefore *A*, *B* measure *C*.

I say next that it is also the least number that they measure.

For, if not, *A*, *B* will measure some number which is less than *C*.

Let them measure *D*.

And, as many times as A measures D, so many units let there be in G,
and, as many times as B measures D, so many units let there be in H.

Therefore A by multiplying G has made D,
and B by multiplying H has made D.

Therefore the product of A, G is equal to the product of B, H;
therefore, as A is to B, so is H to G. [VII. 19]

But, as A is to B, so is F to E.

Therefore also, as F is to E, so is H to G.

But F, E are least,
and the least measure the numbers which have the same ratio the same number
of times, the greater the greater and the less the less; [VII. 20]
therefore E measures G.

And, since A by multiplying E, G has made C, D,
therefore, as E is to G, so is C to D. [VII. 17]

But E measures G;
therefore C also measures D, the greater the less:
which is impossible.

Therefore A, B will not measure any number which is less than C.

Therefore C is the least that is measured by A, B.

Q.E.D.

Proposition 35

*If two numbers measure any number, the least number measured by them will
also measure the same.*

For let the two numbers A, B measure any number CD,
and let E be the least that they measure;
I say that E also measures CD.

For, if E does not measure CD, let E, measuring DF, leave CF less than itself.

Now, since A, B measure E,
and E measures DF,
therefore A, B will also measure DF.

But they also measure the whole CD;
therefore they will also measure the remainder CF which is less than E:
which is impossible.

Therefore E cannot fail to measure CD;
therefore it measures it.

Q.E.D.

Proposition 36

Given three numbers, to find the least number which they measure.

Let A, B, C be the three given numbers; thus it is required to find the least number which they measure.

Let D, the least number measured by the two numbers A, B, be taken. **(VII. 34)**

Then C either measures, or does not measure, D.

First, let it measure it.

But A, B also measure D;
therefore A, B, C measure D.

I say next that it is also the least that they measure.

For, if not, A, B, C will measure some number which is less than D.

Let them measure E.

Since A, B, C measure E,
therefore also A, B measure E.

Therefore the least number measured by A, B will also measure E. **[VII. 35]**

But D is the least number measured by A, B;
therefore D will measure E, the greater the less: which is impossible.

Therefore A, B, C will not measure any number which is less than D;
therefore D is the least that A, B, C measure.

Again, let C not measure D,
and let E, the least number measured by C, D, be taken. **[VII. 34]**

Since A, B measure D,
and D measures E,
therefore also A, B measure E.

But C also measures E;
therefore also A, B, C measure E.

I say next that it is also the least that they measure.

For, if not, A, B, C will measure some number which is less than E.

Let them measure F.

Since A, B, C measure F,
therefore also A, B measure F;
therefore the least number measured by A, B will also measure F. **[VII. 35]**

But D is the least number measured by A, B;
therefore D measures F.

But C also measures F;
therefore D, C measure F,
so that the least number measured by D, C will also measure F.

But *E* is the least number measured by *C, D*;
therefore *E* measures *F*, the greater the less:
which is impossible.

Therefore *A, B, C* will not measure any number which is less than *E*.

Therefore *E* is the least that is measured by *A, B, C*.

Q.E.D.

Proposition 37

If a number be measured by any number, the number which is measured will have a part called by the same name as the measuring number.

For let the number *A* be measured by any number *B*;
I say that *A* has a part called by the same name as *B*.

For, as many times as *B* measures *A*, so many units let there be in *C*.

Since *B* measures *A* according to the units in *C*,
and the unit *D* also measures the number *C* according to the units in it,
therefore the unit *D* measures the number *C* the same number of times as *B* measures *A*.

Therefore, alternately, the unit *D* measures the number *B* the same number of times as *C* measures *A*; **[VII. 15]**
therefore, whatever part the unit *D* is of the number *B*, the same part is *C* of *A* also.

But the unit *D* is a part of the number *B* called by the same name as it;
therefore *C* is also a part of *A* called by the same name as *B*,
so that *A* has a part *C* which is called by the same name as *B*.

Q.E.D.

Proposition 38

If a number have any part whatever, it will be measured by a number called by the same name as the part.

For let the number *A* have any part whatever, *B*,
and let *C* be a number called by the same name as the part *B*;
I say that *C* measures *A*.

For, since *B* is a part of *A* called by the same name as *C*,
and the unit *D* is also a part of *C* called by the same name as it,
therefore, whatever part the unit *D* is of the number *C*,
the same part is *B* of *A* also;

therefore the unit D measures the number C the same number of times that B measures A.

Therefore, alternately, the unit D measures the number B the same number of times that C measures A. [VII. 15]

Therefore C measures A.

Q.E.D.

Proposition 39

To find the number which is the least that will have given parts.

Let A, B, C be the given parts;
thus it is required to find the number which is the least that will have the parts A, B, C.

Let D, E, F be numbers called by the same name as the parts A, B, C,
and let G, the least number measured by D, E, F, be taken. [VII. 36]

Therefore G has parts called by the same name as D, E, F. [VII. 37]

But A, B, C are parts called by the same name as D, E, F;
therefore G has the parts A, B, C.

I say next that it is also the least number that has.

For, if not, there will be some number less than G which will have the parts A, B, C.

Let it be H.

Since H has the parts A, B, C,
therefore H will be measured by numbers called by the same name as the parts A, B, C. [VII. 38]

But D, E, F are numbers called by the same name as the parts A, B, C;
therefore H is measured by D, E, F.

And it is less than G: which is impossible.

Therefore there will be no number less than G that will have the parts A, B, C.

Q.E.D.

Book VIII

Proposition 1

If there be as many numbers as we please in continued proportion, and the extremes of them be prime to one another, the numbers are the least of those which have the same ratio with them.

Let there be as many numbers as we please, A, B, C, D, in continued proportion, and let the extremes of them A, D be prime to one another;
I say that A, B, C, D are the least of those which have the same ratio with them.

A ——— B ——— C ——— D ——— E — F — G ——— H ———

For, if not, let E, F, G, H be less than A, B, C, D, and in the same ratio with them.

Now, since A, B, C, D are in the same ratio with E, F, G, H, and the multitude of the numbers A, B, C, D is equal to the multitude of the numbers E, F, G, H, therefore, *ex aequali*,

$$\text{as } A \text{ is to } D, \text{ so is } E \text{ to } H. \qquad \text{[VII. 14]}$$

But A, D are prime,
primes are also least, [VII. 21]
and the least numbers measure those which have the same ratio the same number of times, the greater the greater and the less the less, that is, the antecedent the antecedent and the consequent the consequent. [VII. 20]

$$\text{Therefore } A \text{ measures } E, \text{ the greater the less:}$$
which is impossible.

Therefore E, F, G, H which are less than A, B, C, D are not in the same ratio with them.

Therefore A, B, C, D are the least of those which have the same ratio with them.

Q.E.D.

Proposition 2

To find numbers in continued proportion, as many as may be prescribed, and the least that are in a given ratio.

—— A ——— C F ———— ———— G
—— B ——— D ———— H
——— E ———— K

Let the ratio of A to B be the given ratio in least numbers;
thus it is required to find numbers in continued proportion, as many as may be prescribed, and the least that are in the ratio of A to B.

Let four be prescribed;
let A by multiplying itself make C, and by multiplying B let it make D;
let B by multiplying itself make E;
further, let A by multiplying C, D, E make F, G, H,
and let B by multiplying E make K.

Now, since A by multiplying itself has made C,
and by multiplying B has made D,
 therefore, as A is to B, so is C to D. [VII. 17]

Again, since A by multiplying B has made D,
and B by multiplying itself has made E,
therefore the numbers A, B by multiplying B have made the numbers D, E respectively.

 Therefore, as A is to B, so is D to E, [VII. 18]

But, as A is to B, so is C to D;
 therefore also, as C is to D, so is D to E.

And, since A by multiplying C, D has made F, G,
 therefore, as C is to D, so is F to G. [VII. 17]

But, as C is to D, so was A to B;
 therefore also, as A is to B, so is F to G.

Again, since A by multiplying D, E has made G, H,
 therefore, as D is to E, so is G to H. [VII. 17]

But, as D is to E, so is A to B.

 Therefore also, as A is to B, so is G to H.

And, since A, B by multiplying E have made H, K,
 therefore, as A is to B, so is H to K. [VII. 18]

But, as A is to B, so is F to G, and G to H.

Therefore also, as F is to G, so is G to H, and H to K;
 therefore C, D, E, and F, G, H, K are proportional in the ratio of A to B.

I say next that they are the least numbers that are so.

For, since A, B are the least of those which have the same ratio with them,
and the least of those which have the same ratio are prime to one another, [VII. 22]
 therefore A, B are prime to one another.

And the numbers A, B by multiplying themselves respectively have made the numbers C, E, and by multiplying the numbers C, E respectively have made the numbers F, K;
 therefore C, E and F, K are prime to one another respectively. [VII. 27]

But, if there be as many numbers as we please in continued proportion, and the extremes of them be prime to one another, they are the least of those which have the same ratio with them. [VIII. 1]

Therefore *C, D, E* and *F, G, H, K* are the least of those which have the same ratio with *A, B*.

<div align="right">Q.E.D.</div>

PORISM. From this it is manifest that, if three numbers in continued proportion be the least of those which have the same ratio with them, the extremes of them are squares, and, if four numbers, cubes.

Proposition 3

If as many numbers as we please in continued proportion be the least of those which have the same ratio with them, the extremes of them are prime to one another.

Let as many numbers as we please, *A, B, C, D*, in continued proportion be the least of those which have the same ratio with them;
I say that the extremes of them *A, D* are prime to one another.

For let two numbers *E, F*, the least that are in the ratio of *A, B, C, D*, be taken,
<div align="right">[VII. 33]</div>
then three others *G, H, K* with the same property;
and others, more by one continually, [VIII. 2]
until the multitude taken becomes equal to the multitude of the numbers *A, B, C, D*.

Let them be taken, and let them be *L, M, N, O*.

Now, since *E, F* are the least of those which have the same ratio with them,
<div align="center">they are prime to one another. [VII. 22]</div>

And, since the numbers *E, F* by multiplying themselves respectively have made the numbers *G, K*, and by multiplying the numbers *G, K* respectively have made the numbers *L, O*,
<div align="right">[VIII. 2, Por.]</div>
<div align="center">therefore both *G, K* and *L, O* are prime to one another. [VII. 27]</div>

And, since *A, B, C, D* are the least of those which have the same ratio with them, while *L, M, N, O* are the least that are in the same ratio with *A, B, C, D*,
and the multitude of the numbers *A, B, C, D* is equal to the multitude of the numbers *L, M, N, O*,
therefore the numbers *A, B, C, D* are equal to the numbers *L, M, N, O* respectively;
<div align="center">therefore *A* is equal to *L*, and *D* to *O*.</div>

And *L, O* are prime to one another.

Therefore *A, D* are also prime to one another.

<div align="right">Q.E.D.</div>

Proposition 4

Given as many ratios as we please in least numbers, to find numbers in continued proportion which are the least in the given ratios.

Let the given ratios in least numbers be that of A to B, that of C to D, and that of E to F;
thus it is required to find numbers in continued proportion which are the least that are in the ratio of A to B, in the ratio of C to D, and in the ratio of E to F.

Let G, the least number measured by B, C, be taken; [VII. 34]

And, as many times as B measures G, so many times also let A measure H, and, as many times as C measures G, so many times also let D measure K.

Now E either measures or does not measure K.

First, let it measure it.

And, as many times as E measures K, so many times let F measure L also.

Now, since A measures H the same number of times that B measures G,

therefore, as A is to B, so is H to G. [VII. Def. 20, VII. 13]

For the same reason also,

as C is to D, so is G to K,

and further, as E is to F, so is K to L;

therefore H, G, K, L are continuously proportional in the ratio of A to B, in the ratio of C to D, and in the ratio of E to F.

I say next that they are also the least that have this property.

For, if H, G, K, L are not the least numbers continuously proportional in the ratios of A to B, of C to D, and of E to F, let them be N, O, M, P.

Then since, as A is to B, so is N to O,
while A, B are least,
and the least numbers measure those which have the same ratio the same number of times, the greater the greater and the less the less, that is, the antecedent the antecedent and the consequent the consequent;

therefore B measures O. [VII. 20]

For the same reason

C also measures O;

therefore B, C measure O;

therefore the least number measured by B, C will also measure O. [VII. 35]

But G is the least number measured by B, C;

therefore G measures O, the greater the less:
which is impossible.

Therefore there will be no numbers less than H, G, K, L which are continuously in the ratio of A to B, of C to D, and of E to F.

Next, let E not measure K.

Let M, the least number measured by E, K, be taken.

And, as many times as K measures M, so many times let H, C measure N, O respectively,
and, as many times as E measures M, so many times let F measure P also.

Since H measures N the same number of times that G measures O,
<div style="text-align:center">therefore, as H is to G, so is N to O.</div> **[VII. 13 and Def. 20]**

But, as H is to G, so is A to B;
<div style="text-align:center">therefore also, as A is to B, so is N to O.</div>

For the same reason also,
<div style="text-align:center">as C is to D, so is O to M.</div>

Again, since E measures M the same number of times that F measures P,
<div style="text-align:center">therefore, as E is to F, so is M to P;</div> **[VII. 13 and Def. 20]**
therefore N, O, M, P are continuously proportional in the ratios of A to B, of C to D, and of E to F.

I say next that they are also the least that are in the ratios $A : B$, $C : D$, $E : F$.

For, if not, there will be some numbers less than N, O, M, P continuously proportional in the ratios $A : B$, $C : D$, $E : F$.

Let them be Q, R, S, T.

Now since, as Q is to R, so is A to B,
while A, B are least,
and the least numbers measure those which have the same ratio with them the same number of times, the antecedent the antecedent and the consequent the consequent, **[VII. 20]**
<div style="text-align:center">therefore B measures R.</div>

For the same reason C also measures R;
<div style="text-align:center">therefore B, C measure R.</div>

Therefore the least number measured by B, C will also measure R. **[VII. 35]**

But G is the least number measured by B, C;
<div style="text-align:center">therefore G measures R.</div>

And, as G is to R, so is K to S: **[VII. 13]**
<div style="text-align:center">therefore K also measures S.</div>

But E also measures S;
<div style="text-align:center">therefore E, K measure S.</div>

Therefore the least number measured by E, K will also measure S. [VII. 35]

But M is the least number measured by E, K;

therefore M measures S, the greater the less:

which is impossible.

Therefore there will not be any numbers less than N, O, M, P continuously proportional in the ratios of A to B, of C to D, and of E to F;

therefore N, O, M, P are the least numbers continuously proportional in the ratios A : B, C : D, E : F.

Q.E.D.

Proposition 5

Plane numbers have to one another the ratio compounded of the ratios of their sides.

Let A, B be plane numbers, and let the numbers C, D be the sides of A, and E, F of B;

I say that A has to B the ratio compounded of the ratios of the sides.

For, the ratios being given which C has to E and D to F, let the least numbers G, H, K that are continuously in the ratios C : E, D : F be taken, so that,

as C is to E, so is G to H,

and, as D is to F, so is H to K. [VIII. 4]

And let D by multiplying E make L.

Now, since D by multiplying C has made A, and by multiplying E has made L,

therefore, as C is to E, so is A to L. [VII. 17]

But, as C is to E, so is G to H;

therefore also, as G is to H, so is A to L.

Again, since E by multiplying D has made L, and further by multiplying F has made B,

therefore, as D is to F, so is L to B. [VII. 17]

But, as D is to F, so is H to K;

therefore also, as H is to K, so is L to B.

But it was also proved that, as G is to H, so is A to L;

therefore, *ex aequali*,

as G is to K, so is A to B. [VII.14]

But *G* has to *K* the ratio compounded of the ratios of the sides;
therefore *A* also has to *B* the ratio compounded of the ratios of the sides.

Q.E.D.

Proposition 6

If there be as many numbers as we please in continued proportion, and the first do not measure the second, neither will any other measure any other.

Let there be as many numbers as we please, *A, B, C, D, E,* in continued proportion, and let *A* not measure *B*;
I say that neither will any other measure any other.

Now it is manifest that *A, B, C, D, E* do not measure one another in order; for *A* does not even measure *B*.

I say, then, that neither will any other measure any other.

For, if possible, let *A* measure *C*.

And, however many *A, B, C* are, let as many numbers *F, G, H,* the least of those which have the same ratio with *A, B, C,* be taken. [VII. 33]

Now, since *F, G, H* are in the same ratio with *A, B, C,* and the multitude of the numbers *A, B, C* is equal to the multitude of the numbers *F, G, H,*
therefore, *ex aequali,* as *A* is to *C,* so is *F* to *H.* [VII. 14]

And since, as *A* is to *B,* so is *F* to *G,*
while *A* does not measure *B,*
therefore neither does *F* measure *G;* [VII. Def. 20]
therefore *F* is not an unit, for the unit measures any number.

Now *F, H* are prime to one another. [VIII. 3]

And, as *F* is to *H,* so is *A* to *C;*
therefore neither does *A* measure *C.*

Similarly we can prove that neither will any other measure any other.

Q.E.D.

Proposition 7

If there be as many numbers as we please in continued proportion, and the first measure the last, it will measure the second also.

191

$$A \text{ ———}$$
$$B \text{ ————}$$
$$C \text{ —————————}$$
$$D \text{ ———————————————}$$

Let there be as many numbers as we please, A, B, C, D, in continued proportion; and let A measure D;
I say that A also measures B.

For, if A does not measure B, neither will any other of the numbers measure any other. [VIII. 6]

But A measures D.

Therefore A also measures B.

Q.E.D.

Proposition 8

If between two numbers there fall numbers in continued proportion with them, then, however many numbers fall between them in continued proportion, so many will also fall in continued proportion between the numbers which have the same ratio with the original numbers.

$$A \text{ ————} \qquad\qquad E \text{ ————}$$
$$C \text{ ————} \qquad\qquad M \text{ —————}$$
$$D \text{ —————} \qquad\qquad N \text{ ———————}$$
$$B \text{ ———————} \qquad F \text{ —————————}$$
$$G \text{ ——}$$
$$H \text{ ———}$$
$$K \text{ ————}$$
$$L \text{ —————}$$

Let the numbers C, D fall between the two numbers A, B in continued proportion with them, and let E be made in the same ratio to F as A is to B;
I say that, as many numbers as have fallen between A, B in continued proportion, so many will also fall between E, F in continued proportion.

For, as many as A, B, C, D are in multitude, let so many numbers G, H, K, L, the least of those which have the same ratio with A, C, D, B, be taken; [VII. 33]
 therefore the extremes of them G, L are prime to one another. [VIII. 3]

Now, since A, C, D, B are in the same ratio with G, H, K, L,
and the multitude of the numbers A, C, D, B is equal to the multitude of the numbers G, H, K, L,
 therefore, *ex aequali*, as A is to B, so is G to L. [VII. 14]

But, as A is to B, so is E to F;
 therefore also, as G is to L, so is E to F.

But G, L are prime,
primes are also least, [VII. 21]
and the least numbers measure those which have the same ratio the same

number of times, the greater the greater and the less the less, that is, the antecedent the antecedent and the consequent the consequent. [VII. 20]

Therefore *G* measures *E* the same number of times as *L* measures *F*.

Next, as many times as *G* measures *E*, so many times let *H*, *K* also measure *M*, *N* respectively;
therefore *G*, *H*, *K*, *L* measure *E*, *M*, *N*, *F* the same number of times.

Therefore *G*, *H*, *K*, *L* are in the same ratio with *E*, *M*, *N*, *F*. [VII. Def. 20]

But *G*, *H*, *K*, *L* are in the same ratio with *A*, *C*, *D*, *B*;
therefore *A*, *C*, *D*, *B* are also in the same ratio with *E*, *M*, *N*, *F*.

But *A*, *C*, *D*, *B* are in continued proportion;
therefore *E*, *M*, *N*, *F* are also in continued proportion.

Therefore, as many numbers as have fallen between *A*, *B* in continued proportion with them, so many numbers have also fallen between *E*, *F* in continued proportion.

Q.E.D.

Proposition 9

If two numbers be prime to one another, and numbers fall between them in continued proportion, then, however many numbers fall between them in continued proportion, so many will also fall between each of them and an unit in continued proportion.

Let *A*, *B* be two numbers prime to one another, and let *C*, *D* fall between them in continued proportion,
and let the unit *E* be set out;
I say that, as many numbers as fall between *A*, *B* in continued proportion, so many will also fall between either of the numbers *A*, *B* and the unit in continued proportion.

For let two numbers *F*, *G*, the least that are in the ratio of *A*, *C*, *D*, *B*, be taken, three numbers *H*, *K*, *L* with the same property,
and others more by one continually, until their multitude is equal to the multitude of *A*, *C*, *D*, *B*. [VIII. 2]

Let them be taken, and let them be *M*, *N*, *O*, *P*.

It is now manifest that *F* by multiplying itself has made *H* and by multiplying *H* has made *M*, while *G* by multiplying itself has made *L* and by multiplying *L* has made *P*. [VIII. 2, Por.]

And, since M, N, O, P are the least of those which have the same ratio with F, G, and A, C, D, B are also the least of those which have the same ratio with F, G,

[VIII.1]

while the multitude of the numbers M, N, O, P is equal to the multitude of the numbers A, C, D, B,

therefore M, N, O, P are equal to A, C, D, B respectively;
therefore M is equal to A, and P to B.

Now, since F by multiplying itself has made H,
therefore F measures H according to the units in F.

But the unit E also measures F according to the units in it;
therefore the unit E measures the number F the same number of times as F measures H.

Therefore, as the unit E is to the number F, so is F to H. [VII. Def. 20]

Again, since F by multiplying H has made M,
therefore H measures M according to the units in F.

But the unit E also measures the number F according to the units in it;
therefore the unit E measures the number F the same number of times as H measures M.

Therefore, as the unit E is to the number F, so is H to M.

But it was also proved that, as the unit E is to the number F, so is F to H;
therefore also, as the unit E is to the number F, so is F to H, and H to M.

But M is equal to A;
therefore, as the unit E is to the number F, so is F to H, and H to A.

For the same reason also,
as the unit E is to the number G, so is G to L and L to B.

Therefore, as many numbers as have fallen between A, B in continued proportion, so many numbers also have fallen between each of the numbers A, B and the unit E in continued proportion.

Q.E.D.

Proposition 10

If numbers fall between each of two numbers and an unit in continued proportion, however many numbers fall between each of them and an unit in continued proportion, so many also will fall between the numbers themselves in continued proportion.

For let the numbers D, E and F, G respectively fall between the two numbers A, B and the unit C in continued proportion;

I say that, as many numbers as have fallen between each of the numbers *A*, *B* and the unit *C* in continued proportion, so many numbers will also fall between *A*, *B* in continued proportion.

For let *D* by multiplying *F* make *H*,
and let the numbers *D*, *F* by multiplying *H* make *K*, *L* respectively.

Now, since, as the unit *C* is to the number *D*, so is *D* to *E*,
therefore the unit *C* measures the number *D* the same number of times as *D* measures *E*. [VII. Def. 20]

But the unit *C* measures the number *D* according to the units in *D*;
therefore the number *D* also measures *E* according to the units in *D*;
therefore *D* by multiplying itself has made *E*.

Again, since, as *C* is to the number *D*, so is *E* to *A*,
therefore the unit *C* measures the number *D* the same number of times as *E* measures *A*.

But the unit *C* measures the number *D* according to the units in *D*;
therefore *E* also measures *A* according to the units in *D*;
therefore *D* by multiplying *E* has made *A*.

For the same reason also
F by multiplying itself has made *G*, and by multiplying *G* has made *B*.

And, since *D* by multiplying itself has made *E* and by multiplying *F* has made *H*,
therefore, as *D* is to *F*, so is *E* to *H*. [VII. 17]

For the same reason also,
as *D* is to *F*, so is *H* to *G*. [VII. 18]
Therefore also, as *E* is to *H*, so is *H* to *G*.

Again, since *D* by multiplying the numbers *E*, *H* has made *A*, *K* respectively,
therefore, as *E* is to *H*, so is *A* to *K*. [VII. 17]

But, as *E* is to *H*, so is *D* to *F*;
therefore also, as *D* is to *F*, so is *A* to *K*.

Again, since the numbers *D*, *F* by multiplying *H* have made *K*, *L* respectively,
therefore, as *D* is to *F*, so is *K* to *L*. [VII. 18]

But, as *D* is to *F*, so is *A* to *K*;
therefore also, as *A* is to *K*, so is *K* to *L*.

Further, since *F* by multiplying the numbers *H*, *G* has made *L*, *B* respectively,
therefore, as *H* is to *G*, so is *L* to *B*. [VII. 17]

But, as *H* is to *G*, so is *D* to *F*;
therefore also, as *D* is to *F*, so is *L* to *B*.

But it was also proved that, as *D* is to *F*, so is *A* to *K* and *K* to *L*;
therefore also, as *A* is to *K*, so is *K* to *L* and *L* to *B*.

Therefore *A, K, L, B* are in continued proportion.

Therefore, as many numbers as fall between each of the numbers *A, B* and the unit *C* in continued proportion, so many also will fall between *A, B* in continued proportion.

Q.E.D.

Proposition 11

Between two square numbers there is one mean proportional number, and the square has to the square the ratio duplicate of that which the side has to the side.

Let *A, B* be square numbers,
and let *C* be the side of *A*, and *D* of *B*;
I say that between *A, B* there is one mean proportional number, and *A* has to *B* the ratio duplicate of that which *C* has to *D*.

For let *C* by multiplying *D* make *E*.

Now, since *A* is a square and *C* is its side,
therefore *C* by multiplying itself has made *A*.

For the same reason also,
D by multiplying itself has made *B*.

Since, then, *C* by multiplying the numbers *C, D* has made *A, E* respectively,
therefore, as *C* is to *D*, so is *A* to *E*. [VII. 17]

For the same reason also,
as *C* is to *D*, so is *E* to *B*. [VII. 18]
Therefore also, as *A* is to *E*, so is *E* to *B*.

Therefore between *A, B* there is one mean proportional number.

I say next that *A* also has to *B* the ratio duplicate of that which *C* has to *D*.

For, since *A, E, B* are three numbers in proportion,
therefore *A* has to *B* the ratio duplicate of that which *A* has to *E*. [V. Def. 9]

But, as *A* is to *E*, so is *C* to *D*.

Therefore *A* has to *B* the ratio duplicate of that which the side *C* has to *D*.

Q.E.D.

Proposition 12

Between two cube numbers there are two mean proportional numbers, and the cube has to the cube the ratio triplicate of that which the side has to the side.

Let *A*, *B* be cube numbers,
and let *C* be the side of *A*, and *D* of *B*;
I say that between *A*, *B* there are two mean proportional numbers, and *A* has to *B* the ratio triplicate of that which *C* has to *D*.

For let *C* by multiplying itself make *E*, and by multiplying *D* let it make *F*;
let *D* by multiplying itself make *G*,
and let the numbers *C*, *D* by multiplying *F* make *H*, *K* respectively.

Now, since *A* is a cube, and *C* its side,
and *C* by multiplying itself has made *E*,
therefore *C* by multiplying itself has made *E* and by multiplying *E* has made *A*.

For the same reason also
 D by multiplying itself has made *G* and by multiplying *G* has made *B*.

And, since *C* by multiplying the numbers *C*, *D* has made *E*, *F* respectively,
 therefore, as *C* is to *D*, so is *E* to *F*. [VII. 17]

For the same reason also,
 as *C* is to *D*, so is *F* to *G*. [VII.18]

Again, since *C* by multiplying the numbers *E*, *F* has made *A*, *H* respectively,
 therefore, as *E* is to *F*, so is *A* to *H*. [VII. 17]

But, as *E* is to *F*, so is *C* to *D*.

 Therefore also, as *C* is to *D*, so is *A* to *H*.

Again, since the numbers *C*, *D* by multiplying *F* have made *H*, *K* respectively,
 therefore, as *C* is to *D*, so is *H* to *K*. [VII. 18]

Again, since *D* by multiplying each of the numbers *F*, *G* has made *K*, *B* respectively,
 therefore, as *F* is to *G*, so is *K* to *B*. [VII. 17]

But, as *F* is to *G*, so is *C* to *D*;
 therefore also, as *C* is to *D*, so is *A* to *H*, *H* to *K*, and *K* to *B*.

Therefore *H*, *K* are two mean proportionals between *A*, *B*.

I say next that *A* also has to *B* the ratio triplicate of that which *C* has to *D*.

For, since *A*, *H*, *K*, *B* are four numbers in proportion,
 therefore *A* has to *B* the ratio triplicate of that which *A* has to *H*.
 [V. Def. 10]

But, as *A* is to *H*, so is *C* to *D*;
 therefore *A* also has to *B* the ratio triplicate of that which *C* has to *D*.

Q.E.D.

Proposition 13

If there be as many numbers as we please in continued proportion, and each by multiplying itself make some number, the products will be proportional; and, if the original numbers by multiplying the products make certain numbers, the latter will also be proportional.

```
A ——           G ——————
B ——           H ————————
C ———          K ——————————
D ————
E —————        M ————
F ———————      N ———————
L ——————       P ————————
O ————————     Q ——————————
```

Let there be as many numbers as we please, *A, B, C*, in continued proportion, so that, as *A* is to *B*, so is *B* to *C*;
let *A, B, C* by multiplying themselves make *D, E, F,*
and by multiplying *D, E, F* let them make *G, H, K*;
I say that *D, E, F* and *G, H, K* are in continued proportion.

For let *A* by multiplying *B* make *L,*
and let the numbers *A, B* by multiplying *L* make *M, N* respectively.

And again let *B* by multiplying *C* make *O,*
and let the numbers *B, C* by multiplying *O* make *P, Q* respectively.

Then, in manner similar to the foregoing, we can prove that
 D, L, E and *G, M, N, H* are continuously proportional in the ratio of *A* to *B,*
and further
 E, O, F and *H, P, Q, K* are continuously proportional in the ratio of *B* to *C.*

Now, as *A* is to *B*, so is *B* to *C*;
 therefore *D, L, E* are also in the same ratio with *E, O, F,*
 and further *G, M, N, H* in the same ratio with *H, P, Q, K.*

And the multitude of *D, L, E* is equal to the multitude of *E, O, F*
and that of *G, M, N, H* to that of *H, P, Q, K*;
therefore, *ex aequali,*

 as *D* is to *E*, so is *E* to *F,*
and,

 as *G* is to *H*, so is *H* to *K.* [VII. 14]

 Q.E.D.

Proposition 14

If a square measure a square, the side will also measure the side; and, if the side measure the side, the square will also measure the square.

Let *A, B* be square numbers, let *C, D* be their sides, and let *A* measure *B*;
I say that *C* also measures *D.*

For let C by multiplying D make E;
therefore A, E, B are continuously proportional in the
ratio of C to D. [VIII. 11]

And, since A, E, B are continuously proportional,
and A measures B,

therefore A also measures E. [VIII. 7]

And, as A is to E, so is C to D;

therefore also C measures D. [VII. Def. 20]

Again, let C measure D;
I say that A also measures B.

For, with the same construction, we can in a similar manner prove that A, E, B
are continuously proportional in the ratio of C to D.

And since, as C is to D, so is A to E,
and C measures D,

therefore A also measures E. [VII. Def. 20]

And A, E, B are continuously proportional;

therefore A also measures B.

Therefore etc.

Q.E.D.

Proposition 15

*If a cube number measure a cube number, the side will also measure the side;
and, if the side measure the side, the cube will also measure the cube.*

For let the cube number A measure the cube B,
and let C be the side of A and D of B;
I say that C measures D.

For let C by multiplying itself make E,
and let D by multiplying itself make G;
further, let C by multiplying D make F,
and let C, D by multiplying F make H, K respectively.

Now it is manifest that E, F, G and A, H, K, B are continuously proportional in
the ratio of C to D. [VIII. 11, 12]

And, since A, H, K, B are continuously proportional,
and A measures B,

therefore it also measures H. [VIII. 7]

And, as A is to H, so is C to D;

therefore C also measures D. [VII. Def. 20]

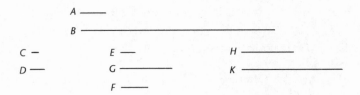

Next, let C measure D;
I say that A will also measure B.

For, with the same construction, we can prove in a similar manner that A, H, K, B are continuously proportional in the ratio of C to D.

And, since C measures D,
and, as C is to D, so is A to H,

<div style="text-align:center">

therefore A also measures H, **[VII. Def. 20]**
so that A measures B also.

</div>

<div style="text-align:right">

Q.E.D.

</div>

Proposition 16

If a square number do not measure a square number, neither will the side measure the side; and, if the side do not measure the side, neither will the square measure the square.

Let A, B be square numbers, and let C, D be their sides; and let A not measure B;
I say that neither does C measure D.

For, if C measures D, A will also measure B. **[VIII. 14]**

But A does not measure B;
therefore neither will C measure D.

Again, let C not measure D;
I say that neither will A measure B.

For, if A measures B, C will also measure D. **[VIII. 14]**

But C does not measure D;
therefore neither will A measure B.

<div style="text-align:right">

Q.E.D.

</div>

Proposition 17

If a cube number do not measure a cube number, neither will the side measure the side; and, if the side do not measure the side, neither will the cube measure the cube.

For let the cube number A not measure the cube number B,

A ————
B ————————————
C —
D ——

and let C be the side of A, and D of B;
I say that C will not measure D.

For if C measures D, A will also measure B. [VIII. 15]

But A does not measure B;
 therefore neither does C measure D.

Again, let C not measure D;
I say that neither will A measure B.

For, if A measures B, C will also measure D. [VIII. 15]

But C does not measure D;
 therefore neither will A measure B.

Q.E.D.

Proposition 18

Between two similar plane numbers there is one mean proportional number; and the plane number has to the plane number the ratio duplicate of that which the corresponding side has to the corresponding side.

A ———— C —
B ———————————— D ——
G —————— E ——
 F ——

Let A, B be two similar plane numbers, and let the numbers C, D be the sides of A, and E, F of B.

Now, since similar plane numbers are those which have their sides proportional,
[VII. Def. 21]
 therefore, as C is to D, so is E to F.

I say then that between A, B there is one mean proportional number, and A has to B the ratio duplicate of that which C has to E, or D to F, that is, of that which the corresponding side has to the corresponding side.

Now since, as C is to D, so is E to F,
 therefore, alternately, as C is to E, so is D to F. [VII. 13]

And, since A is plane, and C, D are its sides,
 therefore D by multiplying C has made A.

For the same reason also
 E by multiplying F has made B.

Now let D by multiplying E make G.

Then, since D by multiplying C has made A, and by multiplying E has made G,
 therefore, as C is to E, so is A to G. [VII. 17]

But, as C is to E, so is D to F;
 therefore also, as D is to F, so is A to G.

Again, since E by multiplying D has made G, and by multiplying F has made B,
 therefore, as D is to F, so is G to B. [VII. 17]

But it was also proved that, as D is to F, so is A to G;
 therefore also, as A is to G, so is G to B.

 Therefore A, G, B are in continued proportion.

 Therefore between A, B there is one mean proportional number.

I say next that A also has to B the ratio duplicate of that which the corresponding side has to the corresponding side, that is, of that which C has to E or D to F.

For, since A, G, B are in continued proportion,
 A has to B the ratio duplicate of that which it has to G. [V. Def. 9]

And, as A is to G, so is C to E, and so is D to F.

Therefore A also has to B the ratio duplicate of that which C has to E or D to F.

 Q.E.D.

Proposition 19

Between two similar solid numbers there fall two mean proportional numbers; and the solid number has to the similar solid number the ratio triplicate of that which the corresponding side has to the corresponding side.

Let A, B be two similar solid numbers, and let C, D, E be the sides of A, and F, G, H of B.

Now, since similar solid numbers are those which have their sides proportional,
 [VII. Def. 21]
 therefore, as C is to D, so is F to G,
 and, as D is to E, so is G to H.

I say that between A, B there fall two mean proportional numbers, and A has to B the ratio triplicate of that which C has to F, D to G, and also E to H.

For let C by multiplying D make K, and let F by multiplying G make L.

Now, since C, D are in the same ratio with F, G,
and K is the product of C, D,
and L the product of F, G,

$\qquad\qquad$ K, L are similar plane numbers; \qquad [VII. Def. 21]
\qquad therefore between K, L there is one mean proportional number. \qquad [VIII. 18]

Let it be M.

Therefore M is the product of D, F, as was proved in the theorem preceding this.
$\qquad\qquad\qquad$ [VIII. 18]

Now, since D by multiplying C has made K, and by multiplying F has made M,
\qquad therefore, as C is to F, so is K to M. \qquad [VII. 17]

But, as K is to M, so is M to L.

\qquad Therefore K, M, L are continuously proportional in the ratio of C to F.

And since, as C is to D, so is F to G,
\qquad alternately therefore, as C is to F, so is D to G. \qquad [VII. 13]

For the same reason also,
$\qquad\qquad$ as D is to G, so is E to H.

Therefore K, M, L are continuously proportional in the ratio of C to F, in the ratio of D to G, and also in the ratio of E to H.

Next, let E, H by multiplying M make N, O respectively.

Now, since A is a solid number, and C, D, E are its sides,
\qquad therefore E by multiplying the product of C, D has made A.

But the product of C, D is K;
$\qquad\qquad$ therefore E by multiplying K has made A.

For the same reason also
$\qquad\qquad$ H by multiplying L has made B.

Now, since E by multiplying K has made A, and further also by multiplying M has made N,
$\qquad\qquad$ therefore, as K is to M, so is A to N. \qquad [VII. 17]

But, as K is to M, so is C to F, D to G, and also E to H;
\qquad therefore also, as C is to F, D to G, and E to H, so is A to N.

Again, since E, H by multiplying M have made N, O respectively,
$\qquad\qquad$ therefore, as E is to H, so is N to O. \qquad [VII. 18]

But, as E is to H, so is C to F and D to G;
\qquad therefore also, as C is to F, D to G, and E to H, so is A to N and N to O.

Again, since H by multiplying M has made O, and further also by multiplying L has made B,
$\qquad\qquad$ therefore, as M is to L, so is O to B. \qquad [VII. 17]

But, as M is to L, so is C to F, D to G, and E to H.

Therefore also, as C is to F, D to G, and E to H, so not only is O to B, but also A to N and N to O.

Therefore A, N, O, B are continuously proportional in the aforesaid ratios of the sides.

I say that A also has to B the ratio triplicate of that which the corresponding side has to the corresponding side, that is, of the ratio which the number C has to F, or D to G, and also E to H.

For, since A, N, O, B are four numbers in continued proportion,

therefore A has to B the ratio triplicate of that which A has to N. [V. Def. 10]

But, as A is to N, so it was proved that C is to F, D to G, and also E to H.

Therefore A also has to B the ratio triplicate of that which the corresponding side has to the corresponding side, that is, of the ratio which the number C has to F, D to G, and also E to H.

Q.E.D.

Proposition 20

If one mean proportional number fall between two numbers, the numbers will be similar plane numbers.

For let one mean proportional number C fall between the two numbers A, B;
I say that A, B are similar plane numbers.

Let D, E, the least numbers of those which have the same ratio with A, C, be taken; [VII. 33]

therefore D measures A the same number of times that E measures C. [VII. 20]

Now, as many times as D measures A, so many units let there be in F;

therefore F by multiplying D has made A,

so that A is plane, and D, F are its sides.

Again, since D, E are the least of the numbers which have the same ratio with C, B,

therefore D measures C the same number of times that E measures B. [VII. 20]

As many times, then, as E measures B, so many units let there be in G;

therefore E measures B according to the units in G;

therefore G by multiplying E has made B.

Therefore B is plane, and E, G are its sides.

Therefore A, B are plane numbers.

I say next that they are also similar.

For, since F by multiplying D has made A, and by multiplying E has made C,[1]
therefore, as D is to E, so is A to C, that is, C to B.[1] [VII. 17]

Again, since E by multiplying F, G has made C, B respectively,
therefore, as F is to G, so is C to B. [VII. 17]

But, as C is to B, so is D to E;
therefore also, as D is to E, so is F to G.

And alternately, as D is to F, so is E to G. [VII. 13]

Therefore A, B are similar plane numbers; for their sides are proportional.

Q.E.D.

Proposition 21

If two mean proportional numbers fall between two numbers, the numbers are similar solid numbers.

For let two mean proportional numbers C, D fall between the two numbers A, B; I say that A, B are similar solid numbers.

For let three numbers E, F, G, the least of those which have the same ratio with A, C, D, be taken; [VII. 33 or VIII. 2]
therefore the extremes of them E, G are prime to one another. [VIII. 3]

Now, since one mean proportional number F has fallen between E, G,
therefore E, G are similar plane numbers. [VIII. 20]

Let, then, H, K be the sides of E, and L, M of G.

Therefore it is manifest from the theorem before this that E, F, G are continuously proportional in the ratio of H to L and that of K to M.

Now, since E, F, G are the least of the numbers which have the same ratio with A, C, D,
and the multitude of the numbers E, F, G is equal to the multitude of the numbers A, C, D,
therefore, *ex aequali*, as E is to G, so is A to D. [VII. 14]

1. Heath regards these lines as defective, for it has nowhere been established that F by multiplying E has made C; and it is not necessary to *infer* that, as D is to E, so is A to C, for this is part of the hypothesis. He offers the following emendation:
 "For, since E measures C the same number of times that D measures A, that is, according to the units in F, therefore F by multiplying E has made C.
 And since E by multiplying F, G," etc. —Ed.

But E, G are prime,

primes are also least, [VII. 21]

and the least measure those which have the same ratio with them the same number of times, the greater the greater and the less the less, that is, the antecedent the antecedent and the consequent the consequent; [VII. 20]

 therefore E measures A the same number of times that G measures D.

Now, as many times as E measures A, so many units let there be in N.

 Therefore N by multiplying E has made A.

But E is the product of H, K;

 therefore N by multiplying the product of H, K has made A.

 Therefore A is solid, and H, K, N are its sides.

Again, since E, F, G are the least of the numbers which have the same ratio as C, D, B,

 therefore E measures C the same number of times that G measures B.

Now, as many times as E measures C, so many units let there be in O.

Therefore G measures B according to the units in O;

 therefore O by multiplying G has made B.

But G is the product of L, M;

 therefore O by multiplying the product of L, M has made B.

Therefore B is solid, and L, M, O are its sides;

 therefore A, B are solid.

I say that they are also similar.

For, since N, O by multiplying E have made A, C,

 therefore, as N is to O, so is A to C, that is, E to F. [VII. 18]

But, as E is to F, so is H to L and K to M;

 therefore also, as H is to L, so is K to M and N to O.

And H, K, N are the sides of A, and O, L, M the sides of B.

Therefore A, B are similar solid numbers.

 Q.E.D.

Proposition 22

If three numbers be in continued proportion, and the first be square, the third will also be square.

Let *A, B, C* be three numbers in continued proportion, and let *A* the first be square;
I say that *C* the third is also square.
For, since between *A, C* there is one mean proportional number, *B*,

$$A \text{——}$$
$$B \text{———}$$
$$C \text{————}$$

therefore *A, C* are similar plane numbers.　　　　　[VIII. 20]
But *A* is square;

therefore *C* is also square.

Q.E.D.

Proposition 23

If four numbers be in continued proportion, and the first be cube, the fourth will also be cube.

$$A \text{——}$$
$$B \text{———}$$
$$C \text{————}$$
$$D \text{—————}$$

Let *A, B, C, D* be four numbers in continued proportion, and let *A* be cube;
I say that *D* is also cube.

For, since between *A, D* there are two mean proportional numbers *B, C*,

therefore *A, D* are similar solid numbers.　　　　　[VIII. 21]
But *A* is cube;

therefore *D* is also cube.

Q.E.D.

Proposition 24

If two numbers have to one another the ratio which a square number has to a square number, and the first be square, the second will also be square.

For let the two numbers *A, B* have to one another the ratio which the square number *C* has to the square number *D*,
and let *A* be square;
I say that *B* is also square.

$$A \text{————}$$
$$B \text{—————}$$
$$C \text{——}$$
$$D \text{——}$$

For, since *C, D* are square,

C, D are similar plane numbers.

Therefore one mean proportional number falls between *C, D*.　　[VIII. 18]
And, as *C* is to *D*, so is *A* to *B*;

therefore one mean proportional number falls between *A, B* also.　　[VIII. 8]
And *A* is square;

therefore *B* is also square.　　　　　[VIII. 22]

Q.E.D.

Proposition 25

If two numbers have to one another the ratio which a cube number has to a cube number, and the first be cube, the second will also be cube.

```
A ——————                              E ——————
B ————————————                        F ——————————
C —————
D ————————————————
```

For let the two numbers *A*, *B* have to one another the ratio which the cube number *C* has to the cube number *D*, and let *A* be cube;
I say that *B* is also cube.

For, since *C*, *D* are cube,

<div align="center">*C*, *D* are similar solid numbers.</div>

<div align="center">Therefore two mean proportional numbers fall between *C*, *D*. [VIII. 19]</div>

And, as many numbers as fall between *C*, *D* in continued proportion, so many will also fall between those which have the same ratio with them; [VIII. 8]

<div align="center">so that two mean proportional numbers fall between *A*, *B* also.</div>

Let *E*, *F* so fall.

Since, then, the four numbers *A*, *E*, *F*, *B* are in continued proportion,
and *A* is cube,

<div align="center">therefore *B* is also cube. [VIII. 23]</div>

<div align="right">Q.E.D.</div>

Proposition 26

Similar plane numbers have to one another the ratio which a square number has to a square number.

```
A ——————        B ————————————————————————
               C ———————————————
D ———       E ——————        F ———————————————
```

Let *A*, *B* be similar plane numbers;
I say that *A* has to *B* the ratio which a square number has to a square number.

For, since *A*, *B* are similar plane numbers,
therefore one mean proportional number falls between *A*, *B*. [VIII. 18]

Let it so fall, and let it be *C*;
and let *D*, *E*, *F*, the least numbers of those which have the same ratio with *A*, *C*, *B*,
be taken; [VII. 33 or VIII. 2]

<div align="center">therefore the extremes of them *D*, *F* are square. [VIII. 2, Por.]</div>

And since, as *D* is to *F*, so is *A* to *B*,

and D, F are square,
therefore A has to B the ratio which a square number has to a square number.

<div align="right">Q.E.D.</div>

Proposition 27

Similar solid numbers have to one another the ratio which a cube number has to a cube number.

Let A, B be similar solid numbers;
I say that A has to B the ratio which a cube number has to a cube number.

For, since A, B are similar solid numbers,
 therefore two mean proportional numbers fall between A, B. **[VIII. 19]**
Let C, D so fall,
and let E, F, G, H, the least numbers of those which have the same ratio with
A, C, D, B, and equal with them in multitude, be taken; **[VII. 33 or VIII. 2]**
 therefore the extremes of them E, H are cube. **[VIII. 2, Por.]**

And, as E is to H, so is A to B;
therefore A also has to B the ratio which a cube number has to a cube number.

<div align="right">Q.E.D.</div>

Book IX

Proposition 1

If two similar plane numbers by multiplying one another make some number, the product will be square.

Let A, B be two similar plane numbers, and let A by multiplying B make C; I say that C is square.

For let A by multiplying itself make D.

Therefore D is square.

Since then A by multiplying itself has made D, and by multiplying B has made C, therefore, as A is to B, so is D to C. [VII. 17]

And, since A, B are similar plane numbers, therefore one mean proportional number falls between A, B. [VIII. 18]

But, if numbers fall between two numbers in continued proportion, as many as fall between them, so many also fall between those which have the same ratio; [VIII. 8]

so that one mean proportional number falls between D, C also.

And D is square;

therefore C is also square. [VIII. 22]

Q.E.D.

Proposition 2

If two numbers by multiplying one another make a square number, they are similar plane numbers.

Let A, B be two numbers, and let A by multiplying B make the square number C; I say that A, B are similar plane numbers.

For let A by multiplying itself make D;

therefore D is square.

Now, since A by multiplying itself has made D, and by multiplying B has made C, therefore, as A is to B, so is D to C. [VII. 17]

And, since D is square, and C is so also, therefore D, C are similar plane numbers.

Therefore one mean proportional number falls between D, C. [VIII. 18]

And, as D is to C, so is A to B;

therefore one mean proportional number falls between A, B also. [VIII. 8]

A ——————
B ——————————
C ————————————————————————————
D ——————————————————————

But, if one mean proportional number fall between two numbers, they are similar plane numbers; [VIII. 20]

therefore A, B are similar plane numbers.

Q.E.D.

Proposition 3

If a cube number by multiplying itself make some number, the product will be cube.

A ——
B ——————
C — D ——

For let the cube number A by multiplying itself make B; I say that B is cube.

For let C, the side of A, be taken, and let C by multiplying itself make D.

It is then manifest that C by multiplying D has made A.

Now, since C by multiplying itself has made D, therefore C measures D according to the units in itself.

But further the unit also measures C according to the units in it; therefore, as the unit is to C, so is C to D. [VII. Def. 20]

Again, since C by multiplying D has made A, therefore D measures A according to the units in C.

But the unit also measures C according to the units in it; therefore, as the unit is to C, so is D to A.

But, as the unit is to C, so is C to D; therefore also, as the unit is to C, so is C to D, and D to A.

Therefore between the unit and the number A two mean proportional numbers C, D have fallen in continued proportion.

Again, since A by multiplying itself has made B, therefore A measures B according to the units in itself.

But the unit also measures A according to the units in it; therefore, as the unit is to A, so is A to B. [VII. Def. 20]

But between the unit and A two mean proportional numbers have fallen; therefore two mean proportional numbers will also fall between A, B. [VIII. 8]

But, if two mean proportional numbers fall between two numbers, and the first be cube, the second will also be cube. [VIII. 23]

And A is cube; therefore B is also cube.

Q.E.D.

Proposition 4

If a cube number by multiplying a cube number make some number, the product will be cube.

For let the cube number A by multiplying the cube number B make C;
I say that C is cube.

A ————————

B ——————————

C ————————————————

D ——————————

For let A by multiplying itself make D;
 therefore D is cube. **[IX. 3]**

And, since A by multiplying itself has made D, and by multiplying B has made C,
 therefore, as A is to B, so is D to C. **[VII. 17]**

And, since A, B are cube numbers,
A, B are similar solid numbers.

Therefore two mean proportional numbers fall between A, B; **[VIII. 19]**
 so that two mean proportional numbers will fall between D, C also. **[VIII. 8]**

And D is cube;
 therefore C is also cube. **[VIII. 23]**

Q.E.D.

Proposition 5

If a cube number by multiplying any number make a cube number, the multiplied number will also be cube.

For let the cube number A by multiplying any number B make the cube number C;
I say that B is cube.

A ————————

B ——————————

C ————————————————

D ——————————

For let A by multiplying itself make D;
 therefore D is cube. **[IX. 3]**

Now, since A by multiplying itself has made D, and by multiplying B has made C,
 therefore, as A is to B, so is D to C. **[VII. 17]**

And since D, C are cube,
 they are similar solid numbers.

Therefore two mean proportional numbers fall between D, C. **[VIII. 19]**

And, as D is to C, so is A to B;
 therefore two mean proportional numbers fall between A, B also. **[VIII. 8]**

And A is cube;
 therefore B is also cube. **[VIII. 23]**

Q.E.D.

Proposition 6

If a number by multiplying itself make a cube number, it will itself also be cube.

A ————

B ————

C ————

For let the number *A* by multiplying itself make the cube number *B*;
I say that *A* is also cube.

For let *A* by multiplying *B* make *C*.

Since, then, *A* by multiplying itself has made *B*, and by multiplying *B* has made *C*,
therefore *C* is cube.

And, since *A* by multiplying itself has made *B*,
therefore *A* measures *B* according to the units in itself.

But the unit also measures *A* according to the units in it.

Therefore, as the unit is to *A*, so is *A* to *B*. [VII. Def. 20]

And, since *A* by multiplying *B* has made *C*,
therefore *B* measures *C* according to the units in *A*.

But the unit also measures *A* according to the units in it.

Therefore, as the unit is to *A*, so is *B* to *C*. [VII. Def. 20]

But, as the unit is to *A*, so is *A* to *B*;
therefore also, as *A* is to *B*, so is *B* to *C*.

And, since *B*, *C* are cube,
they are similar solid numbers.

Therefore there are two mean proportional numbers between *B*, *C*. [VIII. 19]

And, as *B* is to *C*, so is *A* to *B*.

Therefore there are two mean proportional numbers between *A*, *B* also. [VIII. 8]

And *B* is cube;

therefore *A* is also cube. [cf. VIII. 23]

Q.E.D.

Proposition 7

If a composite number by multiplying any number make some number, the product will be solid.

A ————————

B ————

C ————————————

D ———— E ————————————

For, let the composite number *A* by multiplying any number *B* make *C*;
I say that *C* is solid.

For, since *A* is composite, it will be measured by some number. [VII. Def. 13]

Let it be measured by *D*;
and, as many times as *D* measures *A*, so many units let there be in *E*.

Since, then, *D* measures *A* according to the units in *E*,
therefore *E* by multiplying *D* has made *A*. [VII. Def. 15]

And, since A by multiplying B has made C,
and A is the product of D, E,
therefore the product of D, E by multiplying B has made C.

Therefore C is solid, and D, E, B are its sides.

Q.E.D.

Proposition 8

If as many numbers as we please beginning from an unit be in continued proportion, the third from the unit will be square, as will also those which successively leave out one; the fourth will be cube, as will also all those which leave out two; and the seventh will be at once cube and square, as will also those which leave out five.

Let there be as many numbers as we please, A, B, C, D, E, F, beginning from an unit and in continued proportion;
I say that B, the third from the unit, is square, as are also all those which leave out one; C, the fourth, is cube, as are also all those which leave out two; and F, the seventh, is at once cube and square, as are also all those which leave out five.

For since, as the unit is to A, so is A to B,
therefore the unit measures the number A the same number of times that A measures B. [VII. Def. 20]

But the unit measures the number A according to the units in it;
therefore A also measures B according to the units in A.

Therefore A by multiplying itself has made B;
therefore B is square.

And, since B, C, D are in continued proportion, and B is square,
therefore D is also square. [VIII. 22]

For the same reason
F is also square.

Similarly we can prove that all those which leave out one are square.

I say next that C, the fourth from the unit, is cube, as are also all those which leave out two.

For since, as the unit is to A, so is B to C,
therefore the unit measures the number A the same number of times that B measures C.

But the unit measures the number A according to the units in A;
therefore B also measures C according to the units in A.

Therefore A by multiplying B has made C.

Since then A by multiplying itself has made B, and by multiplying B has made C,
therefore C is cube.

And, since C, D, E, F are in continued proportion, and C is cube,

A ——————
B ——————
C ——————
D ——————
E ——————
F ——————

therefore *F* is also cube. [VIII. 23]

But it was also proved square;
therefore the seventh from the unit is both cube and square.

Similarly we can prove that all the numbers which leave
out five are also both cube and square.

Q.E.D.

Proposition 9

*If as many numbers as we please beginning from an unit be in continued
proportion, and the number after the unit be square, all the rest will also be square.
And, if the number after the unit be cube, all the rest will also be cube.*

A ——————
B ——————
C ——————
D ——————
E ——————
F ——————

Let there be as many numbers as we please, *A, B, C, D, E, F,*
beginning from an unit and in continued proportion,
and let *A*, the number after the unit, be square;
I say that all the rest will also be square.

Now it has been proved that *B*, the third from the unit, is
square, as are also all those which leave out one; [IX. 8]
I say that all the rest are also square.

For, since *A, B, C* are in continued proportion,
and *A* is square,

therefore *C* is also square. [VIII. 22]

Again, since *B, C, D* are in continued proportion,
and *B* is square,

D is also square. [VIII. 22]

Similarly we can prove that all the rest are also square.

Next, let *A* be cube;
I say that all the rest are also cube.

Now it has been proved that *C*, the fourth from the unit, is cube, as also are all
those which leave out two; [IX. 8]
I say that all the rest are also cube.

For, since, as the unit is to *A*, so is *A* to *B*,
therefore the unit measures *A* the same number of times as *A* measures *B*.

But the unit measures *A* according to the units in it;
therefore *A* also measures *B* according to the units in itself;
therefore *A* by multiplying itself has made *B*.

And *A* is cube.

But, if a cube number by multiplying itself make some number, the product is
cube. [IX. 3]

Therefore *B* is also cube.

And, since the four numbers *A, B, C, D* are in continued proportion,
and *A* is cube,

D also is cube. [VIII. 23]

For the same reason
E is also cube, and similarly all the rest are cube.

Q.E.D.

Proposition 10

If as many numbers as we please beginning from an unit be in continued proportion, and the number after the unit be not square, neither will any other be square except the third from the unit and all those which leave out one. And, if the number after the unit be not cube, neither will any other be cube except the fourth from the unit and all those which leave out two.

Let there be as many numbers as we please, A, B, C, D, E, F, beginning from an unit and in continued proportion,

| A ——— |
| B ——— |
| C ———— |
| D ————— |
| E —————— |
| F ———————— |

and let A, the number after the unit, not be square;
I say that neither will any other be square except the third from the unit and those which leave out one.

For, if possible, let C be square.

But B is also square; [IX. 8]
therefore B, C have to one another the ratio which a square number has to a square number.

And, as B is to C, so is A to B;
therefore A, B have to one another the ratio which a square number has to a square number;
so that A, B are similar plane numbers. [VIII. 26, converse]

And B is square;
therefore A is also square:
which is contrary to the hypothesis.

Therefore C is not square.

Similarly we can prove that neither is any other of the numbers square except the third from the unit and those which leave out one.

Next, let A not be cube.

I say that neither will any other be cube except the fourth from the unit and those which leave out two.

For, if possible, let D be cube.

Now C is also cube; for it is fourth from the unit. [IX. 8]

And, as C is to D, so is B to C;
therefore B also has to C the ratio which a cube has to a cube.
And C is cube;
therefore B is also cube. [VIII. 25]

And since, as the unit is to A so is A to B,
and the unit measures A according to the units in it;
therefore A also measures B according to the units in itself;

217

A ——

B ———

C ————

D —————

E ——————

F ———————

therefore A by multiplying itself has made the cube number B.

But, if a number by multiplying itself make a cube number, it is also itself cube. [IX .6]

Therefore A is also cube:
which is contrary to the hypothesis.

Therefore D is not cube.

Similarly we can prove that neither is any other of the numbers cube except the fourth from the unit and those which leave out two.

Q.E.D.

Proposition 11

If as many numbers as we please beginning from an unit be in continued proportion, the less measures the greater according to some one of the numbers which have place among the proportional numbers.

A –

B ——

C ———

D ————

E —————

Let there be as many numbers as we please, B, C, D, E, beginning from the unit A and in continued proportion;
I say that B, the least of the numbers B, C, D, E, measures E according to some one of the numbers C, D.

For since, as the unit A is to B, so is D to E,
therefore the unit A measures the number B the same number of times as D measures E;
therefore, alternately, the unit A measures D the same number of times as B measures E. [VII. 15]

But the unit A measures D according to the units in it;
therefore B also measures E according to the units in D;
so that B the less measures E the greater according to some number of those which have place among the proportional numbers.

Q.E.D.

PORISM. And it is manifest that, whatever place the measuring number has, reckoned from the unit, the same place also has the number according to which it measures, reckoned from the number measured, in the direction of the number before it.

Proposition 12

If as many numbers as we please beginning from an unit be in continued proportion, by however many prime numbers the last is measured, the next to the unit will also be measured by the same.

A ———
B ———
C ———
D ———
E —

F ————————
G ——————
H ———

Let there be as many numbers as we please, *A, B, C, D,* beginning from an unit, and in continued proportion;
I say that, by however many prime numbers *D* is measured, *A* will also be measured by the same.

For let *D* be measured by any prime number *E;*
I say that *E* measures *A.*

For suppose it does not;
now *E* is prime, and any prime number is prime to any which it does not measure; [VII. 29]
therefore *E, A* are prime to one another.

And, since *E* measures *D,* let it measure it according to *F,*
therefore *E* by multiplying *F* has made *D.*

Again, since *A* measures *D* according to the units in *C,* [IX. 11 and Por.]
therefore *A* by multiplying *C* has made *D.*

But, further, *E* has also by multiplying *F* made *D;*
therefore the product of *A, C* is equal to the product of *E, F.*

Therefore, as *A* is to *E,* so is *F* to *C.* [VII. 19]

But *A, E* are prime,
primes are also least, [VII. 21]
and the least measure those which have the same ratio the same number of times,
the antecedent the antecedent and the consequent the consequent; [VII. 20]
therefore *E* measures *C.*

Let it measure it according to *G;*
therefore *E* by multiplying *G* has made *C.*

But, further, by the theorem before this,
A has also by multiplying *B* made *C.* [IX. 11 and Por.]

Therefore the product of *A, B* is equal to the product of *E, G.*

Therefore, as *A* is to *E,* so is *G* to *B.* [VII. 19]

But *A, E* are prime,
primes are also least, [VII. 21]
and the least numbers measure those which have the same ratio with them the same number of times, the antecedent the antecedent and the consequent the consequent; [VII.20]
therefore *E* measures *B.*

Let it measure it according to *H;*
therefore *E* by multiplying *H* has made *B.*

But, further, *A* has also by multiplying itself made *B;* [IX. 8]
therefore the product of *E, H* is equal to the square on *A.*

Therefore, as *E* is to *A,* so is *A* to *H.* [VII. 19]

```
A ——            F ————————
B ———           G ——————
C ————          H ————
D —————
E —
```

But *A*, *E* are prime,

primes are also least, [VII. 21]

and the least measure those which have the same ratio the same number of times,

the antecedent the antecedent and the consequent the consequent; [VII. 20]

therefore *E* measures *A*, as antecedent antecedent.

But, again, it also does not measure it:

which is impossible.

Therefore *E*, *A* are not prime to one another.

Therefore they are composite to one another.

But numbers composite to one another are measured by some number.

[VII. Def. 14]

And, since *E* is by hypothesis prime,

and the prime is not measured by any number other than itself,

therefore *E* measures *A*, *E*,

so that *E* measures *A*.

Similarly we can prove that, by however many prime numbers *D* is measured, *A* will also be measured by the same.

Q.E.D.

Proposition 13

If as many numbers as we please beginning from an unit be in continued proportion, and the number after the unit be prime, the greatest will not be measured by any except those which have a place among the proportional numbers.

```
A ———————        E ————
B ————————       F ——————————
C —————————      G ————
D ———————————    H ————————
```

Let there be as many numbers as we please, *A*, *B*, *C*, *D*, beginning from an unit and in continued proportion, and let *A*, the number after the unit, be prime;

I say that *D*, the greatest of them, will not be measured by any other number except *A*, *B*, *C*.

For, if possible, let it be measured by *E*, and let *E* not be the same with any of the numbers *A*, *B*, *C*.

It is then manifest that *E* is not prime.

For, if E is prime and measures D,
it will also measure A, [IX. 12] which is prime, though it is not the same with it:
which is impossible.

<p align="center">Therefore E is not prime.</p>

<p align="center">Therefore it is composite.</p>

But any composite number is measured by some prime number; . [VII. 31]

<p align="center">therefore E is measured by some prime number.</p>

I say next that it will not be measured by any other prime except A.

For, if E is measured by another,
and E measures D,
that other will also measure D;
so that it will also measure A [IX. 12], which is prime, though it is not the same with it:
which is impossible.

<p align="center">Therefore A measures E.</p>

And, since E measures D, let it measure it according to F.

I say that F is not the same with any of the numbers A, B, C.

For, if F is the same with one of the numbers A, B, C,
and measures D according to E,

<p align="center">therefore one of the numbers A, B, C also measures D according to E.</p>

But one of the numbers A, B, C measures D according to some one of the numbers A, B, C; [IX. 11]

<p align="center">therefore E is also the same with one of the numbers A, B, C:</p>

which is contrary to the hypothesis.

<p align="center">Therefore F is not the same as any one of the numbers A, B, C.</p>

Similarly we can prove that F is measured by A, by proving again that F is not prime.

For, if it is, and measures D,
it will also measure A [IX. 12], which is prime, though it is not the same with it:
which is impossible;

<p align="center">therefore F is not prime.</p>

<p align="center">Therefore it is composite.</p>

But any composite number is measured by some prime number; [VII. 31]
therefore F is measured by some prime number.

I say next that it will not be measured by any other prime except A.

For, if any other prime number measures F,
and F measures D,
that other will also measure D;
so that it will also measure A [IX. 12], which is prime, though it is not the same with it:
which is impossible.

<p align="center">Therefore A measures F.</p>

And, since E measures D according to F,

<p align="center">therefore E by multiplying F has made D.</p>

A —————— E ————
B —————— F ——————
C —————— G ———
D —————————— H ————

But, further, A has also by multiplying C made D; [IX. 11]
therefore the product of A, C is equal to the product of E, F.

Therefore, proportionally, as A is to E, so is F to C. [VII. 19]

But A measures E;
therefore F also measures C.

Let it measure it according to G.

Similarly, then, we can prove that G is not the same with any of the numbers A, B, and that it is measured by A.

And, since F measures C according to G,
therefore F by multiplying G has made C.

But, further, A has also by multiplying B made C; [IX. 11]
therefore the product of A, B is equal to the product of F, G.

Therefore, proportionally, as A is to F, so is G to B. [VII. 19]

But A measures F;
therefore G also measures B.

Let it measure it according to H.

Similarly then we can prove that H is not the same with A.

And, since G measures B according to H,
therefore G by multiplying H has made B.

But, further, A has also by multiplying itself made B; [IX. 8]
therefore the product of H, G is equal to the square on A.

Therefore, as H is to A, so is A to G. [VII. 19]

But A measures G;
therefore H also measures A, which is prime, though it is not the same with it: which is absurd.

Therefore D the greatest will not be measured by any other number except A, B, C.

Q.E.D.

Proposition 14

If a number be the least that is measured by prime numbers, it will not be measured by any other prime number except those originally measuring it.

For let the number A be the least that is measured by the prime numbers B, C, D;

I say that A will not be measured by any other prime number except B, C, D.

For, if possible, let it be measured by the prime number E, and let E not be the same with any one of the numbers B, C, D.

Now, since *E* measures *A*, let it measure it according to *F*;

therefore *E* by multiplying *F* has made *A*.

And *A* is measured by the prime numbers *B, C, D*.

But, if two numbers by multiplying one another make some number, and any prime number measure the product, it will also measure one of the original numbers; [VII. 30]

therefore *B, C, D* will measure one of the numbers *E, F*.

Now they will not measure *E*;

for *E* is prime and not the same with any one of the numbers *B, C, D*.

Therefore they will measure *F*, which is less than *A*:

which is impossible, for *A* is by hypothesis the least number measured by *B, C, D*.

Therefore no prime number will measure *A* except *B, C, D*.

<div align="right">Q.E.D.</div>

Proposition 15

If three numbers in continued proportion be the least of those which have the same ratio with them, any two whatever added together will be prime to the remaining number.

Let *A, B, C*, three numbers in continued proportion, be the least of those which have the same ratio with them;

I say that any two of the numbers *A, B, C* whatever added together are prime to the remaining number, namely *A, B* to *C*; *B, C* to *A*; and further, *A, C* to *B*.

For let two numbers *DE, EF*, the least of those which have the same ratio with *A, B, C*, be taken. [VIII. 2]

It is then manifest that *DE* by multiplying itself has made *A*, and by multiplying *EF* has made *B*, and, further, *EF* by multiplying itself has made *C*. [VIII. 2]

Now, since *DE, EF* are least,

they are prime to one another. [VII. 22]

But, if two numbers be prime to one another, their sum is also prime to each; [VII. 28]

therefore *DF* is also prime to each of the numbers *DE, EF*.

But, further, *DE* is also prime to *EF*;

therefore *DF, DE* are prime to *EF*.

But, if two numbers be prime to any number, their product is also prime to the other; [VII. 24]

so that the product of *FD, DE* is prime to *EF*;

hence the product of *FD, DE* is also prime to the square on *EF*. [VII. 25]

But the product of *FD, DE* is the square on *DE* together with the product of *DE, EF*; [II. 3]

therefore the square on *DE* together with the product of *DE, EF* is prime to the square on *EF*.

And the square on *DE* is *A*,
the product of *DE, EF* is *B*,
and the square on *EF* is *C*;

therefore *A, B* added together are prime to *C*.

Similarly we can prove that *B, C* added together are prime to *A*.

I say next that *A, C* added together are also prime to *B*.

For, since *DF* is prime to each of the numbers *DE, EF*,
the square on *DF* is also prime to the product of *DE, EF*. [VII. 24, 25]

But the squares on *DE, EF* together with twice the product of *DE, EF* are equal to the square on *DF*; [II. 4]
therefore the squares on *DE, EF* together with twice the product of *DE, EF* are prime to the product of *DE, EF*.

Separando, the squares on *DE, EF* together with once the product of *DE, EF* are prime to the product of *DE, EF*.

Therefore, *separando* again, the squares on *DE, EF* are prime to the product of *DE, EF*.

And the square on *DE* is *A*,
the product of *DE, EF* is *B*,
and the square on *EF* is *C*.

Therefore *A, C* added together are prime to *B*.

 Q.E.D.

Proposition 16

If two numbers be prime to one another, the second will not be to any other number as the first is to the second.

For let the two numbers *A, B* be prime to one another;
I say that *B* is not to any other number as *A* is to *B*.

For, if possible, as *A* is to *B*, so let *B* be to *C*.

Now *A, B* are prime,
primes are also least, [VII. 21]
and the least numbers measure those which have the same ratio the same number of times, the antecedent the antecedent and the consequent the consequent; [VII. 20]

therefore *A* measures *B* as antecedent antecedent.

But it also measures itself;

therefore A measures A, B which are prime to one another:
which is absurd.

Therefore B will not be to C, as A is to B.

Q.E.D.

Proposition 17

If there be as many numbers as we please in continued proportion, and the extremes of them be prime to one another, the last will not be to any other number as the first to the second.

For let there be as many numbers as we please,
A, B, C, D, in continued proportion,
and let the extremes of them, A, D, be prime to
one another;
I say that D is not to any other number as A is to B.

For, if possible, as A is to B, so let D be to E;

therefore, alternately, as A is to D, so is B to E. [VII. 13]

But A, D are prime,
primes are also least, [VII. 21]
and the least numbers measure those which have the same ratio the same
number of times, the antecedent the antecedent and the consequent the
consequent. [VII. 20]

Therefore A measures B.

And, as A is to B, so is B to C.

Therefore B also measures C;
so that A also measures C.

And since, as B is to C, so is C to D,
and B measures C,

therefore C also measures D.

But A measured C;

so that A also measures D.

But it also measures itself;

therefore A measures A, D which are prime to one another:
which is impossible.

Therefore D will not be to any other number as A is to B.

Q.E.D.

Proposition 18

Given two numbers, to investigate whether it is possible to find a third proportional to them.

A ———

D ———

B ———

C ————————

Let A, B be the given two numbers, and let it be required to investigate whether it is possible to find a third proportional to them.

Now A, B are either prime to one another or not.

And, if they are prime to one another, it has been proved that it is impossible to find a third proportional to them. [IX. 16]

Next, let A, B not be prime to one another,
and let B by multiplying itself make C.

Then A either measures C or does not measure it.

First, let it measure it according to D;
 therefore A by multiplying D has made C.

But, further, B has also by multiplying itself made C;
 therefore the product of A, D is equal to the square on B.

Therefore, as A is to B, so is B to D; [VII. 19]
 therefore a third proportional number D has been found to A, B.

Next, let A not measure C;
I say that it is impossible to find a third proportional number to A, B.

For, if possible, let D, such third proportional, have been found.
 Therefore the product of A, D is equal to the square on B.

But the square on B is C;
 therefore the product of A, D is equal to C.

Hence A by multiplying D has made C;
 therefore A measures C according to D.

But, by hypothesis, it also does not measure it:
which is absurd.

Therefore it is not possible to find a third proportional number to A, B when A does not measure C.

Q.E.D.

Proposition 19

Given three numbers, to investigate when it is possible to find a fourth proportional to them.

Let A, B, C be the given three numbers, and let it be required to investigate when it is possible to find a fourth proportional to them...[1]

———————————

1. The Greek text of this proposition is corrupt, and the intact portion of the proof is erroneous, according to Heath. Nevertheless, analogously to Proposition 18, the condition that a fourth proportional to A, B, C exists is that A measure the product of B and C. —Ed.

Proposition 20

Prime numbers are more than any assigned multitude of prime numbers.

Let *A, B, C* be the assigned prime numbers;
I say that there are more prime numbers
than *A, B, C*.

For let the least number measured by *A, B, C*
be taken, [VII. 36]
and let it be *DE*;
let the unit *DF* be added to *DE*.

A —

B — G ————

C ——

E ————————————— ·— F
 D

Then *EF* is either prime or not.

First, let it be prime;
then the prime numbers *A, B, C, EF* have been found which are more than *A ,B, C*.

Next, let *EF* not be prime;
 therefore it is measured by some prime number: [VII. 31]
Let it be measured by the prime number *G*.

I say that *G* is not the same with any of the numbers *A, B, C*.

For, if possible, let it be so.

Now *A, B, C* measure *DE*;
 therefore *G* also will measure *DE*.

But it also measures *EF*.

 Therefore *G*, being a number, will measure the remainder, the unit *DF*:
which is absurd.

Therefore *G* is not the same with any one of the numbers *A, B, C*.

And by hypothesis it is prime.

Therefore the prime numbers *A, B, C, G* have been found which are more than
the assigned multitude of *A, B, C*.

 Q.E.D.

Proposition 21

If as many even numbers as we please be added together, the whole is even.

For let as many even numbers as we please, *AB, BC, CD, DE,* be added together;
I say that the whole *AE* is even.

For, since each of the numbers *AB, BC, CD, DE* is even, it has a half part;
 [VII. Def. 6]
 so that the whole *AE* also has a half part.

But an even number is that which is divisible into two equal parts; [VII. Def. 6]
 therefore *AE* is even.

 Q.E.D.

Proposition 22

If as many odd numbers as we please be added together, and their multitude be even, the whole will be even.

For let as many odd numbers as we please, *AB, BC, CD, DE,* even in multitude, be added together;
I say that the whole *AE* is even.

For, since each of the numbers *AB, BC, CD, DE* is odd, if an unit be subtracted from each, each of the remainders will be even; [VII. Def. 7]
 so that the sum of them will be even. [IX. 21]

But the multitude of the units is also even.

Therefore the whole *AE* is also even. [IX. 21]

 Q.E.D.

Proposition 23

If as many odd numbers as we please be added together, and their multitude be odd, the whole will also be odd.

For let as many odd numbers as we please, *AB, BC, CD,* the multitude of which is odd, be added together;
I say that the whole *AD* is also odd.

Let the unit *DE* be subtracted from *CD*;
 therefore the remainder *CE* is even. [VII. Def. 7]

But *CA* is also even; [IX. 22]
 therefore the whole *AE* is also even. [IX. 21]

And *DE* is an unit.

Therefore *AD* is odd. [VII. Def. 7]

 Q.E.D.

Proposition 24

If from an even number an even number be subtracted, the remainder will be even.

For from the even number *AB* let the even number *BC* be subtracted:
I say that the remainder *CA* is even.

For, since AB is even, it has a half part. [VII. Def. 6]

For the same reason BC also has a half part;
so that the remainder CA also has a half part, and AC is therefore even.

Q.E.D.

Proposition 25

If from an even number an odd number be subtracted, the remainder will be odd.

For from the even number AB let the odd number BC be subtracted;
I say that the remainder CA is odd.

For let the unit CD be subtracted from BC;
therefore DB is even. [VII. Def. 7]

But AB is also even;
therefore the remainder AD is also even. [IX. 24]

And CD is an unit;
therefore CA is odd. [VII. Def. 7]

Q.E.D.

Proposition 26

If from an odd number an odd number be subtracted, the remainder will be even.

For from the odd number AB let the odd number BC be subtracted;
I say that the remainder CA is even.

For, since AB is odd, let the unit BD be subtracted;
therefore the remainder AD is even. [VII. Def. 7]

For the same reason CD is also even; [VII. Def. 7]
so that the remainder CA is also even. [IX. 24]

Q.E.D.

Proposition 27

If from an odd number an even number be subtracted, the remainder will be odd.

A D C B

For from the odd number AB let the even number BC be subtracted; I say that the remainder CA is odd.

Let the unit AD be subtracted;

therefore DB is even. **[VII. Def. 7]**

But BC is also even;

therefore the remainder CD is even. **[IX. 24]**

Therefore CA is odd. **[VII. Def. 7]**

Q.E.D.

Proposition 28

If an odd number by multiplying an even number make some number, the product will be even.

For let the odd number A by multiplying the even number B make C;
I say that C is even.

A ——
B ———
C ———————————

For, since A by multiplying B has made C, therefore C is made up of as many numbers equal to B as there are units in A. **[VII. Def. 15]**

And B is even;

therefore C is made up of even numbers.

But, if as many even numbers as we please be added together, the whole is even.
 [IX. 21]

Therefore C is even.

Q.E.D.

Proposition 29

If an odd number by multiplying an odd number make some number, the product will be odd.

A ——
B ———
C ———————————

For let the odd number A by multiplying the odd number B make C;
I say that C is odd.

For, since A by multiplying B has made C,

therefore C is made up of as many numbers equal to B as there are units in A.

<div align="right">[VII. Def. 15]</div>

And each of the numbers A, B is odd;

therefore C is made up of odd numbers the multitude of which is odd.

Thus C is odd. <div align="right">[IX. 23]</div>

<div align="right">Q.E.D.</div>

Proposition 30

If an odd number measure an even number, it will also measure the half of it.

For let the odd number A measure the even number B;
I say that it will also measure the half of it.

For, since A measures B,
let it measure it according to C;
I say that C is not odd.

For, if possible, let it be so.

Then, since A measures B according to C,
therefore A by multiplying C has made B.

Therefore B is made up of odd numbers the multitude of which is odd.

Therefore B is odd: <div align="right">[IX. 23]</div>
which is absurd, for by hypothesis it is even.

Therefore C is not odd;
therefore C is even.

Thus A measures B an even number of times.

For this reason then it also measures the half of it.

<div align="right">Q.E.D.</div>

Proposition 31

If an odd number be prime to any number, it will also be prime to the double of it.

For let the odd number A be prime to any number B,
and let C be double of B;
I say that A is prime to C.

For, if they are not prime to one another, some number will measure them.

Let a number measure them, and let it be D.

Now A is odd;
therefore D is also odd.

And since D which is odd measures C,
and C is even,
therefore D will measure the half of C also. <div align="right">[IX. 30]</div>

A ———

B ————

C ————————

D ——

But B is half of C;
 therefore D measures B.

But it also measures A;
therefore D measures A, B which are prime to one another:
which is impossible.

Therefore A cannot but be prime to C.

Therefore A, C are prime to one another.

 Q.E.D.

Proposition 32

Each of the numbers which are continually doubled beginning from a dyad is even-times even only.

A ——

B ————

C ——————

D ————————

For let as many numbers as we please, B, C, D, have been continually doubled beginning from the dyad A;
I say that B, C, D are even-times even only.

Now that each of the numbers B, C, D is even-times even is manifest; for it is doubled from a dyad.

I say that it is also even-times even only.

For let an unit be set out.

Since then as many numbers as we please beginning from an unit are in continued proportion,
and the number A after the unit is prime,
therefore D, the greatest of the numbers A, B, C, D, will not be measured by any other number except A, B, C. **[IX. 13]**

And each of the numbers A, B, C is even;
 therefore D is even-times even only. **[VII. Def. 8]**

Similarly we can prove that each of the numbers B, C is even-times even only.

 Q.E.D.

Proposition 33

If a number have its half odd, it is even-times odd only.

————————
 A

For let the number A have its half odd;
I say that A is even-times odd only.

Now that it is even-times odd is manifest; for the half of it, being odd, measures it an even number of times. **[VII. Def. 9]**

I say next that it is also even-times odd only.

For, if *A* is even-times even also,
 it will be measured by an even number according to an even number;
[VII. Def. 8]
so that the half of it will also be measured by an even number though it is odd:
which is absurd.

Therefore *A* is even-times odd only.

Q.E.D.

Proposition 34

*If a number neither be one of those which are continually doubled from a dyad,
nor have its half odd, it is both even-times even and even-times odd.*

For let the number *A* neither be one of those doubled from
a dyad, nor have its half odd;
I say that *A* is both even-times even and even-times odd.

Now that *A* is even-times even is manifest;
for it has not its half odd.

[VII. Def. 8]

I say next that it is also even-times odd.

For, if we bisect *A*, then bisect its half, and do this continually, we shall come
upon some odd number which will measure *A* according to an even number.

For, if not, we shall come upon a dyad,
and *A* will be among those which are doubled from a dyad:
which is contrary to the hypothesis.

Thus *A* is even-times odd.

But it was also proved even-times even.

Therefore *A* is both even-times even and even-times odd.

Q.E.D.

Proposition 35

*If as many numbers as we please be in continued proportion, and there be
subtracted, from the second and the last, numbers equal to the first, then, as the
excess of the second is to the first, so will the excess of the last be to all those
before it.*

Let there be as many numbers as we please in
continued proportion, *A, BC, D, EF,* beginning
from *A* as least,
and let there be subtracted from *BC* and *EF*
the numbers *BG, FH,* each equal to *A*;
I say that, as *GC* is to *A,* so is *EH* to *A, BC, D.*

For let *FK* be made equal to *BC,* and *FL* equal
to *D.*

Then, since *FK* is equal to *BC*,
and of these the part *FH* is equal to the part *BG*,
therefore the remainder *HK* is equal to the remainder *GC*.

And since, as *EF* is to *D*, so is *D* to *BC*, and *BC* to *A*,
while *D* is equal to *FL*, *BC* to *FK*, and *A* to *FH*,
therefore, as *EF* is to *FL*, so is *LF* to *FK*, and *FK* to *FH*.

Separando, as *EL* is to *LF*, so is *LK* to *FK*, and *KH* to *FH*. [VII. 11, 13]

Therefore also, as one of the antecedents is to one of the consequents, so are all the antecedents to all the consequents; [VII. 12]
therefore, as *KH* is to *FH*, so are *EL*, *LK*, *KH* to *LF*, *FK*, *HF*.

But *KH* is equal to *CG*, *FH* to *A*, and *LF*, *FK*, *HF* to *D*, *BC*, *A*;
therefore, as *CG* is to *A*, so is *EH* to *D*, *BC*, *A*.

Therefore, as the excess of the second is to the first, so is the excess of the last to all those before it.

Q.E.D.

Proposition 36

If as many numbers as we please beginning from an unit be set out continuously in double proportion, until the sum of all becomes prime, and if the sum multiplied into the last make some number, the product will be perfect.

For let as many numbers as we please, *A*, *B*, *C*, *D*, beginning from an unit be set out in double proportion, until the sum of all becomes prime,
let *E* be equal to the sum, and let *E* by multiplying *D* make *FG*;
I say that *FG* is perfect.

For, however many *A*, *B*, *C*, *D* are in multitude, let so many *E*, *HK*, *L*, *M* be taken in double proportion beginning from *E*;
therefore, *ex aequali*, as *A* is to *D*, so is *E* to *M*. [VII. 14]

Therefore the product of *E*, *D* is equal to the product of *A*, *M*. [VII. 19]

And the product of *E*, *D* is *FG*;
therefore the product of *A*, *M* is also *FG*.

Therefore *A* by multiplying *M* has made *FG*;
therefore *M* measures *FG* according to the units in *A*.

And A is a dyad;
<div align="center">therefore FG is double of M.</div>

But M, L, HK, E are continuously double of each other;
therefore E, HK, L, M, FG are continuously proportional in double proportion.

Now let there be subtracted from the second HK and the last FG the numbers HN, FO, each equal to the first E;
therefore, as the excess of the second is to the first, so is the excess of the last to all those before it. [IX. 35]

<div align="center">Therefore, as NK is to E, so is OG to M, L, KH, E.</div>

And NK is equal to E;
<div align="center">therefore OG is also equal to M, L, HK, E.</div>

But FO is also equal to E,
and E is equal to A, B, C, D and the unit.

Therefore the whole FG is equal to E, HK, L, M and A, B, C, D and the unit; and it is measured by them.

I say also that FG will not be measured by any other number except A, B, C, D, E, HK, L, M and the unit.

For, if possible, let some number P measure FG,
and let P not be the same with any of the numbers A, B, C, D, E, HK, L, M.

And, as many times as P measures FG, so many units let there be in Q;
<div align="center">therefore Q by multiplying P has made FG.</div>

But, further, E has also by multiplying D made FG;
<div align="center">therefore, as E is to Q, so is P to D. [VII. 19]</div>

And, since A, B, C, D are continuously proportional beginning from an unit,
therefore D will not be measured by any other number except A, B, C. [IX. 13]

And, by hypothesis, P is not the same with any of the numbers A, B, C;
<div align="center">therefore P will not measure D.</div>

But, as P is to D, so is E to Q;
<div align="center">therefore neither does E measure Q. [VII. Def. 20]</div>

And E is prime;
and any prime number is prime to any number which it does not measure.
<div align="right">[VII. 29]</div>
<div align="center">Therefore E, Q are prime to one another.</div>

But primes are also least, [VII. 21]
and the least numbers measure those which have the same ratio the same number of times, the antecedent the antecedent and the consequent the consequent; [VII. 20]
and, as E is to Q, so is P to D;
therefore E measures P the same number of times that Q measures D.

But D is not measured by any other number except A, B, C;
<div align="center">therefore Q is the same with one of the numbers A, B, C.</div>

Let it be the same with B.

And, however many B, C, D are in multitude, let so many E, HK, L be taken beginning from E.

Now E, HK, L are in the same ratio with B, C, D;

therefore, *ex aequali*, as B is to D, so is E to L. [VII. 14]

Therefore the product of B, L is equal to the product of D, E. [VII. 19]

But the product of D, E is equal to the product of Q, P;

therefore the product of Q, P is also equal to the product of B, L.

Therefore, as Q is to B, so is L to P. [VII. 19]

And Q is the same with B;

therefore L is also the same with P:

which is impossible, for by hypothesis P is not the same with any of the numbers set out.

Therefore no number will measure FG except A, B, C, D, E, HK, L, M and the unit.

And FG was proved equal to A, B, C, D, E, HK, L, M and the unit;

and a perfect number is that which is equal to its own parts; [VII. Def. 22]

therefore FG is perfect.

Q.E.D.

Book X

Definitions I[1]

1. Those magnitudes are said to be *commensurable* which are measured by the same measure, and those *incommensurable* which cannot have any common measure.

2. Straight lines are *commensurable in square* when the squares on them are measured by the same area, and *incommensurable in square* when the squares on them cannot possibly have any area as a common measure.

3. With these hypotheses, it is proved that there exist straight lines infinite in multitude which are commensurable and incommensurable respectively, some in length only, and others in square also, with an assigned straight line. Let then the assigned straight line be called *rational,* and those straight lines which are commensurable with it, whether in length and in square or in square only, *rational,* but those which are incommensurable with it *irrational.*

4. And let the square on the assigned straight line be called *rational* and those areas which are commensurable with it *rational,* but those which are incommensurable with it *irrational,* and the straight lines which produce them *irrational,* that is, in case the areas are squares, the sides themselves, but in case they are any other rectilineal figures, the straight lines on which are described squares equal to them.

Proposition 1

Two unequal magnitudes being set out, if from the greater there be subtracted a magnitude greater than its half, and from that which is left a magnitude greater than its half, and if this process be repeated continually, there will be left some magnitude which will be less than the lesser magnitude set out.

Let *AB, C* be two unequal magnitudes of which *AB* is the greater:
I say that, if from *AB* there be subtracted a magnitude greater than its half, and from that which is left a magnitude greater than its half, and if this process be repeated continually, there will be left some magnitude which will be less than the magnitude *C.*

For *C* if multiplied will sometime be greater than *AB.* [*cf.* V. Def. 4]

Let it be multiplied, and let *DE* be a multiple of *C,* and greater than *AB*;
let *DE* be divided into the parts *DF, FG, GE* equal to *C,*
from *AB* let there be subtracted *BH* greater than its half,

1. A second set of Definitions for Book X begins on page 282, a third on page 323 —Ed.

and, from *AH*, *HK* greater than its half,
and let this process be repeated continually until
the divisions in *AB* are equal in multitude with
the divisions in *DE*.

Let, then, *AK*, *KH*, *HB* be divisions which are
equal in multitude with *DF*, *FG*, *GE*.

Now, since *DE* is greater than *AB*,
and from *DE* there has been subtracted *EG* less than its half,
and, from *AB*, *BH* greater than its half,
 therefore the remainder *GD* is greater than the remainder *HA*.

And, since *GD* is greater than *HA*,
and there has been subtracted, from *GD*, the half *GF*,
and, from *HA*, *HK* greater than its half,
 therefore the remainder *DF* is greater than the remainder *AK*.

But *DF* is equal to *C*;
 therefore *C* is also greater than *AK*.

Therefore *AK* is less than *C*.

Therefore there is left of the magnitude *AB* the magnitude *AK* which is less than
the lesser magnitude set out, namely *C*.

 Q.E.D.

PORISM. And the theorem can be similarly proved even if the parts subtracted
be halves.

Proposition 2

*If, when the less of two unequal magnitudes is continually subtracted in turn from
the greater, that which is left never measures the one before it, the magnitudes will
be incommensurable.*

For, there being two unequal magnitudes *AB*, *CD*, and *AB* being the less, when
the less is continually subtracted in turn from the greater, let that which is left
over never measure the one before it;
I say that the magnitudes *AB*, *CD* are incommensurable.

For, if they are commensurable, some magnitude will measure them.

Let a magnitude measure them, if possible, and let it be *E*;
let *AB*, measuring *FD*, leave *CF* less than itself,
let *CF*, measuring *BG*, leave *AG* less than itself,
and let this process be repeated continually, until there is left some magnitude
which is less than *E*.

Suppose this done, and let there be left *AG* less than *E*.

Then, since *E* measures *AB*,
while *AB* measures *DF*,

therefore *E* will also measure *FD*.

But it measures the whole *CD* also;

therefore it will also measure the remainder *CF*.

But *CF* measures *BG*;

therefore *E* also measures *BG*.

But it measures the whole *AB* also;

therefore it will also measure the remainder *AG*, the greater the less:
which is impossible.

Therefore no magnitude will measure the magnitudes *AB, CD*;

therefore the magnitudes *AB, CD* are incommensurable. **[X. Def. 1]**

Therefore etc.

Q.E.D.

Proposition 3

Given two commensurable magnitudes, to find their greatest common measure.

Let the two given commensurable magnitudes be *AB, CD* of which *AB* is the less;
thus it is required to find the greatest common measure of *AB, CD*.

Now the magnitude *AB* either measures *CD* or it does not.

If then it measures it—and it measures itself also—*AB* is a common measure
of *AB, CD*.

And it is manifest that it is also the greatest;
for a greater magnitude than the magnitude *AB* will not measure *AB*.

Next, let *AB* not measure *CD*.

Then, if the less be continually subtracted in turn from the greater, that which is
left over will sometime measure the one before it, because *AB, CD* are not
incommensurable; **[cf. X. 2]**
let *AB*, measuring *ED*, leave *EC* less than itself,
let *EC*, measuring *FB*, leave *AF* less than itself,
and let *AF* measure *CE*.

Since, then, *AF* measures *CE*,
while *CE* measures *FB*,

therefore *AF* will also measure *FB*.

But it measures itself also;

therefore *AF* will also measure the whole *AB*.

But *AB* measures *DE*;

therefore *AF* will also measure *ED*.

But it measures *CE* also;
> therefore it also measures the whole *CD*.

Therefore *AF* is a common measure of *AB, CD*.

I say next that it is also the greatest.

For, if not, there will be some magnitude greater than *AF* which will measure *AB, CD*.

Let it be *G*.

Since then *G* measures *AB*,
while *AB* measures *ED*,
> therefore *G* will also measure *ED*.

But it measures the whole *CD* also;
> therefore *G* will also measure the remainder *CE*.

But *CE* measures *FB*;
> therefore *G* will also measure *FB*.

But it measures the whole *AB* also,
and it will therefore measure the remainder *AF*, the greater the less:
> which is impossible.

Therefore no magnitude greater than *AF* will measure *AB, CD*;
> therefore *AF* is the greatest common measure of *AB, CD*.

Therefore the greatest common measure of the two given commensurable magnitudes *AB, CD* has been found.

> Q.E.D.

PORISM. From this it is manifest that, if a magnitude measure two magnitudes, it will also measure their greatest common measure.

Proposition 4

Given three commensurable magnitudes, to find their greatest common measure.

Let *A, B, C* be the three given commensurable magnitudes; thus it is required to find the greatest common measure of *A, B, C*.

Let the greatest common measure of the two magnitudes *A, B* be taken, and let it be *D*; [X. 3]
then *D* either measures *C*, or does not measure it.

First, let it measure it.

Since then *D* measures *C*,
while it also measures *A, B*,
> therefore *D* is a common measure of *A, B, C*.

And it is manifest that it is also the greatest;
for a greater magnitude than the magnitude D does not measure A, B.

Next, let D not measure C.

I say first that C, D are commensurable.

For, since A, B, C are commensurable,
 some magnitude will measure them,
and this will of course measure A, B also;
so that it will also measure the greatest common measure of A, B, namely D.

<div align="right">[X. 3, Por.]</div>

But it also measures C;
 so that the said magnitude will measure C, D;
 therefore C, D are commensurable.

Now let their greatest common measure be taken, and let it be E. [X. 3]

Since then E measures D,
while D measures A, B,
 therefore E will also measure A, B.

But it measures C also;
 therefore E measures A, B, C;
 therefore E is a common measure of A, B, C.

I say next that it is also the greatest.

For, if possible, let there be some magnitude F greater than E, and let it measure A, B, C.

Now, since F measures A, B, C,
 it will also measure A, B,
 and will measure the greatest common measure of A, B. [X. 3, Por.]

But the greatest common measure of A, B is D;
 therefore F measures D.

But it measures C also;
 therefore F measures C, D;
therefore F will also measure the greatest common measure of C, D. [X. 3, Por.]

But that is E;
 therefore F will measure E, the greater the less:
which is impossible.

Therefore no magnitude greater than the magnitude E will measure A, B, C;
therefore E is the greatest common measure of A, B, C if D do not measure C,
and, if it measure it, D is itself the greatest common measure.

Therefore the greatest common measure of the three given commensurable magnitudes has been found.

PORISM. From this it is manifest that, if a magnitude measure three magnitudes, it will also measure their greatest common measure.

Similarly too, with more magnitudes, the greatest common measure can be found, and the porism can be extended.

<div align="right">Q.E.D.</div>

Proposition 5

Commensurable magnitudes have to one another the ratio which a number has to a number.

$$\overline{\hspace{2cm}A\hspace{2cm}} \qquad \overline{\hspace{1cm}B\hspace{1cm}} \qquad \overline{\hspace{0.6cm}C\hspace{0.6cm}}$$
$$\overline{\hspace{1.5cm}D\hspace{1.5cm}}$$
$$\overline{\hspace{0.8cm}E\hspace{0.8cm}}$$

Let *A*, *B* be commensurable magnitudes;
I say that *A* has to *B* the ratio which a number has to a number.

For, since *A*, *B* are commensurable, some magnitude will measure them.

Let it measure them, and let it be *C*.

And, as many times as *C* measures *A*, so many units let there be in *D*;
and, as many times as *C* measures *B*, so many units let there be in *E*.

Since then *C* measures *A* according to the units in *D*,
while the unit also measures *D* according to the units in it,
therefore the unit measures the number *D* the same number of times as the magnitude *C* measures *A*;

therefore as *C* is to *A*, so is the unit to *D*; [V. 15]
therefore, inversely, as *A* is to *C*, so is *D* to the unit. [cf. V. 7, Por.]

Again, since *C* measures *B* according to the units in *E*,
while the unit also measures *E* according to the units in it,
therefore the unit measures *E* the same number of times as *C* measures *B*;
therefore, as *C* is to *B*, so is the unit to *E*.

But it was also proved that,
as *A* is to *C*, so is *D* to the unit;
therefore, *ex aequali*,
as *A* is to *B*, so is the number *D* to *E*. [V. 22]

Therefore the commensurable magnitudes *A*, *B* have to one another the ratio which the number *D* has to the number *E*.

Q.E.D.

Proposition 6

If two magnitudes have to one another the ratio which a number has to a number, the magnitudes will be commensurable.

For let the two magnitudes *A*, *B* have to one another the ratio which the number *D* has to the number *E*;
I say that the magnitudes *A*, *B* are commensurable.

For let *A* be divided into as many equal parts as there are units in *D*,

and let C be equal to one of them;
and let F be made up of as many magnitudes equal to C as there are units in E.

Since then there are in A as many magnitudes equal to C as there are units in D,
whatever part the unit is of D, the same part is C of A also;
therefore, as C is to A, so is the unit to D. [VII. Def. 20]

But the unit measures the number D;
therefore C also measures A.

And since, as C is to A, so is the unit to D,
therefore, inversely, as A is to C, so is the number D to the unit. [cf. V. 7, Por.]

Again, since there are in F as many magnitudes equal to C as there are units in E,
therefore, as C is to F, so is the unit to E. [VII. Def. 20]

But it was also proved that,
as A is to C, so is D to the unit;
therefore, *ex aequali*,
as A is to F, so is D to E. [V. 22]

But, as D is to E, so is A to B;
therefore also, as A is to B, so is it to F also. [V. 11]

Therefore A has the same ratio to each of the magnitudes B, F;
therefore B is equal to F. [V. 9]

But C measures F;
therefore it measures B also.

Further it measures A also;
therefore C measures A, B.

Therefore A is commensurable with B.

Therefore etc.

Q.E.D.

PORISM. From this it is manifest that, if there be two numbers, as D, E, and a straight line, as A, it is possible to make a straight line [F] such that the given straight line is to it as the number D is to the number E.

And, if a mean proportional be also taken between A, F, as B,
as A is to F, so will the square on A be to the square on B, that is, as the first is to the third, so is the figure on the first to that which is similar and similarly described on the second. [VI. 19, Por.]

But, as A is to F, so is the number D to the number E;
therefore it has been contrived that, as the number D is to the number E, so also is the figure on the straight line A to the figure on the straight line B.

Proposition 7

Incommensurable magnitudes have not to one another the ratio which a number has to a number.

Let A, B be incommensurable magnitudes;
I say that A has not to B the ratio which a number has to a number.

A
B

For, if *A* has to *B* the ratio which a number has to a number, *A* will be commensurable with *B*. [X. 6]

But it is not;

therefore *A* has not to *B* the ratio which a number has to a number.

Therefore etc.

Q.E.D.

Proposition 8

If two magnitudes have not to one another the ratio which a number has to a number, the magnitudes will be incommensurable.

A
B

For let the two magnitudes *A*, *B* not have to one another the ratio which a number has to a number;

I say that the magnitudes *A*, *B* are incommensurable.

For, if they are commensurable, *A* will have to *B* the ratio which a number has to a number. [X. 5]

But it has not;

therefore the magnitudes *A*, *B* are incommensurable.

Therefore etc.

Q.E.D.

Proposition 9

The squares on straight lines commensurable in length have to one another the ratio which a square number has to a square number; and squares which have to one another the ratio which a square number has to a square number will also have their sides commensurable in length. But the squares on straight lines incommensurable in length have not to one another the ratio which a square number has to a square number; and squares which have not to one another the ratio which a square number has to a square number will not have their sides commensurable in length either.

A B
C
D

For let *A*, *B* be commensurable in length;

I say that the square on *A* has to the square on *B* the ratio which a square number has to a square number.

For, since *A* is commensurable in length with *B*,

therefore *A* has to *B* the ratio which a number has to a number. [X. 5]

Let it have to it the ratio which *C* has to *D*.

Since then, as *A* is to *B*, so is *C* to *D*,

while the ratio of the square on *A* to the square on *B* is duplicate of the ratio of *A* to B,

for similar figures are in the duplicate ratio of their corresponding sides; [VI. 20, Por.]

and the ratio of the square on *C* to the square on *D* is duplicate of the ratio of *C* to *D,*

for between two square numbers there is one mean proportional number, and the square number has to the square number the ratio duplicate of that which the side has to the side; [VIII. 11]

therefore also, as the square on *A* is to the square on *B,* so is the square on *C* to the square on *D.*

Next, as the square on *A* is to the square on *B,* so let the square on *C* be to the square on *D;*

I say that *A* is commensurable in length with *B.*

For since, as the square on *A* is to the square on *B,* so is the square on *C* to the square on *D,*

while the ratio of the square on *A* to the square on *B* is duplicate of the ratio of *A* to B,

and the ratio of the square on *C* to the square on *D* is duplicate of the ratio of *C* to *D,*

therefore also, as *A* is to *B,* so is *C* to *D.*

Therefore *A* has to *B* the ratio which the number *C* has to the number *D;*

therefore *A* is commensurable in length with *B.* [X. 6]

Next, let *A* be incommensurable in length with *B;*

I say that the square on *A* has not to the square on *B* the ratio which a square number has to a square number.

For, if the square on *A* has to the square on *B* the ratio which a square number has to a square number, *A* will be commensurable with *B.*

But it is not;

therefore the square on *A* has not to the square on *B* the ratio which a square number has to a square number.

Again, let the square on *A* not have to the square on *B* the ratio which a square number has to a square number;

I say that *A* is incommensurable in length with *B.*

For, if *A* is commensurable with *B,* the square on *A* will have to the square on *B* the ratio which a square number has to a square number.

But it has not;

therefore *A* is not commensurable in length with *B.*

Therefore etc.

Q.E.D.

PORISM. And it is manifest from what has been proved that straight lines commensurable in length are always commensurable in square also, but those commensurable in square are not always commensurable in length also.

⟨LEMMA. It has been proved in the arithmetical books that similar plane numbers have to one another the ratio which a square number has to a square number, [VIII. 26]

and that, if two numbers have to one another the ratio which a square number has to a square number, they are similar plane numbers. **[Converse of VIII. 26]**

And it is manifest from these propositions that numbers which are not similar plane numbers, that is, those which have not their sides proportional, have not to one another the ratio which a square number has to a square number.

For, if they have, they will be similar plane numbers: which is contrary to the hypothesis.

Therefore numbers which are not similar plane numbers have not to one another the ratio which a square number has to a square number.)[2]

⟨Proposition 10

To find two straight lines incommensurable, the one in length only, and the other in square also, with an assigned straight line.

A ————————

D ——————

E —————

B ———

C ————

Let A be the assigned straight line;
thus it is required to find two straight lines incommensurable, the one in length only, and the other in square also, with A.

Let two numbers B, C be set out which have not to one another the ratio which a square number has to a square number, that is, which are not similar plane numbers;
and let it be contrived that, as B is to C, so is the square on A to the square on D—
for we have learnt how to do this— **[X. 6, Por.]**
therefore the square on A is commensurable with the square on D. **[X. 6]**

And, since B has not to C the ratio which a square number has to a square number,
therefore neither has the square on A to the square on D the ratio which a square number has to a square number;
therefore A is incommensurable in length with D. **[X. 9]**

Let E be taken a mean proportional between A, D;
therefore, as A is to D, so is the square on A to the square on E. **[V. Def. 9]**

But A is incommensurable in length with D;
therefore the square on A is also incommensurable with the square on E; **[X. 11]**
therefore A is incommensurable in square with E.

Therefore two straight lines D, E have been found incommensurable, D in length only, and E in square and of course in length also, with the assigned straight line A.

Q.E.D.⟩

2. Heath comments that a scholium to this proposition "says categorically that the theorem proved in it was the discovery of Theaetetus." The Lemma, however, is thought to be an interpolation; for it has reference to the next proposition, X. 10, and "there are so many objections to X. 10 that it can hardly be accepted as genuine." —Ed.

Proposition 11

If four magnitudes be proportional, and the first be commensurable with the second, the third will also be commensurable with the fourth; and, if the first be incommensurable with the second, the third will also be incommensurable with the fourth.

Let A, B, C, D be four magnitudes in proportion, so that, as A is to B, so is C to D, and let A be commensurable with B;
I say that C will also be commensurable with D.

For, since A is commensurable with B, therefore A has to B the ratio which a number has to a number. [X. 5]

And, as A is to B, so is C to D;
 therefore C also has to D the ratio which a number has to a number;
 therefore C is commensurable with D. [X. 6]

Next, let A be incommensurable with B;
I say that C will also be incommensurable with D.

For, since A is incommensurable with B,
 therefore A has not to B the ratio which a number has to a number. [X. 7]

And, as A is to B, so is C to D;
 therefore neither has C to D the ratio which a number has to a number;
 therefore C is incommensurable with D. [X. 8]

Therefore etc.

Q.E.D.

Proposition 12

Magnitudes commensurable with the same magnitude are commensurable with one another also.

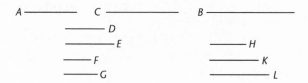

For let each of the magnitudes A, B be commensurable with C;
I say that A is also commensurable with B.

For, since A is commensurable with C,
 therefore A has to C the ratio which a number has to a number. [X. 5]

Let it have the ratio which D has to E.

Again, since C is commensurable with B,
 therefore C has to B the ratio which a number has to a number. [X. 5]

Let it have the ratio which F has to G.

And, given any number of ratios we please, namely the ratio which D has to E and that which F has to G,
let the numbers H, K, L be taken continuously in the given ratios; [cf. VIII. 4]
so that, as D is to E, so is H to K,
and, as F is to G, so is K to L.

Since, then, as A is to C, so is D to E,
while, as D is to E, so is H to K,
 therefore also, as A is to C, so is H to K. [V. 11]

Again, since, as C is to B, so is F to G,
while, as F is to G, so is K to L,
 therefore also, as C is to B, so is K to L. [V. 11]

But also, as A is to C, so is H to K;
therefore, *ex aequali*,
 as A is to B, so is H to L. [V. 22]

Therefore A has to B the ratio which a number has to a number;
 therefore A is commensurable with B. [X. 6]

Therefore etc.

<div align="right">Q.E.D.</div>

Proposition 13

If two magnitudes be commensurable, and the one of them be incommensurable with any magnitude, the remaining one will also be incommensurable with the same.

Let A, B be two commensurable magnitudes, and let one of them, A, be incommensurable with any other magnitude C;
I say that the remaining one, B, will also be incommensurable with C.

For, if B is commensurable with C,
while A is also commensurable with B,
 A is also commensurable with C. [X. 12]

But it is also incommensurable with it:
which is impossible.

Therefore B is not commensurable with C;
 therefore it is incommensurable with it.

Therefore etc.

<div align="right">Q.E.D.</div>

LEMMA. *Given two unequal straight lines, to find by what square the square on the greater is greater than the square on the less.*

Let *AB*, *C* be the given two unequal straight lines, and let *AB* be the greater of them;
thus it is required to find by what square the square on *AB* is greater than the square on *C*.

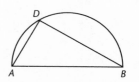

Let the semicircle *ADB* be described on *AB*,
and let *AD* be fitted into it equal to *C*; [IV. 1]
let *DB* be joined.

It is then manifest that the angle *ADB* is right, [III. 31]
and that the square on *AB* is greater than the square on *AD*, that is, *C*, by the square on *DB*. [I. 47]

Similarly also, if two straight lines be given, the straight line the square on which is equal to the sum of the squares on them is found in this manner:

Let *AD*, *DB* be the given two straight lines, and let it be required to find the straight line the square on which is equal to the sum of the squares on them.

Let them be placed so as to contain a right angle, that formed by *AD*, *DB*;
and let *AB* be joined.

It is again manifest that the straight line the square on which is equal to the sum of the squares on *AD*, *DB* is *AB*. [I. 47]

Q.E.D.

Proposition 14

If four straight lines be proportional, and the square on the first be greater than the square on the second by the square on a straight line commensurable with the first, the square on the third will also be greater than the square on the fourth by the square on a straight line commensurable with the third.

And, if the square on the first be greater than the square on the second by the square on a straight line incommensurable with the first, the square on the third will also be greater than the square on the fourth by the square on a straight line incommensurable with the third.

Let *A*, *B*, *C*, *D* be four straight lines in proportion,
so that, as *A* is to *B*, so is *C* to *D*;
and let the square on *A* be greater than the square on *B* by the square on *E*,
and let the square on *C* be greater than the square on *D* by the square on *F*;
I say that, if *A* is commensurable with *E*, *C* is also commensurable with *F*,
and, if *A* is incommensurable with *E*, *C* is also incommensurable with *F*.

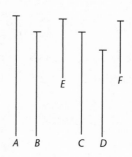

For since, as *A* is to *B*, so is *C* to *D*,
therefore also, as the square on *A* is to the square on *B*, so is the square on *C* to the square on *D*. [VI. 22]

But the squares on E, B are equal to the square on A, and the squares on D, F are equal to the square on C.

Therefore, as the squares on E, B are to the square on B, so are the squares on D, F to the square on D;
therefore, *separando*, as the square on E is to the square on B, so is the square on F to the square on D; [V. 17]
therefore also, as E is to B, so is F to D; [VI. 22]
therefore, inversely, as B is to E, so is D to F.

But, as A is to B, so also is C to D;
therefore, *ex aequali*, as A is to E, so is C to F. [V. 22]

Therefore, if A is commensurable with E, C is also commensurable with F, and, if A is incommensurable with E, C is also incommensurable with F. [X. 11]

Therefore etc.

Q.E.D.

Proposition 15

If two commensurable magnitudes be added together, the whole will also be commensurable with each of them; and, if the whole be commensurable with one of them, the original magnitudes will also be commensurable.

For let the two commensurable magnitudes AB, BC be added together;
I say that the whole AC is also commensurable with each of the magnitudes AB, BC.

For, since AB, BC are commensurable, some magnitude will measure them.

Let it measure them, and let it be D.

Since then D measures AB, BC, it will also measure the whole AC.

But it measures AB, BC also;
therefore D measures AB, BC, AC;
therefore AC is commensurable with each of the magnitudes AB, BC. [X. Def. 1]

Next, let AC be commensurable with AB;
I say that AB, BC are also commensurable.

For, since AC, AB are commensurable, some magnitude will measure them.

Let it measure them, and let it be D.

Since then D measures CA, AB, it will also measure the remainder BC.

But it measures AB also;
therefore D will measure AB, BC;
therefore AB, BC are commensurable. [X. Def. 1]

Therefore etc.

Q.E.D.

Proposition 16

If two incommensurable magnitudes be added together, the whole will also be incommensurable with each of them; and, if the whole be incommensurable with one of them, the original magnitudes will also be incommensurable.

For let the two incommensurable magnitudes AB, BC be added together;

I say that the whole AC is also incommensurable with each of the magnitudes AB, BC.

For, if CA, AB are not incommensurable, some magnitude will measure them.

Let it measure them, if possible, and let it be D.

Since then D measures CA, AB,

therefore it will also measure the remainder BC.

But it measures AB also;

therefore D measures AB, BC.

Therefore AB, BC are commensurable;
but they were also, by hypothesis, incommensurable:
which is impossible.

Therefore no magnitude will measure CA, AB;

therefore CA, AB are incommensurable. [X. Def. 1]

Similarly we can prove that AC, CB are also incommensurable.

Therefore AC is incommensurable with each of the magnitudes AB, BC.

Next, let AC be incommensurable with one of the magnitudes AB, BC.

First, let it be incommensurable with AB;
I say that AB, BC are also incommensurable.

For, if they are commensurable, some magnitude will measure them.

Let it measure them, and let it be D.

Since, then, D measures AB, BC,

therefore it will also measure the whole AC.

But it measures AB also;

therefore D measures CA, AB.

Therefore CA, AB are commensurable;
but they were also, by hypothesis, incommensurable:
which is impossible.

Therefore no magnitude will measure AB, BC;

therefore AB, BC are incommensurable. [X. Def. 1]

Therefore etc.

Q.E.D.

LEMMA. *If to any straight line there be applied a parallelogram deficient by a square figure, the applied parallelogram is equal to the rectangle contained by the segments of the straight line resulting from the application.*

251

For let there be applied to the straight line AB the parallelogram AD deficient by the square figure DB; I say that AD is equal to the rectangle contained by AC, CB.

This is indeed at once manifest;
for, since DB is a square,

$$DC \text{ is equal to } CB;$$

and AD is the rectangle AC, CD, that is, the rectangle AC, CB.

Therefore etc.

<div align="right">Q.E.D.</div>

Proposition 17

If there be two unequal straight lines, and to the greater there be applied a parallelogram equal to the fourth part of the square on the less and deficient by a square figure, and if it divide it into parts which are commensurable in length, then the square on the greater will be greater than the square on the less by the square on a straight line commensurable with the greater.

And, if the square on the greater be greater than the square on the less by the square on a straight line commensurable with the greater, and if there be applied to the greater a parallelogram equal to the fourth part of the square on the less and deficient by a square figure, it will divide it into parts which are commensurable in length.

Let A, BC be two unequal straight lines, of which BC is the greater,
and let there be applied to BC a parallelogram equal to the fourth part of the square on the less, A, that is, equal to the square on the half of A, and deficient by a square figure.

Let this be the rectangle BD, DC, **[cf. Lemma]**
and let BD be commensurable in length with DC;
I say that the square on BC is greater than the square on A by the square on a straight line commensurable with BC.

For let BC be bisected at the point E,
and let EF be made equal to DE.

Therefore the remainder DC is equal to BF.

And, since the straight line BC has been cut into equal parts at E, and into unequal parts at D,
therefore the rectangle contained by BD, DC, together with the square on ED, is equal to the square on EC; **[II. 5]**

And the same is true of their quadruples;
therefore four times the rectangle BD, DC, together with four times the square on DE, is equal to four times the square on EC.

But the square on A is equal to four times the rectangle BD, DC;

and the square on *DF* is equal to four times the square on *DE*, for *DF* is double of *DE*.

And the square on *BC* is equal to four times the square on *EC*, for again *BC* is double of *CE*.

Therefore the squares on *A, DF* are equal to the square on *BC*,
so that the square on *BC* is greater than the square on *A* by the square on *DF*.

It is to be proved that *BC* is also commensurable with *DF*.

Since *BD* is commensurable in length with *DC*,
therefore *BC* is also commensurable in length with *CD*. [X. 15]

But *CD* is commensurable in length with *CD, BF*, for *CD* is equal to *BF*. [X. 6]

Therefore *BC* is also commensurable in length with *BF, CD*, [X. 12]
so that *BC* is also commensurable in length with the remainder *FD*; [X. 15]
therefore the square on *BC* is greater than the square on *A* by the square on a straight line commensurable with *BC*.

Next, let the square on *BC* be greater than the square on *A* by the square on a straight line commensurable with *BC*,
let a parallelogram be applied to *BC* equal to the fourth part of the square on *A* and deficient by a square figure, and let it be the rectangle *BD, DC*.

It is to be proved that *BD* is commensurable in length with *DC*.

With the same construction, we can prove similarly that the square on *BC* is greater than the square on *A* by the square on *FD*.

But the square on *BC* is greater than the square on *A* by the square on a straight line commensurable with *BC*.

Therefore *BC* is commensurable in length with *FD*,
so that *BC* is also commensurable in length with the remainder, the sum of *BF, DC*. [X. 15]

But the sum of *BF, DC* is commensurable with *DC*, [X. 6]
so that *BC* is also commensurable in length with *CD*; [X. 12]
and therefore, *separando*, *BD* is commensurable in length with *DC*. [X. 15]

Therefore etc.

Q.E.D.

Proposition 18

If there be two unequal straight lines, and to the greater there be applied a parallelogram equal to the fourth part of the square on the less and deficient by a square figure, and if it divide it into parts which are incommensurable, the square on the greater will be greater than the square on the less by the square on a straight line incommensurable with the greater.

And, if the square on the greater be greater than the square on the less by the square on a straight line incommensurable with the greater, and if there be applied to the greater a parallelogram equal to the fourth part of the square on the less and deficient by a square figure, it divides it into parts which are incommensurable.

Let A, BC be two unequal straight lines, of which BC is the greater, and to BC let there be applied a parallelogram equal to the fourth part of the square on the less, A, and deficient by a square figure.

Let this be the rectangle BD, DC, [*cf.* Lemma before X. 17]
and let BD be incommensurable in length with DC;
I say that the square on BC is greater than the square on A by the square on a straight line incommensurable with BC.

For, with the same construction as before, we can prove similarly that the square on BC is greater than the square on A by the square on FD.

It is to be proved that BC is incommensurable in length with DF.

Since BD is incommensurable in length with DC,
 therefore BC is also incommensurable in length with CD. [X. 16]

But DC is commensurable with the sum of BF, DC; [X. 6]
 therefore BC is also incommensurable with the sum of BF, DC; [X. 13]
 so that BC is also incommensurable in length with the remainder FD. [X. 16]

And the square on BC is greater than the square on A by the square on FD;
therefore the square on BC is greater than the square on A by the square on a straight line incommensurable with BC.

Again, let the square on BC be greater than the square on A by the square on a straight line incommensurable with BC,
and let there be applied to BC a parallelogram equal to the fourth part of the square on A and deficient by a square figure.

Let this be the rectangle BD, DC.

It is to be proved that BD is incommensurable in length with DC.

For, with the same construction, we can prove similarly that the square on BC is greater than the square on A by the square on FD.

But the square on BC is greater than the square on A by the square on a straight line incommensurable with BC;
 therefore BC is incommensurable in length with FD,
 so that BC is also incommensurable with the remainder, the sum of BF, DC.
 [X. 16]

But the sum of BF, DC is commensurable in length with DC; [X. 6]
 therefore BC is also incommensurable in length with DC, [X. 13]
 so that, *separando*, BD is also incommensurable in length with DC. [X. 16]

Therefore etc.

Q.E.D.

⟨**LEMMA.** Since it has been proved that straight lines commensurable in length are always commensurable in square also, while those commensurable in square are not always commensurable in length also, but can of course be either commensurable or incommensurable in length, it is manifest that, if any straight line be commensurable in length with a given rational straight line, it is called rational and commensurable with the other not only in length but in square also, since straight lines commensurable in length are always commensurable in square also.

But, if any straight line be commensurable in square with a given rational straight line, then, if it is also commensurable in length with it, it is called in this case also rational and commensurable with it both in length and in square; but, if again any straight line, being commensurable in square with a given rational straight line, be incommensurable in length with it, it is called in this case also rational but commensurable in square only.〉

Proposition 19

The rectangle contained by rational straight lines commensurable in length is rational.

For let the rectangle *AC* be contained by the rational straight lines *AB, BC* commensurable in length;
 I say that *AC* is rational.

For on *AB* let the square *AD* be described;
 therefore *AD* is rational. [X. Def. 4]

And, since *AB* is commensurable in length with *BC*, while *AB* is equal to *BD*,
 therefore *BD* is commensurable in length with *BC*.

And, as *BD* is to *BC*, so is *DA* to *AC*. [VI. 1]

 Therefore *DA* is commensurable with *AC*. [X. 11]

But *DA* is rational;
 therefore *AC* is also rational. [X. Def. 4]

Therefore etc.

 Q.E.D.

Proposition 20

If a rational area be applied to a rational straight line, it produces as breadth a straight line rational and commensurable in length with the straight line to which it is applied.

For let the rational area *AC* be applied to *AB*, a straight line once more rational in any of the aforesaid ways, producing *BC* as breadth;
I say that *BC* is rational and commensurable in length with *BA*.

For on *AB* let the square *AD* be described;
 therefore *AD* is rational. [X. Def. 4]

But *AC* is also rational;
 therefore *DA* is commensurable with *AC*.

And, as *DA* is to *AC*, so is *DB* to *BC*. [VI. 1]

Therefore DB is also commensurable with BC; [X. 11]
and DB is equal to BA;
> therefore AB is also commensurable with BC.

But AB is rational;
therefore BC is also rational and commensurable in length with AB.

Therefore etc.

Q.E.D.

Proposition 21

The rectangle contained by rational straight lines commensurable in square only is irrational, and the side of the square equal to it is irrational. Let the latter be called medial.

For let the rectangle AC be contained by the rational straight lines AB, BC commensurable in square only;
I say that AC is irrational, and the side of the square equal to it is irrational;
and let the latter be called *medial*.

For on AB let the square AD be described;
> therefore AD is rational. [X. Def. 4]

And, since AB is incommensurable in length with BC,
for by hypothesis they are commensurable in square only,
while AB is equal to BD,
> therefore DB is also incommensurable in length with BC.

And, as DB is to BC, so is AD to AC; [VI. 1]
> therefore DA is incommensurable with AC. [X. 11]

But DA is rational;
> therefore AC is irrational,
> so that the side of the square equal to AC is also irrational. [X. Def. 4]

And let the latter be called *medial*.

Q.E.D.

LEMMA. If there be two straight lines, then, as the first is to the second, so is the square on the first to the rectangle contained by the two straight lines.

Let FE, EG be two straight lines.

I say that, as FE is to EG, so is the square on FE to the rectangle FE, EG.

For on FE let the square DF be described,
and let GD be completed.

Since then, as *FE* is to *EG*, so is *FD* to *DG*, [VI. 1]
and *FD* is the square on *FE*,
and *DG* the rectangle *DE*, *EG*, that is, the rectangle *FE*, *EG*,
 therefore, as *FE* is to *EG*, so is the square on *FE* to the rectangle *FE*, *EG*.

Similarly also, as the rectangle *GE*, *EF* is to the square on *EF*, that is, as *GD* is to *FD*, so is *GE* to *EF*.

<div align="right">Q.E.D.</div>

Proposition 22

The square on a medial straight line, if applied to a rational straight line, produces as breadth a straight line rational and incommensurable in length with that to which it is applied.

Let *A* be medial and *CB* rational,
and let a rectangular area *BD* equal to the square on *A* be applied to *BC*, producing *CD* as breadth;
I say that *CD* is rational and incommensurable in length with *CB*.

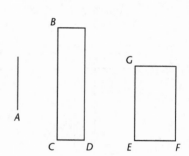

For, since *A* is medial, the square on it is equal to a rectangular area contained by rational straight lines commensurable in square only. [X. 21]

Let the square on it be equal to *GF*.

But the square on it is also equal to *BD*;
 therefore *BD* is equal to *GF*.

But it is also equiangular with it;
and in equal and equiangular parallelograms the sides about the equal angles are reciprocally proportional; [VI. 14]
 therefore, proportionally, as *BC* is to *EG*, so is *EF* to *CD*.

Therefore also, as the square on *BC* is to the square on *EG*, so is the square on *EF* to the square on *CD*. [VI. 22]

But the square on *CB* is commensurable with the square on *EG*, for each of these straight lines is rational;
therefore the square on *EF* is also commensurable with the square on *CD*. [X. 11]

But the square on *EF* is rational;
 therefore the square on *CD* is also rational; [X. Def. 4]
 therefore *CD* is rational.

And since *EF* is incommensurable in length with *EG*,
for they are commensurable in square only,
and, as *EF* is to *EG*, so is the square on *EF* to the rectangle *FE*, *EG*, [Lemma]
 therefore the square on *EF* is incommensurable with the rectangle *FE*, *EG*.
 [X. 11]

But the square on *CD* is commensurable with the square on *EF*, for the straight lines are rational in square;
and the rectangle *DC*, *CB* is commensurable with the rectangle *FE*, *EG*, for they are equal to the square on *A*;

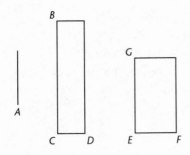

therefore the square on *CD* is also incommensurable with the rectangle *DC, CB*. [X. 13]

But, as the square on *CD* is to the rectangle *DC, CB*, so is *DC* to *CB*; [Lemma] therefore *DC* is incommensurable in length with *CB*. [X. 11]

Therefore *CD* is rational and incommensurable in length with *CB*.

Q.E.D.

Proposition 23

A straight line commensurable with a medial straight line is medial.

Let *A* be medial, and let *B* be commensurable with *A*; I say that *B* is also medial.

For let a rational straight line *CD* be set out, and to *CD* let the rectangular area *CE* equal to the square on *A* be applied, producing *ED* as breadth; therefore *ED* is rational and incommensurable in length with *CD*. [X. 22]

And let the rectangular area *CF* equal to the square on *B* be applied to *CD*, producing *DF* as breadth.

Since, then, *A* is commensurable with *B*, the square on *A* is also commensurable with the square on *B*.

But *EC* is equal to the square on *A*, and *CF* is equal to the square on *B*; therefore *EC* is commensurable with *CF*.

And, as *EC* is to *CF*, so is *ED* to *DF*; [VI. 1] therefore *ED* is commensurable in length with *DF*. [X. 11]

But *ED* is rational and incommensurable in length with *DC*; therefore *DF* is also rational [X. Def. 3] and incommensurable in length with *DC*. [X. 13]

Therefore *CD, DF* are rational and commensurable in square only.

But the straight line the square on which is equal to the rectangle contained by rational straight lines commensurable in square only is medial; [X. 21] therefore the side of the square equal to the rectangle *CD, DF* is medial.

And *B* is the side of the square equal to the rectangle *CD, DF*; therefore *B* is medial.

Q.E.D.

PORISM. From this it is manifest that an area commensurable with a medial area is medial.

And in the same way as was explained in the case of rationals [Lemma following X. 18]

it follows, as regards medials, that a straight line commensurable in length with a medial straight line is called *medial* and commensurable with it not only in length but in square also, since, in general, straight lines commensurable in length are always commensurable in square also.

But, if any straight line be commensurable in square with a medial straight line, then, if it is also commensurable in length with it, the straight lines are called, in this case too, medial and commensurable in length and in square, but, if in square only, they are called medial straight lines commensurable in square only.

Proposition 24

The rectangle contained by medial straight lines commensurable in length is medial.

For let the rectangle *AC* be contained by the medial straight lines *AB*, *BC* which are commensurable in length;
I say that *AC* is medial.

For on *AB* let the square *AD* be described;
therefore *AD* is medial.

And, since *AB* is commensurable in length with *BC*,
while *AB* is equal to *BD*,
therefore *DB* is also commensurable in length with *BC*;
so that *DA* is also commensurable with *AC*. [VI. 1, X. 11]

But *DA* is medial;
therefore *AC* is also medial. [X. 23, Por.]

Q.E.D.

Proposition 25

The rectangle contained by medial straight lines commensurable in square only is either rational or medial.

For let the rectangle *AC* be contained by the medial straight lines *AB*, *BC* which are commensurable in square only;
I say that *AC* is either rational or medial.

For on *AB*, *BC* let the squares *AD*, *BE* be described;
therefore each of the squares *AD*, *BE* is medial.

Let a rational straight line *FG* be set out,
to *FG* let there be applied the rectangular parallelogram *GH* equal to *AD*, producing *FH* as breadth,

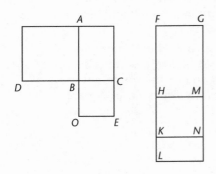

to *HM* let there be applied the rectangular parallelogram *MK* equal to *AC*, producing *HK* as breadth,

and further to *KN* let there be similarly applied *NL* equal to *BE*, producing *KL* as breadth;

therefore *FH, HK, KL* are in a straight line.

Since then each of the squares *AD, BE* is medial,

and *AD* is equal to *GH*,

and *BE* to *NL*,

therefore each of the rectangles *GH, NL* is also medial.

And they are applied to the rational straight line *FG*;

therefore each of the straight lines *FH, KL* is rational and incommensurable in length with *FG*. [X. 22]

And, since *AD* is commensurable with *BE*,

therefore *GH* is also commensurable with *NL*.

And, as *GH* is to *NL*, so is *FH* to *KL*; [VI. 1]

therefore *FH* is commensurable in length with *KL*. [X. 11]

Therefore *FH, KL* are rational straight lines commensurable in length;

therefore the rectangle *FH, KL* is rational. [X. 19]

And, since *DB* is equal to *BA*, and *OB* to *BC*,

therefore; as *DB* is to *BC*, so is *AB* to *BO*.

But, as *DB* is to *BC*, so is *DA* to *AC*, [VI. 1]

and, as *AB* is to *BO*, so is *AC* to *CO*; [VI. 1]

therefore, as *DA* is to *AC*, so is *AC* to *CO*.

But *AD* is equal to *GH*, *AC* to *MK* and *CO* to *NL*;

therefore, as *GH* is to *MK*, so is *MK* to *NL*;

therefore also, as *FH* is to *HK*, so is *HK* to *KL*; [VI. 1, V. 11]

therefore the rectangle *FH, KL* is equal to the square on *HK*. [VI. 17]

But the rectangle *FH, KL* is rational;

therefore the square on *HK* is also rational.

Therefore *HK* is rational.

And, if it is commensurable in length with *FG*,

HN is rational; [X. 19]

but, if it is incommensurable in length with *FG*,

KH, HM are rational straight lines commensurable in square only,

and therefore *HN* is medial. [X. 21]

Therefore *HN* is either rational or medial.

But *HN* is equal to *AC*;

therefore *AC* is either rational or medial.

Therefore etc.

Q.E.D.

Proposition 26

A medial area does not exceed a medial area by a rational area.

For, if possible, let the medial area *AB*
exceed the medial area *AC* by the rational
area *DB*,
and let a rational straight line *EF* be set out;
to *EF* let there be applied the rectangular
parallelogram *FH* equal to *AB*, producing
EH as breadth,
and let the rectangle *FG* equal to *AC* be
subtracted;
therefore the remainder *BD* is equal to the
remainder *KH*.

But *DB* is rational;

therefore *KH* is also rational.

Since, then, each of the rectangles *AB*, *AC* is medial,
and *AB* is equal to *FH*, and *AC* to *FG*,

therefore each of the rectangles *FH*, *FG* is also medial.

And they are applied to the rational straight line *EF*;
therefore each of the straight lines *HE*, *EG* is rational and incommensurable in
length with *EF*. [X. 22]

And, since *KH* is rational;
and it is applied to the rational straight line *EF*;

therefore *GH* is rational and commensurable in length with *EF*. [X. 20]

But *EG* is also rational, and is incommensurable in length with *EF*;

therefore *EG* is incommensurable in length with *GH*. [X. 13]

And, as *EG* is to *GH*, so is the square on *EG* to the rectangle *EG*, *GH*;

therefore the square on *EG* is incommensurable with the rectangle *EG*, *GH*.

[X. 11]

But the squares on *EG*, *GH* are commensurable with the square on *EG*, for both
are rational;
and twice the rectangle *EG*, *GH* is commensurable with the rectangle *EG*, *GH*,
for it is double of it; [X. 6]
therefore the squares on *EG*, *GH* are incommensurable with twice the rectangle
EG, *GH*; [X. 13]
therefore also the sum of the squares on *EG*, *GH* and twice the rectangle *EG*, *GH*,
that is, the square on *EH* [II. 4] is incommensurable with the squares on *EG*, *GH*.

[X. 16]

But the squares on *EG*, *GH* are rational;

therefore the square on *EH* is irrational. [X. Def. 4]

Therefore *EH* is irrational.

But it is also rational: which is impossible.

Therefore etc.

Q.E.D.

Proposition 27

To find medial straight lines commensurable in square only which contain a rational rectangle.

Let two rational straight lines A, B commensurable in square only be set out;

let C be taken a mean proportional between A, B, **[VI. 13]**

and let it be contrived that, as A is to B, so is C to D. **[VI. 12]**

Then, since A, B are rational and commensurable in square only,

the rectangle A, B, that is, the square on C **[VI. 17]**, is medial. **[X. 21]**

Therefore C is medial. **[X. 21]**

And since, as A is to B, so is C to D,

and A, B are commensurable in square only,

therefore C, D are also commensurable in square only. **[X. 11]**

And C is medial;

therefore D is also medial. **[X. 23, addition]**

Therefore C, D are medial and commensurable in square only.

I say that they also contain a rational rectangle.

For since, as A is to B, so is C to D,

therefore, alternately, as A is to C, so is B to D. **[V. 16]**

But, as A is to C, so is C to B;

therefore also, as C is to B, so is B to D;

therefore the rectangle C, D is equal to the square on B.

But the square on B is rational,

therefore the rectangle C, D is also rational.

Therefore medial straight lines commensurable in square only have been found which contain a rational rectangle.

Q.E.D.

Proposition 28

To find medial straight lines commensurable in square only, which contain a medial rectangle.

Let the rational straight lines A, B, C commensurable in square only be set out;

let D be taken a mean proportional between A, B, **[VI. 13]**

and let it be contrived that, as B is to C, so is D to E. **[VI. 12]**

Since A, B are rational straight lines commensurable in square only,

therefore the rectangle A, B, that is, the square on D **[VI. 17]**, is medial. **[X. 21]**

Therefore D is medial.

And since B, C are commensurable in square only,
and, as B is to C, so is D to E,
therefore D, E are also commensurable in square only. [X. 11]

But D is medial;
therefore E is also medial. [X. 23, addition]

Therefore D, E are medial straight lines commensurable in square only.

I say next that they also contain a medial rectangle.

For since, as B is to C, so is D to E,
therefore, alternately, as B is to D, so is C to E. [V. 16]

But, as B is to D, so is D to A;
therefore also, as D is to A, so is C to E;
therefore the rectangle A, C is equal to the rectangle D, E. [VI. 16]

But the rectangle A, C is medial; [X. 21]
therefore the rectangle D, E is also medial.

Therefore medial straight lines commensurable in square only have been found which contain a medial rectangle.

<div align="right">Q.E.D.</div>

LEMMA 1. *To find two square numbers such that their sum is also square.*

Let two numbers AB, BC be set out, and let them be either both even or both odd.

Then since, whether an even number is subtracted from an even number, or an odd number from an odd number, the remainder is even, [IX. 24, 26]
therefore the remainder AC is even.

Let AC be bisected at D.

Let AB, BC also be either similar plane numbers, or square numbers, which are themselves also similar plane numbers.

Now the product of AB, BC together with the square on CD is equal to the square on BD. [II. 6]

And the product of AB, BC is square, inasmuch as it was proved that, if two similar plane numbers by multiplying one another make some number, the product is square. [IX. 1]

Therefore two square numbers, the product of AB, BC, and the square on CD, have been found which, when added together, make the square on BD.

And it is manifest that two square numbers, the square on BD and the square on CD, have again been found such that their difference, the product of AB, BC, is a square, whenever AB, BC are similar plane numbers.

But when they are not similar plane numbers, two square numbers, the square on BD and the square on DC, have been found such that their difference, the product of AB, BC, is not square.

<div align="right">Q.E.D.</div>

LEMMA 2. *To find two square numbers such that their sum is not square.*

For let the product of *AB, BC*, as we said, be square,
and *CA* even,
and let *CA* be bisected by *D*.

It is then manifest that the square product of *AB, BC* together with the square on *CD* is equal to the square on *BD*. **[See Lemma 1]**

Let the unit *DE* be subtracted;
therefore the product of *AB, BC* together with the square on *CE* is less than the square on *BD*.

I say then that the square product of *AB, BC* together with the square on *CE* will not be square.

For, if it is square, it is either equal to the square on *BE*, or less than the square on *BE*, but cannot any more be greater, lest the unit be divided.

First, if possible, let the product of *AB, BC* together with the square on *CE* be equal to the square on *BE*,
and let *GA* be double of the unit *DE*.

Since then the whole *AC* is double of the whole *CD*,
and in them *AG* is double of *DE*,
> therefore the remainder *GC* is also double of the remainder *EC*;
>> therefore *GC* is bisected by *E*.

Therefore the product of *GB, BC* together with the square on *CE* is equal to the square on *BE*. **[II. 6]**

But the product of *AB, BC* together with the square on *CE* is also, by hypothesis, equal to the square on *BE*;
therefore the product of *GB, BC* together with the square on *CE* is equal to the product of *AB, BC* together with the square on *CE*.

And, if the common square on *CE* be subtracted,
> it follows that *AB* is equal to *GB*:
which is absurd.

Therefore the product of *AB, BC* together with the square on *CE* is not equal to the square on *BE*.

I say next that neither is it less than the square on *BE*.

For, if possible, let it be equal to the square on *BF*,
and let *HA* be double of *DF*.

Now it will again follow that *HC* is double of *CF*;
> so that *CH* has also been bisected at *F*,
and for this reason the product of *HB, BC* together with the square on *FC* is equal to the square on *BF*. **[II. 6]**

But, by hypothesis, the product of *AB, BC* together with the square on *CE* is also equal to the square on *BF*.

Thus the product of *HB, BC* together with the square on *CF* will also be equal to the product of *AB, BC* together with the square on *CE*:

which is absurd.

Therefore the product of *AB, BC* together with the square on *CE* is not less than the square on *BE*.

And it was proved that neither is it equal to the square on *BE*.

Therefore the product of *AB, BC* together with the square on *CE* is not square.

Q.E.D.

Proposition 29

To find two rational straight lines commensurable in square only and such that the square on the greater is greater than the square on the less by the square on a straight line commensurable in length with the greater.

For let there be set out any rational straight line *AB*, and two square numbers *CD, DE* such that their difference *CE* is not square; **[Lemma 1]**
let there be described on *AB* the semicircle *AFB*,
and let it be contrived that, as *DC* is to *CE*, so is the square on *BA* to the square on *AF*. **[X. 6, Por.]**

Let *FB* be joined.

Since, as the square on *BA* is to the square on *AF*, so is *DC* to *CE*,
therefore the square on *BA* has to the square on *AF* the ratio which the number *DC* has to the number *CE*;
therefore the square on *BA* is commensurable with the square on *AF*. **[X. 6]**

But the square on *AB* is rational; **[X. Def. 4]**
therefore the square on *AF* is also rational; **[X. Def. 4]**
therefore *AF* is also rational.

And, since *DC* has not to *CE* the ratio which a square number has to a square number,
neither has the square on *BA* to the square on *AF* the ratio which a square number has to a square number;
therefore *AB* is incommensurable in length with *AF*. **[X. 9]**

Therefore *BA, AF* are rational straight lines commensurable in square only.

And since, as *DC* is to *CE*, so is the square on *BA* to the square on *AF*,
therefore, *convertendo,*
as *CD* is to *DE*, so is the square on *AB* to the square on *BF*.
[V. 19, Por., III. 31, I. 47]

But *CD* has to *DE* the ratio which a square number has to a square number:
therefore also the square on *AB* has to the square on *BF* the ratio which a square number has to a square number;
therefore *AB* is commensurable in length with *BF*. **[X. 9]**

And the square on *AB* is equal to the squares on *AF, FB*;
therefore the square on *AB* is greater than the square on *AF* by the square on *BF* commensurable with *AB*.

Therefore there have been found two rational straight lines BA, AF commensurable in square only and such that the square on the greater AB is greater than the square on the less AF by the square on BF commensurable in length with AB.

Q.E.D.

Proposition 30

To find two rational straight lines commensurable in square only and such that the square on the greater is greater than the square on the less by the square on a straight line incommensurable in length with the greater.

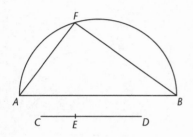

Let there be set out a rational straight line AB, and two square numbers CE, ED such that their sum CD is not square; **[Lemma 2]**
let there be described on AB the semicircle AFB, let it be contrived that, as DC is to CE, so is the square on BA to the square on AF, **[X. 6, Por.]**
and let FB be joined.

Then, in a similar manner to the preceding, we can prove that BA, AF are rational straight lines commensurable in square only.

And since, as DC is to CE, so is the square on BA to the square on AF,
therefore, *convertendo*,
 as CD is to DE, so is the square on AB to the square on BF.
 [V. 19, Por., III. 31, I. 47]

But CD has not to DE the ratio which a square number has to a square number; therefore neither has the square on AB to the square on BF the ratio which a square number has to a square number;
 therefore AB is incommensurable in length with BF. **[X. 9]**

And the square on AB is greater than the square on AF by the square on FB incommensurable with AB.

Therefore AB, AF are rational straight lines commensurable in square only, and the square on AB is greater than the square on AF by the square on FB incommensurable in length with AB.

Q.E.D.

Proposition 31

To find two medial straight lines commensurable in square only, containing a rational rectangle, and such that the square on the greater is greater than the square on the less by the square on a straight line commensurable in length with the greater.

Let there be set out two rational straight lines A, B commensurable in square only and such that the square on A, being the greater, is greater than the square

on B the less by the square on a straight line commensurable in length with A. [X. 29]

And let the square on C be equal to the rectangle A, B.

Now the rectangle A, B is medial; [X. 21]
therefore the square on C is also medial;
therefore C is also medial. [X. 21]

Let the rectangle C, D be equal to the square on B.

Now the square on B is rational;
therefore the rectangle C, D is also rational.

And since, as A is to B, so is the rectangle A, B to the square on B,
while the square on C is equal to the rectangle A, B,
and the rectangle C, D is equal to the square on B,
therefore, as A is to B, so is the square on C to the rectangle C, D.

But, as the square on C is to the rectangle C, D, so is C to D;
therefore also, as A is to B, so is C to D.

But A is commensurable with B in square only;
therefore C is also commensurable with D in square only. [X. 11]

And C is medial;
therefore D is also medial. [X. 23, addition]

And since, as A is to B, so is C to D,
and the square on A is greater than the square on B by the square on a straight line commensurable with A,
therefore also the square on C is greater than the square on D by the square on a straight line commensurable with C. [X. 14]

Therefore two medial straight lines C, D, commensurable in square only and containing a rational rectangle, have been found, and the square on C is greater than the square on D by the square on a straight line commensurable in length with C.

Similarly also it can be proved that the square on C exceeds the square on D by the square on a straight line incommensurable with C, when the square on A is greater than the square on B by the square on a straight line incommensurable with A. [X. 30]

Q.E.D.

Proposition 32

To find two medial straight lines commensurable in square only, containing a medial rectangle, and such that the square on the greater is greater than the square on the less by the square on a straight line commensurable with the greater.

Let there be set out three rational straight lines A, B, C commensurable in square only, and such that the square on A is greater than the square on C by the square on a straight line commensurable with A, [X. 29]

A ─────────────

B ────────── D ───────────

 E ─────────
C ──────────

and let the square on *D* be equal to the rectangle *A, B*.

Therefore the square on *D* is medial;

therefore *D* is also medial. [X. 21]

Let the rectangle *D, E* be equal to the rectangle *B, C*.

Then since, as the rectangle *A, B* is to the rectangle *B, C*, so is *A* to *C*;
while the square on *D* is equal to the rectangle *A, B*,
and the rectangle *D, E* is equal to the rectangle *B, C*,

therefore, as *A* is to *C*, so is the square on *D* to the rectangle *D, E*.

But, as the square on *D* is to the rectangle *D, E*, so is *D* to *E*;

therefore also, as *A* is to *C*, so is *D* to *E*.

But *A* is commensurable with *C* in square only;

therefore *D* is also commensurable with *E* in square only. [X. 11]

But *D* is medial;

therefore *E* is also medial. [X. 23, addition]

And, since, as *A* is to *C*, so is *D* to *E*,
while the square on *A* is greater than the square on *C* by the square on a straight line commensurable with *A*,
therefore also the square on *D* will be greater than the square on *E* by the square on a straight line commensurable with *D*. [X. 14]

I say next that the rectangle *D, E* is also medial.

For, since the rectangle *B, C* is equal to the rectangle *D, E*, while the rectangle *B, C* is medial, [X. 21]

therefore the rectangle *D, E* is also medial.

Therefore two medial straight lines *D, E*, commensurable in square only, and containing a medial rectangle, have been found such that the square on the greater is greater than the square on the less by the square on a straight line commensurable with the greater.

Similarly again it can be proved that the square on *D* is greater than the square on *E* by the square on a straight line incommensurable with *D*, when the square on *A* is greater than the square on *C* by the square on a straight line incommensurable with *A*. [X. 30]

Q.E.D.

LEMMA. Let *ABC* be a right-angled triangle having the angle *A* right, and let the perpendicular *AD* be drawn;
I say that the rectangle *CB, BD* is equal to the square on *BA*,
the rectangle *BC, CD* equal to the square on *CA*,
the rectangle *BD, DC* equal to the square on *AD*,
and, further, the rectangle *BC, AD* equal to the rectangle *BA, AC*.

And first that the rectangle *CB, BD* is equal to the square on *BA*.

For, since in a right-angled triangle *AD* has been drawn from the right angle perpendicular to the base,
therefore the triangles *ABD, ADC* are similar both to the whole *ABC* and to one another. [VI. 8]

And since the triangle *ABC* is similar to the triangle *ABD*,
therefore, as *CB* is to *BA*, so is *BA* to *BD*; [VI. 4]
therefore the rectangle *CB, BD* is equal to the square on *AB*. [VI. 17]

For the same reason the rectangle *BC, CD* is also equal to the square on *AC*.

And since, if in a right-angled triangle a perpendicular be drawn from the right angle to the base, the perpendicular so drawn is a mean proportional between the segments of the base, [VI. 8, Por.]
therefore, as *BD* is to *DA*, so is *AD* to *DC*;
therefore the rectangle *BD, DC* is equal to the square on *AD*. [VI. 17]

I say that the rectangle *BC, AD* is also equal to the rectangle *BA, AC*.

For since, as we said, *ABC* is similar to *ABD*,
therefore, as *BC* is to *CA*, so is *BA* to *AD*. [VI. 4]

Therefore the rectangle *BC, AD* is equal to the rectangle *BA, AC*. [VI. 16]

Q.E.D.

Proposition 33

To find two straight lines incommensurable in square which make the sum of the squares on them rational but the rectangle contained by them medial.

Let there be set out two rational straight lines *AB, BC* commensurable in square only and such that the square on the greater *AB* is greater than the square on the less *BC* by the square on a straight line incommensurable with *AB*, [X. 30]

let *BC* be bisected at *D*,
let there be applied to *AB* a parallelogram equal to the square on either of the straight lines *BD, DC* and deficient by a square figure,
and let it be the rectangle *AE, EB*; [VI. 28]
let the semicircle *AFB* be described on *AB*,
let *EF* be drawn at right angles to *AB*,
and let *AF, FB* be joined.

Then, since *AB, BC* are unequal straight lines,
and the square on *AB* is greater than the square on *BC* by the square on a straight line incommensurable with *AB*,
while there has been applied to *AB* a parallelogram equal to the fourth part of the square on *BC*, that is, to the square on half of it, and deficient by a square figure, making the rectangle *AE, EB*,
therefore *AE* is incommensurable with *EB*. [X. 18]

And, as *AE* is to *EB*, so is the rectangle *BA, AE* to the rectangle *AB, BE*,

269

while the rectangle *BA, AE* is equal to the square on *AF*,
and the rectangle *AB, BE* to the square on *BF*;
 therefore the square on *AF* is incommensurable with the square on *FB*;
 therefore *AF, FB* are incommensurable in square.

And, since *AB* is rational,
 therefore the square on *AB* is also rational;
 so that the sum of the squares on *AF, FB* is also rational. **[I. 47]**

And since, again, the rectangle *AE, EB* is equal to the square on *EF*,
and, by hypothesis, the rectangle *AB, EB* is also equal to the square on *BD*,
 therefore *FE* is equal to *BD*;
 therefore *BC* is double of *FE*,
so that the rectangle *AB, BC* is also commensurable with the rectangle *AB, EF*.

But the rectangle *AB, BC* is medial; **[X. 21]**
 therefore the rectangle *AB, EF* is also medial. **[X. 23, Por.]**

But the rectangle *AB, EF* is equal to the rectangle *AF, FB*; **[Lemma]**
 therefore the rectangle *AF, FB* is also medial.

But it was also proved that the sum of the squares on these straight lines is rational.

Therefore two straight lines *AF, FB* incommensurable in square have been found which make the sum of the squares on them rational, but the rectangle contained by them medial.

 Q.E.D.

Proposition 34

To find two straight lines incommensurable in square which make the sum of the squares on them medial but the rectangle contained by them rational.

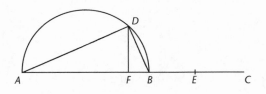

Let there be set out two medial straight lines *AB, BC*, commensurable in square only, such that the rectangle which they contain is rational, and the square on *AB* is greater than the square on *BC* by the square on a straight line incommensurable with *AB*; **[X. 31, *ad fin.*]**

let the semicircle *ADB* be described on *AB*,
let *BC* be bisected at *E*,
let there be applied to *AB* a parallelogram equal to the square on *BE* and deficient by a square figure, namely the rectangle *AF, FB*; **[VI. 28]**
 therefore *AF* is incommensurable in length with *FB*. **[X. 18]**

Let *FD* be drawn from *F* at right angles to *AB*,
and let *AD*, *DB* be joined.

Since *AF* is incommensurable in length with *FB*,
therefore the rectangle *BA*, *AF* is also incommensurable with the rectangle *AB*, *BF*.

[X. 11]

But the rectangle *BA*, *AF* is equal to the square on *AD*, and the rectangle *AB*, *BF* to the square on *DB*;
 therefore the square on *AD* is also incommensurable with the square on *DB*.

And, since the square on *AB* is medial,
 therefore the sum of the squares on *AD*, *DB* is also medial. [III. 31, I. 47]

And, since *BC* is double of *DF*,
 therefore the rectangle *AB*, *BC* is also double of the rectangle *AB*, *FD*.

But the rectangle *AB*, *BC* is rational;
 therefore the rectangle *AB*, *FD* is also rational. [X. 6]

But the rectangle *AB*, *FD* is equal to the rectangle *AD*, *DB*; [Lemma]
 so that the rectangle *AD*, *DB* is also rational.

Therefore two straight lines *AD*, *DB* incommensurable in square have been found which make the sum of the squares on them medial, but the rectangle contained by them rational.

Q.E.D.

Proposition 35

To find two straight lines incommensurable in square which make the sum of the squares on them medial and the rectangle contained by them medial and moreover incommensurable with the sum of the squares on them.

Let there be set out two medial straight lines *AB*, *BC* commensurable in square only, containing a medial rectangle, and such that the square on *AB* is greater than the square on *BC* by the square on a straight line incommensurable with *AB*;

[X. 32, ad fin.]

let the semicircle *ADB* be described on *AB*,
and let the rest of the construction be as above.

Then, since *AF* is incommensurable in length with *FB*, [X. 18]
 AD is also incommensurable in square with *DB*. [X. 11]

And, since the square on *AB* is medial,
 therefore the sum of the squares on *AD*, *DB* is also medial.

[III. 31, I. 47]

And, since the rectangle *AF, FB* is equal to the square on each of the straight lines *BE, DF,*

<div style="text-align:center">therefore BE is equal to DF;</div>
<div style="text-align:center">therefore BC is double of FD,</div>

so that the rectangle *AB, BC* is also double of the rectangle *AB, FD.*

But the rectangle *AB, BC* is medial;

<div style="text-align:center">therefore the rectangle AB, FD is also medial.</div> [X. 32, Por.]

And it is equal to the rectangle *AD, DB*; **[Lemma after X. 32]**

<div style="text-align:center">therefore the rectangle AD, DB is also medial.</div>

And, since *AB* is incommensurable in length with *BC*,
while *CB* is commensurable with *BE*,

<div style="text-align:center">therefore AB is also incommensurable in length with BE, [X. 13]</div>

so that the square on *AB* is also incommensurable with the rectangle *AB, BE.*
[X. 11]

But the squares on *AD, DB* are equal to the square on *AB*, [I. 47]
and the rectangle *AB, FD*, that is, the rectangle *AD, DB*, is equal to the rectangle *AB, BE*;
therefore the sum of the squares on *AD, DB* is incommensurable with the rectangle *AD, DB.*

Therefore two straight lines *AD, DB* incommensurable in square have been found which make the sum of the squares on them medial and the rectangle contained by them medial and moreover incommensurable with the sum of the squares on them.

<div style="text-align:right">Q.E.D.</div>

Proposition 36

If two rational straight lines commensurable in square only be added together, the whole is irrational; and let it be called binomial.

For let two rational straight lines *AB, BC* commensurable in square only be added together;
I say that the whole *AC* is irrational.

For, since *AB* is incommensurable in length with *BC*—
for they are commensurable in square only—
and, as *AB* is to *BC*, so is the rectangle *AB, BC* to the square on *BC*,

<div style="text-align:center">therefore the rectangle AB, BC is incommensurable with the square on BC.</div>

<div style="text-align:right">[X. 11]</div>

But twice the rectangle *AB, BC* is commensurable with the rectangle *AB, BC,*

[X. 6]

and the squares on *AB, BC* are commensurable with the square on *BC*—
for *AB, BC* are rational straight lines commensurable in square only— [X. 15]
therefore twice the rectangle *AB, BC* is incommensurable with the squares on *AB, BC.*

[X. 13]

And, *componendo,* twice the rectangle *AB, BC* together with the squares on *AB, BC,* that is, the square on *AC* [II. 4], is incommensurable with the sum of the squares on *AB, BC.*

[X. 16]

But the sum of the squares on *AB, BC* is rational;

therefore the square on *AC* is irrational,

so that *AC* is also irrational. [X. Def. 4]

And let it be called *binomial.*

Q.E.D.

Proposition 37

If two medial straight lines commensurable in square only and containing a rational rectangle be added together, the whole is irrational; and let it be called a first bimedial *straight line.*

For let two medial straight lines *AB, BC* commensurable in square only and containing a rational rectangle be added together;
I say that the whole *AC* is irrational.

For, since *AB* is incommensurable in length with *BC,*
therefore the squares on *AB, BC* are also incommensurable with twice the rectangle *AB, BC;* [cf. X. 36, lines 9–19]
and, *componendo,* the squares on *AB, BC* together with twice the rectangle *AB, BC,* that is, the square on *AC* [II. 4], is incommensurable with the rectangle *AB, BC.*

[X. 16]

But the rectangle *AB, BC* is rational, for, by hypothesis, *AB, BC* are straight lines containing a rational rectangle;

therefore the square on *AC* is irrational;

therefore *AC* is irrational. [X. Def. 4]

And let it be called a *first bimedial* straight line.

Q.E.D.

Proposition 38

If two medial straight lines commensurable in square only and containing a medial rectangle be added together, the whole is irrational; and let it be called a second bimedial *straight line.*

For let two medial straight lines *AB*, *BC* commensurable in square only and containing a medial rectangle be added together;
I say that *AC* is irrational.

For let a rational straight line *DE* be set out, and let the parallelogram *DF* equal to the square on *AC* be applied to *DE*, producing *DG* as breadth. [I. 44]

Then, since the square on *AC* is equal to the squares on *AB*, *BC* and twice the rectangle *AB*, *BC*, [II. 4]
let *EH*, equal to the squares on *AB*, *BC*, be applied to *DE*;
 therefore the remainder *HF* is equal to twice the rectangle *AB*, *BC*.

And, since each of the straight lines *AB*, *BC* is medial,
 therefore the squares on *AB*, *BC* are also medial.

But, by hypothesis, twice the rectangle *AB*, *BC* is also medial.

And *EH* is equal to the squares on *AB*, *BC*,
while *FH* is equal to twice the rectangle *AB*, *BC*;
 therefore each of the rectangles *EH*, *HF* is medial.

And they are applied to the rational straight line *DE*;
therefore each of the straight lines *DH*, *HG* is rational and incommensurable in length with *DE*. [X. 22]

Since then *AB* is incommensurable in length with *BC*,
and, as *AB* is to *BC*, so is the square on *AB* to the rectangle *AB*, *BC*,
 therefore the square on *AB* is incommensurable with the rectangle *AB*, *BC*.
 [X. 11]

But the sum of the squares on *AB*, *BC* is commensurable with the square on *AB*,
 [X. 15]
and twice the rectangle *AB*, *BC* is commensurable with the rectangle *AB*, *BC*.
 [X. 6]

Therefore the sum of the squares on *AB*, *BC* is incommensurable with twice the rectangle *AB*, *BC*. [X. 13]

But *EH* is equal to the squares on *AB*, *BC*,
and *HF* is equal to twice the rectangle *AB*, *BC*.

Therefore *EH* is incommensurable with *HF*,
 so that *DH* is also incommensurable in length with *HG*. [VI. 1, X. 11]

Therefore *DH*, *HG* are rational straight lines commensurable in square only;
 so that *DG* is irrational. [X. 36]

But *DE* is rational;
and the rectangle contained by an irrational and a rational straight line is irrational; [cf. X. 20]
 therefore the area *DF* is irrational,
 and the side of the square equal to it is irrational, [X. Def. 4]

But *AC* is the side of the square equal to *DF*;
 therefore *AC* is irrational.

And let it be called a *second bimedial* straight line.

 Q.E.D.

Proposition 39

If two straight lines incommensurable in square which make the sum of the squares on them rational, but the rectangle contained by them medial, be added together, the whole straight line is irrational: and let it be called major.

For let two straight lines *AB, BC* incommensurable in square, and fulfilling the given conditions [x. 33], be added together;
I say. that *AC* is irrational.

For, since the rectangle *AB, BC* is medial,
twice the rectangle *AB, BC* is also medial.　　　[X. 6 and 23, Por.]

But the sum of the squares on *AB, BC* is rational;
therefore twice the rectangle *AB, BC* is incommensurable with the sum of the squares on *AB, BC,*
so that the squares on *AB, BC* together with twice the rectangle *AB, BC,* that is, the square on *AC,* is also incommensurable with the sum of the squares on *AB, BC;*
　　　　　　　　　　　　　　　　　　　　　　[X. 16]
therefore the square on *AC* is irrational,
so that *AC* is also irrational.　　　　[X. Def. 4]

And let it be called *major.*

Q.E.D.

Proposition 40

If two straight lines incommensurable in square which make the sum of the squares on them medial, but the rectangle contained by them rational, be added together, the whole straight line is irrational; and let it be called the side of a rational plus a medial area.

For let two straight lines *AB, BC* incommensurable in square, and fulfilling the given conditions [x. 34], be added together;
I say that *AC* is irrational.

For, since the sum of the squares on *AB, BC* is medial, while twice the rectangle *AB, BC* is rational,
therefore the sum of the squares on *AB, BC* is incommensurable with twice the rectangle *AB, BC;*
so that the square on *AC* is also incommensurable with twice the rectangle *AB, BC.*
　　　　　　　　　　　　　　　　　　　　[X. 16]

But twice the rectangle *AB, BC* is rational;
therefore the square on *AC* is irrational.

Therefore *AC* is irrational.　　　　[X. Def. 4]

And let it be called the *side of a rational plus a medial area.*

Q.E.D.

Proposition 41

If two straight lines incommensurable in square which make the sum of the squares on them medial, and the rectangle contained by them medial and also incommensurable with the sum of the squares on them, be added together, the whole straight line is irrational; and let it be called the side of the sum of two medial areas.

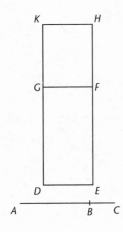

For let two straight lines *AB, BC* incommensurable in square and satisfying the given conditions [x. 35] be added together; I say that *AC* is irrational.

Let a rational straight line *DE* be set out, and let there be applied to *DE* the rectangle *DF* equal to the squares on *AB, BC*, and the rectangle *GH* equal to twice the rectangle *AB, BC*;
therefore the whole *DH* is equal to the square on *AC*. [II. 4]

Now, since the sum of the squares on *AB, BC* is medial, and is equal to *DF*,
therefore *DF* is also medial.

And it is applied to the rational straight line *DE*;
therefore *DG* is rational and incommensurable in length with *DE*. [X. 22]

For the same reason *GK* is also rational and incommensurable in length with *GF*, that is, *DE*.

And, since the squares on *AB, BC* are incommensurable with twice the rectangle *AB, BC*,
DF is incommensurable with *GH*;
so that *DG* is also incommensurable with *GK*. [VI. 1, X. 11]

And they are rational;
therefore *DG, GK* are rational straight lines commensurable in square only;
therefore *DK* is irrational and what is called binomial. [X. 36]

But *DE* is rational;
therefore *DH* is irrational, and the side of the square which is equal to it is irrational. [X. Def. 4]

But *AC* is the side of the square equal to *HD*;
therefore *AC* is irrational.

And let it be called the *side of the sum of two medial areas*.

Q.E.D.

LEMMA. And that the aforesaid irrational straight lines are divided only in one way into the straight lines of which they are the sum and which produce the types in question, we will now prove after premising the following lemma.

Let the straight line *AB* be set out, let the whole be cut into unequal parts at each of the points *C, D*,
and let *AC* be supposed greater than *DB*;

I say that the squares on *AC, CB* are greater than the squares on *AD, DB*.

For let *AB* be bisected at *E*.

Then, since *AC* is greater than *DB*,
let *DC* be subtracted from each;
 therefore the remainder *AD* is greater than the remainder *CB*.

But *AE* is equal to *EB*;
 therefore *DE* is less than *EC*;
 therefore the points *C, D* are not equidistant from the point of bisection.

And, since the rectangle *AC, CB* together with the square on *EC* is equal to the square on *EB*, **[II. 5]**
and, further, the rectangle *AD, DB* together with the square on *DE* is equal to the square on *EB*, **[II. 5]**
therefore the rectangle *AC, CB* together with the square on *EC* is equal to the rectangle *AD, DB* together with the square on *DE*.

And of these the square on *DE* is less than the square on *EC*;
therefore the remainder, the rectangle *AC, CB*, is also less than the rectangle *AD, DB*,
so that twice the rectangle *AC, CB* is also less than twice the rectangle *AD, DB*.

Therefore also the remainder, the sum of the squares on *AC, CB*, is greater than the sum of the squares on *AD, DB*.

 Q.E.D.

Proposition 42

A binomial straight line is divided into its terms at one point only.

Let *AB* be a binomial straight line divided
into its terms at *C*;
therefore *AC, CB* are rational straight lines
commensurable in square only.

I say that *AB* is not divided at another point into two rational straight lines commensurable in square only.

For, if possible, let it be divided at *D* also, so that *AD, DB* are also rational straight lines commensurable in square only.

It is then manifest that *AC* is not the same with *DB*.

For, if possible, let it be so.

Then *AD* will also be the same as *CB*,
 and, as *AC* is to *CB*, so will *BD* be to *DA*;
thus *AB* will be divided at *D* also in the same way as by the division at *C*:
which is contrary to the hypothesis.

Therefore *AC* is not the same with *DB*.

For this reason also the points *C, D* are not equidistant from the point of bisection.

Therefore that by which the squares on *AC, CB* differ from the squares on *AD, DB* is also that by which twice the rectangle *AD, DB* differs from twice the rectangle *AC, CB*,

because both the squares on *AC, CB* together with twice the rectangle *AC, CB,* and the squares on *AD, DB* together with twice the rectangle *AD, DB,* are equal to the square on *AB.* [II. 4]

But the squares on *AC, CB* differ from the squares on *AD, DB* by a rational area, for both are rational;
therefore twice the rectangle *AD, DB* also differs from twice the rectangle *AC, CB* by a rational area, though they are medial: [X. 21]
which is absurd, for a medial area does not exceed a medial by a rational area.
 [X. 26]

Therefore a binomial straight line is not divided at different points;
therefore it is divided at one point only.

 Q.E.D.

Proposition 43

A first bimedial straight line is divided at one point only.

Let *AB* be a first bimedial straight line divided at *C,* so that *AC, CB* are medial straight lines commensurable in square only and containing a rational rectangle;
 [X. 37]
I say that *AB* is not so divided at another point.

For, if possible, let it be divided at *D* also, so that *AD, DB* are also medial straight lines commensurable in square only and containing a rational rectangle.

Since, then, that by which twice the rectangle *AD, DB* differs from twice the rectangle *AC, CB* is that by which the squares on *AC, CB* differ from the squares on *AD, DB,*
while twice the rectangle *AD, DB* differs from twice the rectangle *AC, CB* by a rational area—for both are rational—
therefore the squares on *AC, CB* also differ from the squares on *AD, DB* by a rational area, though they are medial:
which is absurd. [X. 26]

Therefore a first bimedial straight line is not divided into its terms at different points;
therefore it is so divided at one point only.

 Q.E.D.

Proposition 44

A second bimedial straight line is divided at one point only.

Let *AB* be a second bimedial straight line divided at *C,* so that *AC, CB* are medial

straight lines commensurable in square only and containing a medial rectangle; [X. 38]

it is then manifest that *C* is not at the point of bisection, because the segments are not commensurable in length.

I say that *AB* is not so divided at another point.

For, if possible, let it be divided at *D* also, so that *AC* is not the same with *DB*, but *AC* is supposed greater;
it is then clear that the squares on *AD, DB* are also, as we proved above [Lemma], less than the squares on *AC, CB*;
and suppose that *AD, DB* are medial straight lines commensurable in square only and containing a medial rectangle.

Now let a rational straight line *EF* be set out,
let there be applied to *EF* the rectangular parallelogram *EK* equal to the square on *AB*,
and let *EG*, equal to the squares on *AC, CB*, be subtracted;
therefore the remainder *HK* is equal to twice the rectangle *AC, CB*. [II. 4]

Again, let there be subtracted *EL,* equal to the squares on *AD, DB*, which were proved less than the squares on *AC, CB*; [Lemma]
therefore the remainder *MK* is also equal to twice the rectangle *AD, DB*.

Now, since the squares on *AC, CB* are medial,
therefore *EG* is medial.

And it is applied to the rational straight line *EF*;
therefore *EH* is rational and incommensurable in length with *EF*. [X. 22]

For the same reason
HN is also rational and incommensurable in length with *EF*.

And, since *AC, CB* are medial straight lines commensurable in square only,
therefore *AC* is incommensurable in length with *CB*.

But, as *AC* is to *CB*, so is the square on *AC* to the rectangle *AC, CB*;
therefore the square on *AC* is incommensurable with the rectangle *AC, CB*. [X. 11]

But the squares on *AC, CB* are commensurable with the square on *AC*; for *AC, CB* are commensurable in square. [X. 15]

And twice the rectangle *AC, CB* is commensurable with the rectangle *AC, CB*. [X. 6]

Therefore the squares on *AC, CB* are also incommensurable with twice the rectangle *AC, CB*. [X. 13]

But *EG* is equal to the squares on *AC, CB*;

and *HK* is equal to twice the rectangle *AC, CB*;

therefore *EG* is incommensurable with *HK*,

so that *EH* is also incommensurable in length with *HN*. [VI. 1, X. 11]

And they are rational;

therefore *EH, HN* are rational straight lines commensurable in square only.

But, if two rational straight lines commensurable in square only be added together, the whole is the irrational which is called binomial. [X. 36]

Therefore *EN* is a binomial straight line divided at *H*.

In the same way *EM, MN* will also be proved to be rational straight lines commensurable in square only;

and *EN* will be a binomial straight line divided at different points, *H* and *M*.

And *EH* is not the same with *MN*.

For the squares on *AC, CB* are greater than the squares on *AD, DB*.

But the squares on *AD, DB* are greater than twice the rectangle *AD, DB*; therefore also the squares on *AC, CB*, that is, *EG*, are much greater than twice the rectangle *AD, DB*, that is, *MK*,

so that *EH* is also greater than *MN*.

Therefore *EH* is not the same with *MN*.

Q.E.D.

Proposition 45

A major straight line is divided at one and the same point only.

Let *AB* be a major straight line divided at *C*, so that *AC, CB* are incommensurable in square and make the sum of the squares on *AC, CB* rational, but the rectangle *AC, CB* medial;

I say that *AB* is not so divided at another point.

For, if possible, let it be divided at *D* also, so that *AD, DB* are also incommensurable in square and make the sum of the squares on *AD, DB* rational, but the rectangle contained by them medial.

Then, since that by which the squares on *AC, CB* differ from the squares on *AD, DB* is also that by which twice the rectangle *AD, DB* differs from twice the rectangle *AC, CB*,

while the squares on *AC, CB* exceed the squares on *AD, DB* by a rational area—
for both are rational—
therefore twice the rectangle *AD, DB* also exceeds twice the rectangle *AC, CB* by
a rational area, though they are medial:
which is impossible. [X. 26]

Therefore a major straight line is not divided at different points;
therefore it is only divided at one and the same point.

Q.E.D.

Proposition 46

The side of a rational plus a medial area is divided at one point only.

Let *AB* be the side of a rational plus a medial area divided at *C*, so that *AC, CB* are
incommensurable in square and make the sum of the squares on *AC, CB* medial,
but twice the rectangle *AC, CB* rational; [X. 40]
I say that *AB* is not so divided at another point.

For, if possible, let it be divided at *D* also, so that *AD, DB* are also incom-
mensurable in square and make the sum of the squares on *AD, DB* medial, but
twice the rectangle *AD, DB* rational.

Since, then, that by which twice the rectangle *AC, CB* differs from twice the
rectangle *AD, DB* is also that by which the squares on *AD, DB* differ from the
squares on *AC, CB,*
while twice the rectangle *AC, CB* exceeds twice the rectangle *AD, DB* by a
rational area,
therefore the squares on *AD, DB* also exceed the squares on *AC, CB* by a rational
area, though they are medial:
which is impossible. [X. 26]

Therefore the side of a rational plus a medial area is not divided at different
points;
therefore it is divided at one point only.

Q.E.D.

Proposition 47

The side of the sum of two medial areas is divided at one point only.

Let *AB* be divided at *C*, so that *AC, CB* are incommensurable in square and make
the sum of the squares on *AC, CB* medial, and the rectangle *AC, CB* medial and
also incommensurable with the sum of the squares on them;
I say that *AB* is not divided at another point so as to fulfil the given conditions.

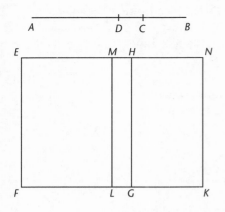

For, if possible, let it be divided at *D*, so that again *AC* is of course not the same as *BD*, but *AC* is supposed greater;
let a rational straight line *EF* be set out,
and let there be applied to *EF* the rectangle *EG* equal to the squares on *AC, CB*,
and the rectangle *HK* equal to twice the rectangle *AC, CB*;
therefore the whole *EK* is equal to the square on *AB*. [II. 4]

Again, let *EL*, equal to the squares on *AD, DB*, be applied to *EF*;
therefore the remainder, twice the rectangle *AD, DB*, is equal to the remainder *MK*.

And since, by hypothesis, the sum of the squares on *AC, CB* is medial,
therefore *EG* is also medial.

And it is applied to the rational straight line *EF*;
therefore *HE* is rational and incommensurable in length with *EF*. [X. 22]

For the same reason
HN is also rational and incommensurable in length with *EF*.

And, since the sum of the squares on *AC, CB* is incommensurable with twice the rectangle *AC, CB*,
therefore *EG* is also incommensurable with *GN*,
so that *EH* is also incommensurable with *HN*. [VI. 1, X. 11]

And they are rational;
therefore *EH, HN* are rational straight lines commensurable in square only;
therefore *EN* is a binomial straight. line divided at *H*. [X. 36]

Similarly we can prove that it is also divided at *M*.

And *EH* is not the same with *MN*;
therefore a binomial has been divided at different points:
which is absurd. [X. 42]

Therefore a side of the sum of two medial areas is not divided at different points;
therefore it is divided at one point only.

Q.E.D.

Definitions II

1. Given a rational straight line and a binomial, divided into its terms, such that the square on the greater term is greater than the square on the lesser by the square on a straight line commensurable in length with the greater, then, if the greater term be commensurable in length with the rational straight line set out, let the whole be called a *first binomial* straight line;

2. but if the lesser term be commensurable in length with the rational straight line set out, let the whole be called a *second binomial*;

3. and if neither of the terms be commensurable in length with the rational straight line set out, let the whole be called a *third binomial*.

4. Again, if the square on the greater term be greater than the square on the lesser by the square on a straight line incommensurable in length with the greater, then, if the greater term be commensurable in length with the rational straight line set out, let the whole be called a *fourth binomial*;

5. if the lesser, a *fifth binomial*;

6. and if neither, a *sixth binomial*.

Proposition 48

To find the first binomial straight line.

Let two numbers *AC*, *CB* be set out such that the sum of them *AB* has to *BC* the ratio which a square number has to a square number, but has not to *CA* the ratio which a square number has to a square number; **[Lemma 1 after X. 28]** let any rational straight line *D* be set out, and let *EF* be commensurable in length with *D*.

Therefore *EF* is also rational.

Let it be contrived that,
as the number *BA* is to *AC*, so is the square on *EF* to the square on *FG*.

[X. 6, Por.]

But *AB* has to *AC* the ratio which a number has to a number;
therefore the square on *EF* also has to the square on *FG* the ratio which a number has to a number,
 so that the square on *EF* is commensurable with the square on *FG*. [X. 6]

And *EF* is rational;
 therefore *FG* is also rational.

And, since *BA* has not to *AC* the ratio which a square number has to a square number,
neither, therefore, has the square on *EF* to the square on *FG* the ratio which a square number has to a square number;
 therefore *EF* is incommensurable in length with *FG*. [X. 9]

Therefore *EF*, *FG* are rational straight lines commensurable in square only;
 therefore *EG* is binomial. [X. 36]

I say that it is also a first binomial straight line.

For since, as the number *BA* is to *AC*, so is the square on *EF* to the square on *FG*, while *BA* is greater than *AC*,
 therefore the square on *EF* is also greater than the square on *FG*.

Let then the squares on *FG*, *H* be equal to the square on *EF*.

Now since, as *BA* is to *AC*, so is the square on *EF* to the square on *FG*,
therefore, *convertendo*,
 as *AB* is to *BC*, so is the square on *EF* to the square on *H*. [V. 19, Por.]

But *AB* has to *BC* the ratio which a square number has to a square number;
therefore the square on *EF* also has to the square on *H* the ratio which a square number has to a square number.

Therefore *EF* is commensurable in length with *H*; [X. 9]

therefore the square on *EF* is greater than the square on *FG* by the square on a straight line commensurable with *EF*.

And *EF, FG* are rational, and *EF* is commensurable in length with *D*.

Therefore *EF* is a first binomial straight line.

Q.E.D.

Proposition 49

To find the second binomial straight line.

Let two numbers *AC, CB* be set out such that the sum of them *AB* has to *BC* the ratio which a square number has to a square number, but has not to *AC* the ratio which a square number has to a square number;
let a rational straight line *D* be set out, and let *EF* be commensurable in length with *D*;
therefore *EF* is rational.

Let it be contrived then that,
as the number *CA* is to *AB*, so also is the square on *EF* to the square on *FG*; [X. 6, Por.]
therefore the square on *EF* is commensurable with the square on *FG*. [X. 6]

Therefore *FG* is also rational.

Now, since the number *CA* has not to *AB* the ratio which a square number has to a square number, neither has the square on *EF* to the square on *FG* the ratio which a square number has to a square number.

Therefore *EF* is incommensurable in length with *FG*; [X. 9]
therefore *EF, FG* are rational straight lines commensurable in square only;
therefore *EG* is binomial. [X. 36]

It is next to be proved that it is also a second binomial straight line.

For since, inversely, as the number *BA* is to *AC*, so is the square on *GF* to the square on *FE*,
while *BA* is greater than *AC*,
therefore the square on *GF* is greater than the square on *FE*.

Let the squares on *EF, H* be equal to the square on *GF*;
therefore, *convertendo*, as *AB* is to *BC*, so is the square on *FG* to the square on *H*.
[IV. 19, Por]

But *AB* has to *BC* the ratio which a square number has to a square number;
therefore the square on *FG* also has to the square on *H* the ratio which a square number has to a square number.

Therefore *FG* is commensurable in length with *H*; [X. 9]
so that the square on *FG* is greater than the square on *FE* by the square on a straight line commensurable with *FG*.

And *FG, FE* are rational straight lines commensurable in square only, and *EF,* the lesser term, is commensurable in length with the rational straight line *D* set out.

Therefore *EG* is a second binomial straight line.

<div align="right">Q.E.D.</div>

Proposition 50

To find the third binomial straight line.

Let two numbers *AC, CB* be set out such that the sum of them *AB* has to *BC* the ratio which a square number has to a square number, but has not to *AC* the ratio which a square number has to a square number.

Let any other number *D*, not square, be set out also, and let it not have to either of the numbers *BA, AC* the ratio which a square number has to a square number.

Let any rational straight line *E* be set out,
and let it be contrived that, as *D* is to *AB*, so is the square on *E* to the square on *FG*; [X. 6, Por.]
 therefore the square on *E* is commensurable with the square on *FG*. [X. 6]

And *E* is rational;
 therefore *FG* is also rational.

And, since *D* has not to *AB* the ratio which a square number has to a square number,
neither has the square on *E* to the square on *FG* the ratio which a square number has to a square number;
 therefore *E* is incommensurable in length with *FG*. [X. 9]

Next let it be contrived that, as the number *BA* is to *AC*, so is the square on *FG* to the square on *GH*; [X. 6, Por.]
 therefore the square on *FG* is commensurable with the square on *GH*. [X. 6]

But *FG* is rational;
 therefore *GH* is also rational.

And, since *BA* has not to *AC* the ratio which a square number has to a square number,
neither has the square on *FG* to the square on *HG* the ratio which a square number has to a square number;
 therefore *FG* is incommensurable in length with *GH*. [X. 9]

Therefore *FG, GH* are rational straight lines commensurable in square only;
 therefore *FH* is binomial. [X. 36]

I say next that it is also a third binomial straight line.

For since, as *D* is to *AB*, so is the square on *E* to the square on *FG*,
and, as *BA* is to *AC*, so is the square on *FG* to the square on *GH*,
therefore, *ex aequali*,
 as *D* is to *AC*, so is the square on *E* to the square on *GH*. [V. 22]

But *D* has not to *AC* the ratio which a square number has to a square numher;
therefore neither has the square on *E* to the square on *GH* the ratio which a
square number has to a square number;
 therefore *E* is incommensurable in length with *GH*. [X. 9]

And since, as *BA* is to *AC*, so is the square on *FG* to the square on *GH*,
 therefore the square on *FG* is greater than the square on *GH*.

Let then the squares on *GH, K* be equal to the square on *FG*;
therefore, *convertendo*,
 as *AB* is to *BC*, so is the square on *FG* to the square on *K*. [V. 19, Por.]

But *AB* has to *BC* the ratio which a square number has to a square number;
therefore the square on *FG* also has to the square on *K* the ratio which a square
number has to a square number;
 therefore *FG* is commensurable in length with *K*. [X. 9]

Therefore the square on *FG* is greater than the square on *GH* by the square on a
straight line commensurable with *FG*.

And *FG, GH* are rational straight lines commensurable in square only, and neither
of them is commensurable in length with *E*.

Therefore *FH* is a third binomial straight line.

Q.E.D.

Proposition 51

To find the fourth binomial straight line.

Let two numbers *AC, CB* be set out such that *AB* neither has to *BC*, nor
yet to *AC*, the ratio which a square number has to a square number.

Let a rational straight line *D* be set out,
and let *EF* be commensurable in length with *D*;
 therefore *EF* is also rational.

Let it be contrived that, as the number *BA* is to *AC*, so is the square
on *EF* to the square on *FG*; [X. 6, Por.]
therefore the square on *EF* is commensurable with the square on *FG*;
 [X. 6]
 therefore *FG* is also rational.

Now, since *BA* has not to *AC* the ratio which a square number has to a square
number, neither has the square on *EF* to the square on *FG* the ratio which a

square number has to a square number;

therefore *EF* is incommensurable in length with *FG*. [X. 9]

Therefore *EF*, *FG* are rational straight lines commensurable in square only;
so that *EG* is binomial.

I say next that it is also a fourth binomial straight line.

For since, as *BA* is to *AC*, so is the square on *EF* to the square on *FG*,
therefore the square on *EF* is greater than the square on *FG*.

Let then the squares on *FG*, *H* be equal to the square on *EF*;
therefore, *convertendo*,
as the number *AB* is to *BC*, so is the square on *EF* to the square on *H*. [V. 19, Por.]

But *AB* has not to *BC* the ratio which a square number has to a square number;
therefore neither has the square on *EF* to the square on *H* the ratio which a
square number has to a square number.

Therefore *EF* is incommensurable in length with *H*; [X. 9]

therefore the square on *EF* is greater than the square on *GF* by the square on a
straight line incommensurable with *EF*.

And *EF*, *FG* are rational straight lines commensurable in square only, and *EF* is
commensurable in length with *D*.

Therefore *EG* is a fourth binomial straight line.

Q.E.D.

Proposition 52

To find the fifth binomial straight line.

Let two numbers *AC*, *CB* be set out such that *AB* has not to either
of them the ratio which a square number has to a square number;
let any rational straight line *D* be set out,
and let *EF* be commensurable with *D*;
therefore *EF* is rational.

Let it be contrived that, as *CA* is to *AB*, so is the square on *EF* to
the square on *FG*. [X. 6, Por.]

But *CA* has not to *AB* the ratio which a square number has to a
square number;
therefore neither has the square on *EF* to the square on *FG* the
ratio which a square number has to a square number.

Therefore *EF*, *FG* are rational straight lines commensurable in square only;
[X. 9]
therefore *EG* is binomial. [X. 36]

I say next that it is also a fifth binomial straight line.

For since, as *CA* is to *AB*, so is the square on *EF* to the square on *FG*,
inversely, as *BA* is to *AC*, so is the square on *FG* to the square on *FE*;
therefore the square on *GF* is greater than the square on *FE*.

Let then the squares on *EF*, *H* be equal to the square on *GF*;

therefore, *convertendo,*

as the number AB is to BC, so is the square on GF to the square on H.

[V. 19, Por.]

But AB has not to BC the ratio which a square number has to a square number;

therefore neither has the square on FG to the square on H the ratio which a square number has to a square number.

Therefore FG is incommensurable in length with H; [X. 9]

so that the square on FG is greater than the square on FE by the square on a straight line incommensurable with FG.

And GF, FE are rational straight lines commensurable in square only, and the lesser term EF is commensurable in length with the rational straight line D set out.

Therefore EG is a fifth binomial straight line.

Q.E.D.

Proposition 53

To find the sixth binomial straight line.

Let two numbers AC, CB be set out such that AB has not to either of them the ratio which a square number has to a square number;

and let there also be another number D which is not square and which has not to either of the numbers BA, AC the ratio which a square number has to a square number.

Let any rational straight line E be set out,

and let it be contrived that, as D is to AB, so is the square on E to the square on FG; [X. 6, Por.]

therefore the square on E is commensurable with the square on FG. [X. 6]

And E is rational;

therefore FG is also rational.

Now, since D has not to AB the ratio which a square number has to a square number,

neither has the square on E to the square on FG the ratio which a square number has to a square number;

therefore E is incommensurable in length with FG. [X. 9]

Again, let it be contrived that, as BA is to AC, so is the square on FG to the square on GH. [X. 6, Por.]

Therefore the square on FG is commensurable with the square on HG. [X. 6]

Therefore the square on HG is rational;

therefore HG is rational.

And, since BA has not to AC the ratio which a square number has to a square number,

neither has the square on *FG* to the square on *GH* the ratio which a square number has to a square number;

therefore *FG* is incommensurable in length with *GH*. [X. 9]

Therefore *FG, GH* are rational straight lines commensurable in square only;

therefore *FH* is binomial. [X. 36]

It is next to be proved that it is also a sixth binomial straight line.

For since, as *D* is to *AB*, so is the square on *E* to the square on *FG*, and also, as *BA* is to *AC*, so is the square on *FG* to the square on *GH*, therefore, *ex aequali*,

as *D* is to *AC*, so is the square on *E* to the square on *GH*. [V. 22]

But *D* has not to *AC* the ratio which a square number has to a square number; therefore neither has the square on *E* to the square on *GH* the ratio which a square number has to a square number;

therefore *E* is incommensurable in length with *GH*. [X. 9]

But it was also proved incommensurable with *FG*; therefore each of the straight lines *FG, GH* is incommensurable in length with *E*.

And, since, as *BA* is to *AC*, so is the square on *FG* to the square on *GH*, therefore the square on *FG* is greater than the square on *GH*.

Let then the squares on *GH, K* be equal to the square on *FG*; therefore, *convertendo*,

as *AB* is to *BC*, so is the square on *FG* to the square on *K*. [V. 19, Por.]

But *AB* has not to *BC* the ratio which a square number has to a square number; so that neither has the square on *FG* to the square on *K* the ratio which a square number has to a square number.

Therefore *FG* is incommensurable in length with *K*; [X. 9] therefore the square on *FG* is greater than the square on *GH* by the square on a straight line incommensurable with *FG*.

And *FG, GH* are rational straight lines commensurable in square only, and neither of them is commensurable in length with the rational straight line *E* set out.

Therefore *FH* is a sixth binomial straight line.

Q.E.D.

LEMMA. Let there be two squares *AB, BC,* and let them be placed so that *DB* is in a straight line with *BE*;

therefore *FB* is also in a straight line with *BG*.

Let the parallelogram *AC* be completed; I say that *AC* is a square, that *DG* is a mean proportional between *AB, BC,* and further that *DC* is a mean proportional between *AC, CB*.

For, since *DB* is equal to *BF*, and *BE* to *BG*,

therefore the whole *DE* is equal to the whole *FG*.

But *DE* is equal to each of the straight lines *AH, KC,*

and *FG* is equal to each of the straight lines *AK, HC*; [I. 34] therefore each of the straight lines *AH, KC* is also equal to each of the straight lines *AK, HC*.

Therefore the parallelogram AC is equilateral.

And it is also rectangular;

therefore AC is a square.

And since, as FB is to BG, so is DB to BE,
while, as FB is to BG, so is AB to DG,
and, as DB is to BE, so is DG to BC, [VI. 1]
therefore also, as AB is to DG, so is DG to BC. [V. 11]

Therefore DG is a mean proportional between AB, BC.

I say next that DC is also a mean proportional between AC, CB.

For since, as AD is to DK, so is KG to GC—
for they are equal respectively—
and, *componendo*,

as AK is to KD, so is KC to CG, [V. 18]
while, as AK is to KD, so is AC to CD,
and, as KC is to CG, so is DC to CB, [VI. 1]
therefore also, as AC is to DC, so is DC to BC. [V. 11]

Therefore DC is a mean proportional between AC, CB.

Being what it was proposed to prove.

Proposition 54

If an area be contained by a rational straight line and the first binomial, the "side" of the area is the irrational straight line which is called binomial.

For let the area AC be contained by the rational straight line AB and the first binomial AD;
I say that the "side" of the area AC is the irrational straight line which is called binomial.

For, since AD is a first binomial straight line, let it be divided into its terms at E, and let AE be the greater term.

It is then manifest that AE, ED are rational straight lines commensurable in square only,
the square on AE is greater than the square on ED by the square on a straight line commensurable with AE,
and AE is commensurable in length with the rational straight line AB set out.
[X. Deff. II. 1]

Let ED be bisected at the point F.

Then, since the square on *AE* is greater than the square on *ED* by the square on a straight line commensurable with *AE*,
therefore, if there be applied to the greater *AE* a parallelogram equal to the fourth part of the square on the less, that is, to the square on *EF*, and deficient by a square figure, it divides it into commensurable parts. [X. 17]

Let then the rectangle *AG*, *GE* equal to the square on *EF* be applied to *AE*;
 therefore *AG* is commensurable in length with *EG*.

Let *GH*, *EK*, *FL* be drawn from *G*, *E*, *F* parallel to either of the straight lines *AB*, *CD*;
let the square *SN* be constructed equal to the. parallelogram *AH*, and the square *NQ* equal to *GK*, [II. 14]
and let them be placed so that *MN* is in a straight line with *NO*;
 therefore *RN* is also in a straight line with *NP*.

And let the parallelogram *SQ* be completed;
 therefore *SQ* is a square. [Lemma]

Now, since the rectangle *AG*, *GE* is equal to the square on *EF*,
 therefore, as *AG* is to *EF*, so is *FE* to *EG*; [VI. 17]
 therefore also, as *AH* is to *EL*, so is *EL* to *KG*; [VI. 1]
 therefore *EL* is a mean proportional between *AH*, *GK*.

But *AH* is equal to *SN*, and *GK* to *NQ*;
 therefore *EL* is a mean proportional between *SN*, *NQ*.

But *MR* is also a mean proportional between the same *SN*, *NQ*; [Lemma]
 therefore *EL* is equal to *MR*,
 so that it is also equal to *PO*.

But *AH*, *GK* are also equal to *SN*, *NQ*;
therefore the whole *AC* is equal to the whole *SQ*, that is, to the square on *MO*;
 therefore *MO* is the "side" of *AC*.

I say next that *MO* is binomial.

For, since *AG* is commensurable with *GE*,
 therefore *AE* is also commensurable with each of the straight lines *AG*, *GE*.
 [X. 15]

But *AE* is also, by hypothesis, commensurable with *AB*;
 therefore *AG*, *GE* are also commensurable with *AB*. [X. 12]

And *AB* is rational;
 therefore each of the straight lines *AG*, *GE* is also rational;
 therefore each of the rectangles *AH*, *GK* is rational, [X. 19]
 and *AH* is commensurable with *GK*.

But *AH* is equal to *SN*, and *GK* to *NQ*;
therefore *SN*, *NQ*, that is, the squares on *MN*, *NO*, are rational and commensurable.

And, since *AE* is incommensurable in length with *ED*,
while *AE* is commensurable with *AG*, and *DE* is commensurable with *EF*,
 therefore *AG* is also incommensurable with *EF*, [X. 13]
 so that *AH* is also incommensurable with *EL*. [VI. 1, X. 11]

But *AH* is equal to *SN*, and *EL* to *MR*;
 therefore *SN* is also incommensurable with *MR*.

But, as *SN* is to *MR*, so is *PN* to *NR*; [VI. 1]
 therefore *PN* is incommensurable with *NR*. [X. 11]

But *PN* is equal to *MN*, and *NR* to *NO*;
 therefore *MN* is incommensurable with *NO*.

And the square on *MN* is commensurable with the square on *NO*,
and each is rational;
 therefore *MN*, *NO* are rational straight lines commensurable in square only.

Therefore *MO* is binomial [X. 36] and the "side" of *AC*.

 Q.E.D.

Proposition 55

If an area be contained by a rational straight line and the second binomial, the
"side" of the area is the irrational straight line which is called a first bimedial.

For let the area *ABCD* be contained by the rational straight line *AB* and the
second binomial *AD*;
I say that the "side" of the area *AC* is a first bimedial straight line.

For, since *AD* is a second binomial straight line, let it be divided into its terms
at *E*, so that *AE* is the greater term;
therefore *AE*, *ED* are rational straight lines commensurable in square only,
the square on *AE* is greater than the square on *ED* by the square on a straight
line commensurable with *AE*,
and the lesser term *ED* is commensurable in length with *AB*. [X. Deff. II. 2]

Let *ED* be bisected at *F*,
and let there be applied to *AE* the rectangle *AG*, *GE* equal to the square on *EF*
and deficient by a square figure;
 therefore *AG* is commensurable in length with *GE*. [X. 17]

Through *G*, *E*, *F* let *GH*, *EK*, *FL* be drawn parallel to *AB*, *CD*,
let the square *SN* be constructed equal to the parallelogram *AH*, and the square
NQ equal to *GK*,

and let them be placed so that *MN* is in a straight line with *NO*;
therefore *RN* is also in a straight line with *NP*.

Let the square *SQ* be completed.

It is then manifest from what was proved before that *MR* is a mean proportional
between *SN, NQ* and is equal to *EL*, and that *MO* is the "side" of the area *AC*.

It is now to be proved that *MO* is a first bimedial straight line.

Since *AE* is commensurable in length with *ED,*
while *ED* is commensurable with *AB,*
therefore *AE* is incommensurable with *AB*. [X. 13]

And, since *AG* is commensurable with *EG,*
AE is also commensurable with each of the straight lines *AG, GE*. [X. 15]

But *AE* is incommensurable in length with *AB*;
therefore *AG, GE* are also incommensurable with *AB*. [X. 13]

Therefore *BA, AG* and *BA, GE* are pairs of rational straight lines commensurable
in square only;
so that each of the rectangles *AH, GK* is medial. [X. 21]

Hence each of the squares *SN, NQ* is medial.

Therefore *MN, NO* are also medial.

And, since *AG* is commensurable in length with *GE,*
AH is also commensurable with *GK,* [VI. 1, X. 11]
that is, *SN* is commensurable with *NQ,*
that is, the square on *MN* with the square on *NO*.

And, since *AE* is incommensurable in length with *ED,*
while *AE* is commensurable with *AG,*
and *ED* is commensurable with *EF,*
therefore *AG* is incommensurable with *EF*; [X. 13]
so that *AH* is also incommensurable with *EL,*
that is, *SN* is incommensurable with *MR,*
that is, *PN* with *NR,* [VI. 1, X. 11]
that is, *MN* is incommensurable in length with *NO*.

But *MN, NO* were proved to be both medial and commensurable in square;
therefore *MN, NO* are medial straight lines commensurable in square only.

I say next that they also contain a rational rectangle.

For, since *DE* is, by hypothesis, commensurable with each of the straight lines
AB, EF,
therefore *EF* is also commensurable with *EK*. [X. 12]

And each of them is rational;
therefore *EL,* that is, *MR* is rational, [X. 19]
and *MR* is the rectangle *MN, NO*.

But, if two medial straight lines commensurable in square only and containing
a rational rectangle be added together, the whole is irrational and is called a first
bimedial straight line. [X. 37]

Therefore *MO* is a first bimedial straight line.

Q.E.D.

Proposition 56

If an area be contained by a rational straight line and the third binomial, the "side" of the area is the irrational straight line called a second bimedial.

For let the area *ABCD* be contained by the rational straight line *AB* and the third binomial *AD* divided into its terms at *E*, of which terms *AE* is the greater;
I say that the "side" of the area *AC* is the irrational straight line called a second bimedial.

For let the same construction be made as before.

Now, since *AD* is a third binomial straight line,
therefore *AE*, *ED* are rational straight lines commensurable in square only,
the square on *AE* is greater than the square on *ED* by the square on a straight line commensurable with *AE*,
and neither of the terms *AE*, *ED* is commensurable in length with *AB*.

[X. Deff. II. 3]

Then, in manner similar to the foregoing, we shall prove that *MO* is the "side" of the area *AC*,
and *MN*, *NO* are medial straight lines commensurable in square only;
so that *MO* is bimedial.

It is next to be proved that it is also a second bimedial straight line.

Since *DE* is incommensurable in length with *AB*, that is, with *EK*,
and *DE* is commensurable with *EF*,
therefore *EF* is incommensurable in length with *EK*. [X. 13]

And they are rational;
therefore *FE*, *EK* are rational straight lines commensurable in square only.

Therefore *EL*, that is, *MR*, is medial. [X. 21]

And it is contained by *MN*, *NO*;
therefore the rectangle *MN*, *NO* is medial.

Therefore *MO* is a second bimedial straight line. [X. 38]

Q.E.D.

Proposition 57

If an area be contained by a rational straight line and the fourth binomial, the "side" of the area is the irrational straight line called major.

For let the area *AC* be contained by the rational straight line *AB* and the fourth binomial *AD* divided into its terms at *E*, of which terms let *AE* be the greater; I say that the "side" of the area *AC* is the irrational straight line called major.

For, since *AD* is a fourth binomial straight line,
therefore *AE*, *ED* are rational straight lines commensurable in square only,
the square on *AE* is greater than the square on *ED* by the square on a straight line incommensurable with *AE*,
and *AE* is commensurable in length with *AB*. [X. Deff. II. 4]

Let *DE* be bisected at *F*,
and let there be applied to *AE* a parallelogram, the rectangle *AG*, *GE*, equal to the square on *EF*;
therefore *AG* is incommensurable in length with *GE*. [X. 18]

Let *GH*, *EK*, *FL* be drawn parallel to *AB*,
and let the rest of the construction be as before;
it is then manifest that *MO* is the "side" of the area *AC*.

It is next to be proved that *MO* is the irrational straight line called major.

Since *AG* is incommensurable with *EG*,
AH is also incommensurable with *GK*, that is, *SN* with *NQ*; [VI. 1, X. 11]
therefore *MN*, *NO* are incommensurable in square.

And, since *AE* is commensurable with *AB*,
AK is rational; [X. 19]
and it is equal to the squares on *MN*, *NO*;
therefore the sum of the squares on *MN*, *NO* is also rational.

And, since *DE* is incommensurable in length with *AB*, that is, with *EK*,
while *DE* is commensurable with *EF*,
therefore *EF* is incommensurable in length with *EK*. [X. 13]

Therefore *EK*, *EF* are rational straight lines commensurable in square only;
therefore *LE*, that is, *MR*, is medial. [X. 21]

And it is contained by *MN*, *NO*;
therefore the rectangle *MN*, *NO* is medial.

And the sum of the squares on *MN*, *NO* is rational,
and *MN*, *NO* are incommensurable in square.

But, if two straight lines incommensurable in square and making the sum of the squares on them rational, but the rectangle contained by them medial, be added together, the whole is irrational and is called major. [X. 39]

Therefore *MO* is the irrational straight line called major and is the "side" of the area *AC*.

Q.E.D.

Proposition 58

If an area be contained by a rational straight line and the fifth binomial, the "side" of the area is the irrational straight line called the side of a rational plus a medial area.

For let the area *AC* be contained by the rational straight line *AB* and the fifth binomial *AD* divided into its terms at *E*, so that *AE* is the greater term;
I say that the "side" of the area *AC* is the irrational straight line called the side of a rational plus a medial area.

For let the same construction be made as before shown;
it is then manifest that *MO* is the "side" of the area *AC*.

It is then to be proved that *MO* is the side of a rational plus a medial area.

For, since *AG* is incommensurable with *GE*, **[X. 18]**
 therefore *AH* is also commensurable with *HE*, **[VI. 1, X. 11]**
that is, the square on *MN* with the square on *NO*;
 therefore *MN*, *NO* are incommensurable in square.

And, since *AD* is a fifth binomial straight line, and *ED* the lesser segment,
 therefore *ED* is commensurable in length with *AB*. **[X. Deff II. 5]**

But *AE* is incommensurable with *ED*;
 therefore *AB* is also incommensurable in length with *AE*. **[X. 13]**

Therefore *AK*, that is, the sum of the squares on *MN*, *NO*, is medial. **[X. 21]**

And, since *DE* is commensurable in length with *AB*, that is, with *EK*,
while *DE* is commensurable with *EF*,
 therefore *EF* is also commensurable with *EK*. **[X. 12]**

And *EK* is rational;
therefore *EL*, that is, *MR*, that is, the rectangle *MN*, *NO*, is also rational. **[X. 19]**

Therefore *MN*, *NO* are straight lines incommensurable in square which make the sum of the squares on them medial, but the rectangle contained by them rational.

Therefore *MO* is the side of a rational plus a medial area **[X. 40]** and is the "side" of the area *AC*.

Q.E.D.

Proposition 59

If an area be contained by a rational straight line and the sixth binomial, the "side" of the area is the irrational straight line called the side of the sum of two medial areas.

For let the area *ABCD* be contained by the rational straight line *AB* and the sixth binomial *AD*, divided into its terms at *E*, so that *AE* is the greater term; I say that the "side" of *AC* is the side of the sum of two medial areas.

Let the same construction be made as before shown.

It is then manifest that *MO* is the "side" of *AC*, and that *MN* is incommensurable in square with *NO*.

Now, since *EA* is incommensurable in length with *AB*,
 therefore *EA*, *AB* are rational straight lines commensurable in square only;
 therefore *AK*, that is, the sum of the squares on *MN*, *NO*, is medial. [X. 21]

Again, since *ED* is incommensurable in length with *AB*,
 therefore *FE* is also incommensurable with *EK*; [X. 13]
 therefore *FE*, *EK* are rational straight lines commensurable in square only;
 therefore *EL*, that is, *MR*, that is, the rectangle *MN*, *NO*, is medial. [X. 21]

And, since *AE* is incommensurable with *EF*,
 AK is also incommensurable with *EL*. [VI. 1, X. 11]

But *AK* is the sum of the squares on *MN*, *NO*,
and *EL* is the rectangle *MN*, *NO*;
therefore the sum of the squares on *MN*, *NO* is incommensurable with the rectangle *MN*, *NO*.

And each of them is medial, and *MN*, *NO* are incommensurable in square.

Therefore *MO* is the side of the sum of two medial areas [X. 41], and is the "side" of *AC*.

 Q.E.D.

⟨**LEMMA.** *If a straight line be cut into unequal parts, the squares on the unequal parts are greater than twice the rectangle contained by the unequal parts.*

Let *AB* be a straight line, and let it be cut into unequal parts at *C*, and let *AC* be the greater;
I say that the squares on *AC*, *CB* are greater than twice the rectangle *AC*, *CB*.

For let *AB* be bisected at *D*.

Since, then, a straight line has been cut into equal parts at *D*, and into unequal parts at *C*,

therefore the rectangle *AC*, *CB* together with the square on *CD* is equal to the square on *AD*, [II. 5]
 so that the rectangle *AC*, *CB* is less than the square on *AD*;
 therefore twice the rectangle *AC*, *CB* is less than double of the square on *AD*.

But the squares on *AC*, *CB* are double of the squares on *AD*, *DC*; [II. 9]
therefore the squares on *AC*, *CB* are greater than twice the rectangle *AC*, *CB*.

 Q.E.D.⟩

Proposition 60

The square on the binomial straight line applied to a rational straight line produces as breadth the first binomial.

Let AB be a binomial straight line divided into its terms at C, so that AC is the greater term;
let a rational straight line DE be set out,
and let $DEFG$ equal to the square on AB be applied to DE producing DG as its breadth;
I say that DG is a first binomial straight line.

For let there be applied to DE the rectangle DH equal to the square on AC, and KL equal to the square on BC;
therefore the remainder, twice the rectangle AC, CB, is equal to MF.

Let MG be bisected at N, and let NO be drawn parallel to ML or GF.

Therefore each of the rectangles MO, NF is equal to once the rectangle AC, CB.

Now, since AB is a binomial divided into its terms at C,
therefore AC, CB are rational straight lines commensurable in square only;
[X. 36]
therefore the squares on AC, CB are rational and commensurable with one another,
so that the sum of the squares on AC, CB is also rational. [X. 15]

And it is equal to DL;
therefore DL is rational.

And it is applied to the rational straight line DE;
therefore DM is rational and commensurable in length with DE. [X. 20]

Again, since AC, CB are rational straight lines commensurable in square only,
therefore twice the rectangle AC, CB, that is MF, is medial. [X. 21]

And it is applied to the rational straight line ML;
therefore MG is also rational and incommensurable in length with ML, that is, DE.
[X. 22]

But MD is also rational and is commensurable in length with DE;
therefore DM is incommensurable in length with MG. [X. 13]

And they are rational;
therefore DM, MG are rational straight lines commensurable in square only;
therefore DG is binomial. [X. 36]

It is next to be proved that it is also a first binomial straight line.
Since the rectangle AC, CB is a mean proportional between the squares on AC, CB,
[*cf.* Lemma after X. 53]
therefore MO is also a mean proportional between DH, KL.

Therefore, as DH is to MO, so is MO to KL,
that is, as DK is to MN, so is MN to MK; [VI. 1]
therefore the rectangle DK, KM is equal to the square on MN. [VI. 17]

And, since the square on AC is commensurable with the square on CB,
DH is also commensurable with KL,

so that DK is also commensurable with KM. [VI. 1, X. 11]

And, since the squares on AC, CB are greater than twice the rectangle AC, CB,
[Lemma]

therefore DL is also greater than MF,
so that DM is also greater than MG. [VI. 1]

And the rectangle DK, KM is equal to the square on MN, that is, to the fourth part of the square on MG,
and DK is commensurable with KM.

But, if there be two unequal straight lines, and to the greater there be applied a parallelogram equal to the fourth part of the square on the less and deficient by a square figure, and if it divide it into commensurable parts, the square on the greater is greater than the square on the less by the square on a straight line commensurable with the greater; [X. 17]
therefore the square on DM is greater than the square on MG by the square on a straight line commensurable with DM.

And DM, MG are rational,
and DM, which is the greater term, is commensurable in length with the rational straight line DE set out.

Therefore DG is a first binomial straight line. [X. Deff. II. 1]

Q.E.D.

Proposition 61

The square on the first bimedial straight line applied to a rational straight line produces as breadth the second binomial.

Let AB be a first bimedial straight line divided into its medials at C, of which medials AC is the greater;
let a rational straight line DE be set out,
and let there be applied to DE the parallelogram DF equal to the square on AB, producing DG as its breadth;
I say that DG is a second binomial straight line.

For let the same construction as before be made.

Then, since AB is a first bimedial divided at C,
therefore AC, CB are medial straight lines commensurable in square only, and containing a rational rectangle, [X. 37]
so that the squares on AC, CB are also medial. [X. 21]

Therefore DL is medial. [X. 15 and 13, Por.]

And it has been applied to the rational straight line DE;
therefore MD is rational and incommensurable in length with DE. [X. 22]

Again, since twice the rectangle AC, CB is rational, MF is also rational.

And it is applied to the rational straight line ML;
therefore MG is also rational and commensurable in length with ML, that is, DE; [X. 20]
therefore DM is incommensurable in length with MG. [X. 13]

And they are rational;
therefore *DM, MG* are rational straight lines commensurable in square only;
therefore *DG* is binomial. [X. 36]

It is next to be proved that it is also a second binomial straight line.

For, since the squares on *AC, CB* are greater than twice the rectangle *AC, CB*,
therefore *DL* is also greater than *MF*,
so that *DM* is also greater than *MG*. [VI. 1]

And, since the square on *AC* is commensurable with the square on *CB*,
DH is also commensurable with *KL*,
so that *DK* is also commensurable with *KM*. [VI. 1, X. 11]

And the rectangle *DK, KM* is equal to the square on *MN*;
therefore the square on *DM* is greater than the square on *MG* by the square on a straight line commensurable with *DM*. [X. 17]

And *MG* is commensurable in length with *DE*.

Therefore *DG* is a second binomial straight line. [X. Deff. II. 2]

Q.E.D.

Proposition 62

The square on the second bimedial straight line applied to a rational straight line produces as breadth the third binomial.

Let *AB* be a second bimedial straight line divided into its medials at C, so that *AC* is the greater segment;
let *DE* be any rational straight line, and to *DE* let there be applied the parallelogram *DF* equal to the square on *AB* and producing *DG* as its breadth;
I say that *DG* is a third binomial straight line.

Let the same construction be made as before shown.

Then, since *AB* is a second bimedial divided at C,
therefore *AC, CB* are medial straight lines commensurable in square only and containing a medial rectangle, [X. 38]
so that the sum of the squares on AC, CB is also medial. [X. 15 and 23 Por.]

And it is equal to *DL*;
therefore *DL* is also medial.

And it is applied to the rational straight line *DE*;
therefore *MD* is also rational and incommensurable in length with *DE*. [X. 22]

For the same reason,
MG is also rational and incommensurable in length with *ML*, that is, with *DE*;
therefore each of the straight lines *DM, MG* is rational and incommensurable in length with *DE*.

And, since AC is incommensurable in length with CB,
and, as AC is to CB, so is the square on AC to the rectangle AC, CB,
therefore the square on AC is also incommensurable with the rectangle AC, CB.

[X. 11]

Hence the sum of the squares on AC, CB is incommensurable with twice the rectangle AC, CB,

[X. 12, 13]

that is, DL is incommensurable with MF,

so that DM is also incommensurable with MG.

[VI. 1, X. 11]

And they are rational;

therefore DG is binomial.

[X. 36]

It is to be proved that it is also a third binomial, straight line.

In manner similar to the foregoing we may conclude that DM is greater than MG, and that DK is commensurable with KM.

And the rectangle DK, KM is equal to the square on MN;
therefore the square on DM is greater than the square on MG by the square on a straight line commensurable with DM.

And neither of the straight lines DM, MG is commensurable in length with DE.

Therefore DG is a third binomial straight line.

[X. Deff. II. 3]

Q.E.D.

Proposition 63

The square on the major straight line applied to a rational straight line produces as breadth the fourth binomial.

Let AB be a major straight line divided at C, so that AC is greater than CB;
let DE be a rational straight line,
and to DE let there be applied the parallelogram DF equal to the square on AB and producing DG as its breadth;
I say that, DG is a fourth binomial straight line.

Let the same construction be made as before shown.

Then, since AB is a major straight line divided at C,
AC, CB are straight lines incommensurable in square which make the sum of the squares on them rational, but the rectangle contained by them medial.

[X. 39]

Since, then, the sum of the squares on AC, CB is rational,

therefore DL is rational;

therefore DM is also rational and commensurable in length with DE. [X. 20]

Again, since twice the rectangle AC, CB, that is, MF, is medial,
and it is applied to the rational straight line ML,

therefore MG is also rational and incommensurable in length with DE; [X. 22]

therefore DM is also incommensurable in length with MG. [X. 13]

Therefore DM, MG are rational straight lines commensurable in square only;

therefore DG is binomial. [X. 36]

It is to be proved that it is also a fourth binomial straight line.

In manner similar to the. foregoing we can prove that DM is greater than MG, and that the rectangle DK, KM is equal to the square on MN.

Since then the square on AC is incommensurable with the square on CB,
therefore DH is also incommensurable with KL,
so that DK is also incommensurable with KM. [VI. 1, X. 11]

But, if there be two unequal straight lines, and to the greater there be applied a parallelogram equal to the fourth part of the square on the less and deficient by a square figure, and if it divide it into incommensurable parts, then the square on the greater will be greater than the square on the less by the square on a straight line incommensurable in length with the greater; [X. 18]
therefore the square on DM is greater than the square on MG by the square on a straight line incommensurable with DM.

And DM, MG are rational straight lines commensurable in square only,
and DM is commensurable with the rational straight line DE set out.

Therefore DG is a fourth binomial straight line. [X. Deff. II. 41]

Q.E.D.

Proposition 64

The square on the side of a rational plus a medial area applied to a rational straight line produces as breadth the fifth binomial.

Let AB be the side of a rational plus a medial area, divided into its straight lines at C, so that AC is the greater;
let a rational straight line DE be set out,
and let there be applied to DE the parallelogram DF equal to the square on AB, producing DG as its breadth;
I say that DG is a fifth binomial straight line.

Let the same construction as before be made.

Since then AB is the side of a rational plus a medial area, divided at C,
therefore AC, CB are straight lines incommensurable in square which make the sum of the squares on them medial, but the rectangle contained by them rational. [X. 40]

Since, then, the sum of the squares on AC, CB is medial,
therefore DL is medial,
so that DM is rational and incommensurable in length with DE. [X. 22]
Again, since twice the rectangle AC, CB, that is MF, is rational,
therefore MG is rational and commensurable with DE. [X. 20]

Therefore DM is incommensurable with MG; [X. 13]
Therefore DM, MG are rational straight lines commensurable in square only;

therefore *DG* is binomial. [X. 36]

I say next that, it is also a fifth binomial straight line.

For it can be proved similarly that the rectangle *DK, KM* is equal to the square on *MN*,
and that *DK* is incommensurable in length with *KM*;
therefore the square on *DM* is greater than the square on *MG* by the square a straight line incommensurable with *DM*. [X. 18]

And *DM, MG* are commensurable in square only, and the less, *MG*, is commensurable in length with *DE*.

Therefore *DG* is a fifth binomial.

Q.E.D.

Proposition 65

The square on the side of the sum of two medial areas applied to a rational straight line produces as breadth the sixth binomial.

Let *AB* be the side of the sum of two medial areas, divided at *C*,
let *DE* be a rational straight line,
and let there be applied to *DE* the parallelogram *DF* equal to the square on *AB*, producing *DG* as its breadth;
I say that *DG* is a sixth binomial straight line.

For let the same construction be made as before.

Then, since *AB* is the side of the sum of two medial areas, divided at *C*,
therefore *AC, CB* are straight lines incommensurable in square which make the sum of the squares on them medial, the rectangle contained by them medial, and moreover the sum of the squares on them incommensurable with the rectangle contained by them, [X. 41]
so that, in accordance with what was before proved, each of the rectangles *DL, MF* is medial.

And they are applied to the rational straight line *DE*;
therefore each of the straight lines *DM, MG* is rational and incommensurable in length with *DE*. [X. 22]

And, since the sum of the squares on *AC, CB* is incommensurable with twice the rectangle *AC, CB*,
therefore *DL* is incommensurable with *MF*.

Therefore *DM* is also incommensurable with *MG*; [VI. 1, X. 11]
therefore *DM, MG* are rational straight lines commensurable in square only;
therefore *DG* is binomial. [X. 36]

I say next that it is also a sixth binomial straight line.

Similarly again we can prove that the rectangle *DK, KM* is equal to the square on *MN*,

and that DK is incommensurable in length with KM; and, for the same reason, the square on DM is greater than the square on MG by the square on a straight line incommensurable in length with DM.

And neither of the straight lines DM, MG is commensurable in length with the rational straight line DE set out.

Therefore DG is a sixth binomial straight line.

Q.E.D.

Proposition 66

A straight line commensurable in length with a binomial straight line is itself also binomial and the same in order.

Let AB be binomial, and let CD be commensurable in length with AB; I say that CD is binomial and the same in order with AB.

For, since AB is binomial, let it be divided into its terms at E, and let AE be the greater term;
 therefore AE, EB are rational straight lines commensurable in square only. [X. 36]

Let it be contrived that, as AB is to CD, so is AE to CF; [VI. 12]
 therefore also the remainder EB is to the remainder FD as AB is to CD. [V. 19]

But AB is commensurable in length with CD;
 therefore AE is also commensurable with CF, and EB with FD. [X. 11]

And AE, EB are rational;
 therefore CF, FD are also rational.

And, as AE is to CF, so is EB to FD. [V. 11]

 Therefore, alternately, as AE is to EB, so is CF to FD. [V. 16]

But AE, EB are commensurable in square only;
 therefore CF, FD are also commensurable in square only. [X. 11]

And they are rational;
 therefore CD is binomial. [X. 36]

I say next that it is the same in order with AB.

For the square on AE is greater than the square on EB either by the square on a straight line commensurable with AE or by the square on a straight line incommensurable with it.

If then the square on AE is greater than the square on EB by the square on a straight line commensurable with AE, the square on CF will also be greater than the square on FD by the square on a straight line commensurable with CF. [X. 14]

And, if *AE* is commensurable with the rational straight line set out, *CF* will also be commensurable with it, [X. 12]
and for this reason each of the straight lines *AB, CD* is a first binomial, that is, the same in order. [X. Deff. II. 1]

But, if *EB* is commensurable with the rational straight line set out, *FD* is also commensurable with it, [X. 12]
 and for this reason again *CD* will be the same in order with *AB*,
for each of them will be a second binomial. [X. Deff. II. 2]

But, if neither of the straight lines *AE, EB* is commensurable with the rational straight line set out, neither of the straight lines *CF, FD* will be commensurable with it, [X. 13]
 and each of the straight lines *AB, CD* is a third binomial. [X. Deff. II. 3]

But, if the square on *AE* is greater than the square on *EB* by the square on a straight line incommensurable with *AE*, the square on *CF* is also greater than the square on *FD* by the square on a straight line incommensurable with *CF*. [X. 14]

And, if *AE* is commensurable with the rational straight line set out, *CF* is also commensurable with it,
 and each of the straight lines *AB, CD* is a fourth binomial. [X. Deff. II. 4]

But, if *EB* is so commensurable, so is *FD* also,
 and each of the straight lines *AB, CD* will be a fifth binomial. [X. Deff. II. 5]

But, if neither of the straight lines *AE, EB* is so commensurable, neither of the straight lines *CF, FD* is commensurable with the rational straight line set out,
 and each of the straight lines *AB, CD* will be a sixth binomial. [X. Deff. II. 6]

Hence a straight line commensurable in length with a binomial straight line is binomial and the same in order.

Q.E.D.

Proposition 67

A straight line commensurable in length with a bimedial straight line is itself also bimedial and the same in order.

Let *AB* be bimedial, and let *CD* be commensurable in length with *AB*;
I say that *CD* is bimedial and the same in order with *AB*.

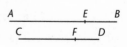

For, since *AB* is bimedial, let it be divided into its medials at *E*;
 therefore *AE, EB* are medial straight lines commensurable in square only.
 [X. 37, 38]

And let it be contrived that, as *AB* is to *CD*, so is *AE* to *CF*;
 therefore also the remainder *EB* is to the remainder *FD* as *AB* is to *CD*. [V. 19]

But *AB* is commensurable in length with *CD*;
 therefore *AE, EB* are also commensurable with *CF, FD* respectively. [X. 11]

But *AE, EB* are medial;
 therefore *CF, FD* are also medial. [X. 23]

And since, as *AE* is to *EB*, so is *CF* to *FD*, [V. 11]
and *AE*, *EB* are commensurable in square only,
CF, FD are also commensurable in square only.
[X. 11]

But they were also proved medial;
therefore *CD* is bimedial.

I say next that it is also the same in order with *AB*.

For since, as *AE* is to *EB*, so is *CF* to *FD*,
therefore also, as the square on *AE* is to the rectangle *AE*, *EB*, so is the square on *CF* to the rectangle *CF*, *FD*;
therefore, alternately, as the square on *AE* is to the square on *CF*, so is the rectangle *AE*, *EB* to the rectangle *CF*, *FD*. [V. 16]

But the square on *AE* is commensurable with the square on *CF*;
therefore the rectangle *AE*, *EB* is also commensurable with the rectangle *CF*, *FD*.

If therefore the rectangle *AE*, *EB* is rational,
the rectangle *CF*, *FD* is also rational,
and for this reason *CD* is a first bimedial; [X. 37]
but if medial, medial, [X. 23, Por.]
and each of the straight lines *AB*, *CD* is a second bimedial. [X. 38]

And for this reason *CD* will be the same in order with *AB*.

Q.E.D.

Proposition 68

A straight line commensurable with a major straight line is itself also major.

Let *AB* be major, and let *CD* be commensurable with *AB*;
I say that *CD* is major.

Let *AB* be divided at *E*;
therefore *AE*, *EB* are straight lines incommensurable in square which make the sum of the squares on them rational, but the rectangle contained by them medial. [X. 39]

Let the same construction be made as before.

Then since, as *AB* is to *CD*, so is *AE* to *CF*, and *EB* to *FD*,
therefore also, as *AE* is to *CF*, so is *EB* to *FD*. [V. 11]

But *AB* is commensurable with *CD*;
therefore *AE*, *EB* are also commensurable with *CF*, *FD* respectively. [X. 11]

And since, as *AE* is to *CF*, so is *EB* to *FD*,
alternately also,
as *AE* is to *EB*, so is *CF* to *FD*; [V. 16]
therefore also, *componendo*,
as *AB* is to *BE*, so is *CD* to *DF*; [V. 18]
therefore also, as the square on *AB* is to the square on *BE*, so is the square on *CD* to the square on *DF*. [VI. 20]

Similarly we can prove that, as the square on AB is to the square on AE, so also is the square on CD to the square on CF.

Therefore also, as the square on AB is ts the squares on AE, EB, so is the square on CD to the squares on CF, FD;
therefore also, alternately, as the square on AB is to the square on CD, so are the squares on AE, EB to the squares on CF, FD. [V. 16]

But the square on AB is commensurable with the square on CD;
therefore the squares on AE, EB are also commensurable with the squares on CF, FD.

And the squares on AE, EB together are rational;
 therefore the squares on CF, FD together are rational.

Similarly also twice the rectangle AE, EB is commensurable with twice the rectangle CF, FD.

And twice the rectangle AE, EB is medial;
 therefore twice the rectangle CF, FD is also medial. [X. 23, Por.]

Therefore CF, FD are straight lines incommensurable in square which make, at the same time, the sum of the squares on them rational, but the rectangle contained by them medial; therefore the whole CD is the irrational straight line called major. [X. 39]

Therefore a straight line commensurable with the major straight line is major.

 Q.E.D.

Proposition 69

A straight line commensurable with the side of a rational plus a medial area is itself also the side of a rational plus a medial area.

Let AB be the side of a rational plus a medial area, and let CD be commensurable with AB;
it is to be proved that CD is also the side of a rational plus a medial area.

Let AB be divided into its straight lines at E;
therefore AE, EB are straight lines incommensurable in square which make the sum of the squares on them medial, but the rectangle contained by them rational. [X. 40]

Let the same construction be made as before.

We can then prove similarly that CF, FD are incommensurable in square, and the sum of the squares on AE, EB is commensurable with the sum of the squares on CF, FD,
and the rectangle AE, EB with the rectangle CE, FD;
so that the sum of the squares on CF, FD is also medial, and the rectangle CF, FD rational.

Therefore CD is the side of a rational plus a medial area.

 Q.E.D.

Proposition 70

A straight line commensurable with the side of the sum of two medial areas is the side of the sum of two medial areas.

Let *AB* be the side of the sum of two medial areas, and *CD* commensurable with *AB*;
it is to be proved that *CD* is also the side of the sum of two medial areas.

For, since *AB* is the side of the sum of two medial areas, let it be divided into its straight lines at *E*;
therefore *AE, EB* are straight lines incommensurable in square which make the sum of the squares on them medial, the rectangle contained by them medial, and furthermore the sum of the squares on *AE, EB* incommensurable with the rectangle *AE, EB*.　　　　[X. 41]

Let the same construction be made as before.

We can then prove similarly that *CF, FD* are also incommensurable in square, the sum of the squares on *AE, EB* is commensurable with the sum of the squares on *CF, FD*,
and the rectangle *AE, EB* with the rectangle *CF, FD*;
　　　so that the sum of the squares on *CF, FD* is also medial,
　　　　　the rectangle *CF, FD* is medial,
and moreover the sum of the squares on *CF, FD* is incommensurable with the rectangle *CF, FD*.

Therefore *CD* is the side of the sum of two medial areas.

Q.E.D.

Proposition 71

If a rational and a medial area be added together, four irrational straight lines arise, namely a binomial or a first bimedial or a major or a side of a rational plus a medial area.

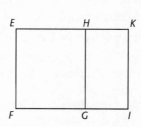

Let *AB* be rational, and *CD* medial;
I say that the "side" of the area *AD* is a binomial or a first bimedial or a major or a side of a rational plus a medial area.

For *AB* is either greater or less than *CD*.

First, let it be greater;
let a rational straight line *EF* be set out,
let there be applied to *EF* the rectangle *EG* equal to *AB*, producing *EH* as breadth,
and let *HI*, equal to *DC*, be applied to *EF*, producing *HK* as breadth.

Then, since *AB* is rational and is equal to *EG*,
<div align="center">therefore *EG* is also rational.</div>

And it has been applied to *EF*, producing *EH* as breadth;
<div align="center">therefore *EH* is rational and commensurable in length with *EF*. [X. 20]</div>

Again, since *CD* is medial and is equal to *HI*,
<div align="center">therefore *HI* is also medial.</div>

And it is applied to the rational straight line *EF*, producing *HK* as breadth;
<div align="center">therefore *HK* is rational and incommensurable in length with *EF*. [X. 22]</div>

And, since *CD* is medial,
while *AB* is rational,
<div align="center">therefore *AB* is incommensurable with *CD*,
so that *EG* is also incommensurable with *HI*.</div>

But, as *EG* is to *HI*, so is *EH* to *HK*; [VI. 1]
<div align="center">therefore *EH* is also incommensurable in length with *HK*. [X. 11]</div>

And both are rational;

<div align="center">therefore *EH*, *HK* are rational straight lines commensurable in square only;
therefore *EK* is a binomial straight line, divided at *H*. [X. 36]</div>

And, since *AB* is greater than *CD*,
while *AB* is equal to *EG* and *CD* to *HI*,
<div align="center">therefore *EG* is also greater than *HI*;
therefore *EH* is also greater than *HK*.</div>

The square, then, on *EH* is greater than the square on *HK* either by the square on a straight line commensurable in length with *EH* or by the square on a straight line incommensurable with it.

First, let the square on it be greater by the square on a straight line commensurable with itself.

Now the greater straight line *HE* is commensurable in length with the rational straight line *EF* set out;
<div align="center">therefore *EK* is a first binomial. [X. Deff. II. 1]</div>

But *EF* is rational;
and, if an area be contained by a rational straight line and the first binomial, the side of the square equal to the area is binomial. [X. 54]

Therefore the "side" of *EI* is binomial;
<div align="center">so that the "side" of *AD* is also binomial.</div>

Next, let the square on *EH* be greater than the square on *HK* by the square on a straight line incommensurable with *EH*.

Now the greater straight line *EH* is commensurable in length with the rational straight line *EF* set out;
<div align="center">therefore *EK* is a fourth binomial. [X. Deff. II. 4]</div>

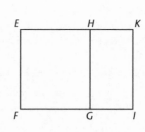

But *EF* is rational;
and, if an area be contained by a rational straight line and the fourth binomial, the "side" of the area is the irrational straight line called major. [X. 57]

Therefore the "side" of the area *EI* is major;
so that the "side" of the area *AD* is also major.

Next, let *AB* be less than *CD*;
therefore *EG* is also less than *HI*,
so that *EH* is also less than *HK*.

Now the square on *HK* is greater than the square on *EH* either by the square on a straight line commensurable with *HK* or by the square on a straight line incommensurable with it.

First, let the square on it be greater by the square on a straight line commensurable in length with itself.

Now the lesser straight line *EH* is commensurable in length with the rational straight line *EF* set out;
therefore *EK* is a second binomial, [X. Deff. II. 2]

But *EF* is rational,
and, if an area be contained by a rational straight line and the second binomial, the side of the square equal to it is a first bimedial; [X. 55]
therefore the "side" of the area *EI* is a first bimedial,
so that the "side" of *AD* is also a first bimedial.

Next, let the square on *HK* be greater than the square on *HE* by the square on a straight line incommensurable with *HK*.

Now the lesser straight line *EH* is commensurable with the rational straight line *EF* set out;
therefore *EK* is a fifth binomial. [X. Deff. II. 5]

But *EF* is rational;
and, if an area be contained by a rational straight line and the fifth binomial, the side of the square equal to the area is a side of a rational plus a medial area. [X. 58]

Therefore the "side" of the area *EI* is a side of a rational plus a medial area,
so that the "side" of the area *AD* is also a side of a rational plus a medial area.

Therefore etc.

Q.E.D.

Proposition 72

If two medial areas incommensurable with one another be added together, the remaining two irrational straight lines arise, namely either a second bimedial or a side of the sum of two medial areas.

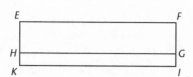

For let two medial areas *AB, CD* incommensurable with one another be added together;

I say that the "side" of the area *AD* is either a second bimedial or a side of the sum of two medial areas.

For *AB* is either greater or less than *CD*.

First, if it so chance, let *AB* be greater than *CD*.

Let the rational straight line *EF* be set out,

and to *EF* let there be applied the rectangle *EG* equal to *AB* and producing *EH* as breadth, and the rectangle *HI* equal to *CD* and producing *HK* as breadth.

Now, since each of the areas *AB, CD* is medial,
therefore each of the areas *EG, HI* is also medial.

And they are applied to the rational straight line *FE*, producing *EH, HK* as breadth;

therefore each of the straight lines *EH, HK* is rational and incommensurable in length with *EF*. [X. 22]

And, since *AB* is incommensurable with *CD*,
and *AB* is equal to *EG*, and *CD* to *HI*,
therefore *EG* is also incommensurable with *HI*.

But, as *EG* is to *HI*, so is *EH* to *HK*; [VI. 1]
therefore *EH* is incommensurable in length with *HK*. [X. 11]

Therefore *EH, HK* are rational straight lines commensurable in square only;
therefore *EK* is binomial. [X. 36]

But the square on *EH* is greater than the square on *HK* either by the square on a straight line commensurable with *EH* or by the square on a straight line incommensurable with it.

First, let the square on it be greater by the square on a straight line commensurable in length with itself.

Now neither of the straight lines *EH, HK* is commensurable in length with the rational straight line *EF* set out;

therefore *EK* is a third binomial. [X. Deff. II. 3]

But *EF* is rational;

and, if an area be contained by a rational straight line and the third binomial, the "side" of the area is a second bimedial; [X. 56]

therefore the "side" of EI, that is, of AD, is a second bimedial.

Next, let the square on EH be greater than the square on HK by the square on a straight line incommensurable in length with EH.

Now each of the straight lines EH, HK is incommensurable in length with EF;

therefore EK is a sixth binomial. [X. Deff. II. 6]

But, if an area be contained by a rational straight line and the sixth binomial, the "side" of the area is the side of the sum of two medial areas; [X. 59]
so that the "side" of the area AD is also the side of the sum of two medial areas.

Therefore etc.

Q.E.D.

The binomial straight line and the irrational straight lines after it are neither the same with the medial nor with one another.

For the square on a medial, if applied to a rational straight line, produces as breadth a straight line rational and incommensurable in length with that to which it is applied. [X. 22]

But the square on the binomial, if applied to a rational straight line, produces as breadth the first binomial. [X. 60]

The square on the first bimedial, if applied to a rational straight line, produces as breadth the second binomial. [X. 61]

The square on the second bimedial, if applied to a rational straight line, produces as breadth the third binomial. [X. 62]

The square on the major, if applied to a rational straight line, produces as breadth the fourth binomial. [X. 63]

The square on the side of a rational plus a medial area, if applied to a rational straight line, produces as breadth the fifth binomial. [X. 64]

The square on the side of the sum of two medial areas, if applied to a rational straight line, produces as breadth the sixth binomial. [X. 65]

And the said breadths differ both from the first and from one another: from the first because it is rational, and from one another because they are not the same in order;
so that the irrational straight lines themselves also differ from one another.

Proposition 73

If from a rational straight line there be subtracted a rational straight line commensurable with the whole in square only, the remainder is irrational; and let it be called an apotome.

For from the rational straight line AB let the rational straight line BC, commensurable with the whole in square only, be subtracted;
I say that the remainder AC is the irrational straight line called *apotome*.

For, since AB is incommensurable in length with BC,
and, as AB is to BC, so is the square on AB to the rectangle AB, BC,
 therefore the square on AB is incommensurable with the rectangle AB, BC.
 [X. 11]

But the squares on AB, BC are commensurable with the square on AB, [X. 15]
and twice the rectangle AB, BC is commensurable with the rectangle AB, BC.
 [X. 6]

And, inasmuch as the squares on AB, BC are equal to twice the rectangle AB, BC
together with the square on CA, [II. 7]
therefore the squares on AB, BC are also incommensurable with the remainder,
the square on AC. [X. 13, 16]

But the squares on AB, BC are rational;
 therefore AC is irrational. [X. Def. 4]

And let it be called an *apotome*.

 Q.E.D.

Proposition 74

If from a medial straight line there be subtracted a medial straight line which is commensurable with the whole in square only, and which contains with the whole a rational rectangle, the remainder is irrational. And let it be called a first apotome of a medial *straight line.*

For from the medial straight line AB let there be subtracted the medial straight line BC which is commensurable with AB in square only and with AB makes the rectangle AB, BC rational;

I say that the remainder AC is irrational; and let it be called a *first apotome of a medial* straight line.

For, since AB, BC are medial,
 the squares on AB, BC are also medial.

But twice the rectangle AB, BC is rational;
therefore the squares on AB, BC are incommensurable with twice the rectangle AB, BC;

therefore twice the rectangle *AB, BC* is also incommensurable with the remainder, the square on *AC*, [*cf.* II. 7]

since, if the whole is incommensurable with one of the magnitudes, the original magnitudes will also be incommensurable, [X. 16]

But twice the rectangle *AB, BC* is rational;

therefore the square on *AC* is irrational;

therefore *AC* is irrational. [X. Def. 4]

And let it be called a *first apotome of a medial* straight line.

Q.E.D.

Proposition 75

If from a medial straight line there be subtracted a medial straight line which is commensurable with the whole in square only, and which contains with the whole a medial rectangle, the remainder is irrational; and let it be called a second apotome of a medial *straight line.*

For from the medial straight line *AB* let there be subtracted the medial straight line *CB* which is commensurable with the whole *AB* in square only and such that the rectangle *AB, BC* which it contains with the whole *AB*, is medial; [X. 28]
I say that the remainder *AC* is irrational; and let it be called a *second apotome of a medial* straight line.

For let a rational straight line *DI* be set out,
let *DE*, equal to the squares on *AB, BC*, be applied to *DI*, producing *DG* as breadth,
and let *DH* equal to twice the rectangle *AB, BC* be applied to *DI*, producing *DF* as breadth;

therefore the remainder *FE* is equal to the square on *AC*. [II. 7]

Now, since the squares on *AB, BC* are medial and commensurable,

therefore *DE* is also medial. [X. 15 and 23, Por.]

And it is applied to the rational straight line *DI*, producing *DG* as breadth;

therefore *DG* is rational and incommensurable in length with *DI*. [X. 22]

Again, since the rectangle *AB, BC* is medial,

therefore twice the rectangle *AB, BC* is also medial. [X. 23, Por.]

And it is equal to *DH*;

therefore *DH* is also medial.

And it has been applied to the rational straight line *DI*, producing *DF* as breadth;

therefore *DF* is rational and incommensurable in length with *DI*. [X. 22]

And, since *AB, BC* are commensurable in square only,

therefore *AB* is incommensurable in length with *BC*;

therefore the square on *AB* is also incommensurable with the rectangle *AB, BC*.
[X. 11]

But the squares on *AB, BC* are commensurable with the square on *AB*, [X. 15]
and twice the rectangle *AB, BC* is commensurable with the rectangle *AB, BC*;
[X. 6]

therefore twice the rectangle *AB, BC* is incommensurable with the squares on *AB, BC*. [X. 13]

But *DE* is equal to the squares on *AB, BC*,
and *DH* to twice the rectangle *AB, BC*;

therefore *DE* is incommensurable with *DH*.

But, as *DE* is to *DH*, so is *GD* to *DF*; [VI. 1]

therefore *GD* is incommensurable with *DF*. [X. 11]

And both are rational;

therefore *GD, DF* are rational straight lines commensurable in square only;

therefore *FG* is an apotome. [X. 73]

But *DI* is rational,

and the rectangle contained by a rational and an irrational straight line is irrational,
[deduction from X. 20]

and its "side" is irrational.

And *AC* is the "side" of *FE*;

therefore *AC* is irrational.

And let it be called a *second apotome of a medial* straight line.

Q.E.D.

Proposition 76

If from a straight line there be subtracted a straight line which is incommensurable in square with the whole and which with the whole makes the squares on them added together rational, but the rectangle contained by them medial, the remainder is irrational; and let it be called minor.

For from the straight line *AB* let there be subtracted
the straight line *BC* which is incommensurable in
square with the whole and fulfils the given conditions.
 [X. 33]

I say that the remainder *AC* is the irrational straight line called *minor*.

For, since the sum of the squares on *AB, BC* is rational, while twice the rectangle *AB, BC* is medial,

therefore the squares on *AB, BC* are incommensurable with twice the rectangle *AB, BC*;

and, *convertendo,* the squares on *AB, BC* are incommensurable with the remainder, the square on *AC*.

[II. 7, X. 16]

But the squares on *AB, BC* are rational;

therefore the square on *AC* is irrational;

therefore *AC* is irrational.

And let it be called *minor.*

Q.E.D.

Proposition 77

If from a straight line there be subtracted a straight line which is incommensurable in square with the whole, and which with the whole makes the sum of the squares on them medial, but twice the rectangle contained by them rational, the remainder is irrational; and let it be called that which produces with a rational area a medial whole.

For from the straight line *AB* let there be subtracted the straight line *BC* which is incommensurable in square with *AB* and fulfils the given conditions; [X. 34]

I say that the remainder *AC* is the irrational straight line aforesaid.

For, since the sum of the squares on *AB, BC* is medial,

while twice the rectangle *AB, BC* is rational,

therefore the squares on *AB, BC* are incommensurable with twice the rectangle *AB, BC*;

therefore the remainder also, the square on *AC,* is incommensurable with twice the rectangle *AB, BC*. [II. 7, X. 16]

And twice the rectangle *AB, BC* is rational;

therefore the square on *AC* is irrational;

therefore *AC* is irrational.

And let it be called *that which produces with a rational area a medial whole.*

Q.E.D.

Proposition 78

If from a straight line there be subtracted a straight line which is incommensurable in square with the whole and which with the whole makes the sum of the squares on them medial, twice the rectangle contained by them medial, and further, the squares on them incommensurable with twice the rectangle contained by them, the remainder is irrational; and let it be called that which produces with a medial area a medial whole.

For from the straight line *AB* let there be subtracted the straight line *BC* incommensurable in square with *AB* and fulfilling the given conditions; [X. 35]

I say that the remainder *AC* is the irrational straight line called *that which produces with a medial area a medial whole.*

For let a rational straight line *DI* be set out,
to *DI* let there be applied *DE* equal to the squares on
AB, BC, producing *DG* as breadth,
and let *DH* equal to twice the rectangle *AB, BC* be
subtracted.

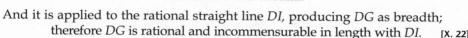

Therefore the remainder *FE* is equal to the square on
AC, [II. 7]
> so that *AC* is the "side" of *FE.*

Now, since the sum of the squares on *AB, BC* is medial
and is equal to *DE,*
> therefore *DE* is medial.

And it is applied to the rational straight line *DI,* producing *DG* as breadth;
> therefore *DG* is rational and incommensurable in length with *DI.* [X. 22]

Again, since twice the rectangle *AB, BC* is medial and is equal to *DH,*
> therefore *DH* is medial.

And it is applied to the rational straight line *DI,* producing *DF* as breadth;
therefore *DF* is also rational and incommensurable in length with *DI.* [X. 22]

And, since the squares on *AB, BC* are incommensurable with twice the rectangle
AB, BC,
> therefore *DE* is also incommensurable with *DH.*

But, as *DE* is to *DH,* so also is *DG* to *DF;* [VI. 1]
> therefore *DG* is incommensurable with *DF.* [X. 11]

And both are rational;
> therefore *GD, DF* are rational straight lines commensurable in square only.

Therefore *FG* is an apotome. [X. 73]

And *FH* is rational;
but the rectangle contained by a rational straight line and an apotome is irrational,
> [deduction from X. 20]
and its "side" is irrational.

And *AC* is the "side" of *FE;*
> therefore *AC* is irrational.

And let it be called *that which produces with a medial area a medial whole.*

> Q.E.D.

Proposition 79

*To an apotome only one rational straight line can be annexed which is
commensurable with the whole in square only.*

Let *AB* be an apotome, and *BC* an annex to it;
therefore *AC, CB* are rational straight lines
commensurable in square only. [X. 73]

I say that no other rational straight line can be annexed to *AB* which is
commensurable with the whole in square only.

For, if possible, let *BD* be so annexed;

therefore *AD, DB* are also rational straight lines commensurable in square only.

<div align="right">[X. 73]</div>

Now, since the excess of the squares on *AD, DB* over twice the rectangle *AD, DB* is also the excess of the squares on *AC, CB* over twice the rectangle *AC, CB*, for both exceed by the same, the square on *AB*, [II. 7]
therefore, alternately, the excess of the squares on *AD, DB* over the squares on *AC, CB* is the excess of twice the rectangle *AD, DB* over twice the rectangle *AC, CB*.

But the squares on *AD, DB* exceed the squares on *AC, CB* by a rational area, for both are rational;
therefore twice the rectangle *AD, DB* also exceeds twice the rectangle *AC, CB* by a rational area:
which is impossible,
for both are medial, [X. 21]
and a medial area does not exceed a medial by a rational area. [X. 26]

Therefore no other rational straight line can be annexed to *AB* which is commensurable with the whole in square only.

Therefore only one rational straight line can be annexed to an apotome which is commensurable with the whole in square only.

<div align="right">Q.E.D.</div>

Proposition 80

To a first apotome of a medial straight line only one medial straight line can be annexed which is commensurable with the whole in square only and which contains with the whole a rational rectangle.

For let *AB* be a first apotome of a medial straight line, and let *BC* be an annex to *AB*;
therefore *AC, CB* are medial straight lines commensurable in square only and such that the rectangle *AC, CB* which they contain is rational; [X. 74]
I say that no other medial straight line can be annexed to *AB* which is commensurable with the whole in square only and which contains with the whole a rational area.

For, if possible, let *DB* also be so annexed;
therefore *AD, DB* are medial straight lines commensurable in square only and such that the rectangle *AD, DB* which they contain is rational. [X. 74]

Now, since the excess of the squares on *AD, DB* over twice the rectangle *AD, DB* is also the excess of the squares on *AC, CB* over twice the rectangle *AC, CB*, for they exceed by the same, the square on *AB*, [II. 7]
therefore, alternately, the excess of the squares on *AD, DB* over the squares on *AC, CB* is also the excess of twice the rectangle *AD, DB* over twice the rectangle *AC, CB*.

But twice the rectangle *AD, DB* exceeds twice the rectangle *AC, CB* by a rational area,
for both are rational.

Therefore the squares on *AD, DB* also exceed the squares on *AC, CB* by a rational area:
which is impossible,
for both are medial, [X. 15 and 23, Por]
and a medial area does not exceed a medial by a rational area. [X. 26]

Therefore etc.

Q.E.D.

Proposition 81

To a second apotome of a medial straight line only one medial straight line can be annexed which is commensurable with the whole in square only and which contains with the whole a medial rectangle.

Let *AB* be a second apotome of a medial straight line and *BC* an annex to *AB*;
therefore *AC, CB* are medial straight lines commensurable in square only and such that the rectangle *AC, CB* which they contain is medial. [X. 75]

I say that no other medial straight line can be annexed to *AB* which is commensurable with the whole in square only and which contains with the whole a medial rectangle.

For, if possible, let *BD* also be so annexed;
therefore *AD, DB* are also medial straight lines commensurable in square only and such that the rectangle *AD, DB* which they contain is medial. [X. 75]

Let a rational straight line *EF* be set out,
let *EG* equal to the squares on *AC, CB* be applied to *EF*, producing *EM* as breadth,
and let *HG* equal to twice the rectangle *AC, CB* be subtracted, producing *HM* as breadth;
 therefore the remainder *EL* is equal to the square on *AB*,
 [II. 7]
 so that *AB* is the "side" of *EL.*

Again, let *EI* equal to the squares on *AD, DB* be applied to *EF*, producing *EN* as breadth.

But *EL* is also equal to the square on *AB*;
therefore the remainder *HI* is equal to twice the rectangle *AD, DB.* [II. 7]

Now, since *AC, CB* are medial straight lines,
 therefore the squares on *AC, CB* are also medial.

And they are equal to *EG*;
 therefore *EG* is also medial, [X. 15 and 23, Por.]

And it is applied to the rational straight line *EF*, producing *EM* as breadth;

319

therefore *EM* is rational and incommensurable in length with *EF*. [X. 22]

Again, since the rectangle *AC*, *CB* is medial,
 twice the rectangle *AC*, *CB* is also medial. [X. 23, Por.]

And it is equal to *HG*;
 therefore *HG* is also medial.

And it is applied to the rational straight line *EF*, producing *HM* as breadth;
therefore *HM* is also rational and incommensurable in length with *EF*. [X. 22]

And, since *AC*, *CB* are commensurable in square only,
 therefore *AC* is incommensurable in length with *CB*.

But, as *AC* is to *CB*, so is the square on *AC* to the rectangle *AC*, *CB*;
therefore the square on *AC* is incommensurable with the rectangle *AC*, *CB*. [X. 11]

But the squares on *AC*, *CB* are commensurable with the square on *AC*,
while twice the rectangle *AC*, *CB* is commensurable with the rectangle *AC*, *CB*; [X. 6]

therefore the squares on *AC*, *CB* are incommensurable with twice the rectangle *AC*, *CB*. [X. 13]

And *EG* is equal to the squares on *AC*, *CB*,
while *GH* is equal to twice the rectangle *AC*, *CB*;
 therefore *EG* is incommensurable with *HG*.

But, as *EG* is to *HG*, so is *EM* to *HM*; [VI. 1]
 therefore *EM* is incommensurable in length with *MH*. [X. 11]

And both are rational;
 therefore *EM*, *MH* are rational straight lines commensurable in square only;
 therefore *EH* is an apotome, and *HM* an annex to it. [X. 73]

Similarly we can prove that *HN* is also an annex to it;
therefore to an apotome different straight lines are annexed which are commensurable with the wholes in square only:
which is impossible. [X. 79]

Therefore etc.

Q.E.D.

Proposition 82

To a minor straight line only one straight line can be annexed which is incommensurable in square with the whole and which makes, with the whole, the sum of the squares on them rational but twice the rectangle contained by them medial.

Let *AB* be the minor straight line, and let *BC* be an annex to *AB*;
therefore *AC*, *CB* are straight lines incommensurable in square which make the

sum of the squares on them rational, but twice the rectangle contained by them medial. [X. 76]

I say that no other straight line can be annexed to AB fulfilling the same conditions.

For, if possible, let BD be so annexed;
therefore AD, DB are also straight lines incommensurable in square which fulfil the aforesaid conditions. [X. 76]

Now, since the excess of the squares on AD, DB over the squares on AC, CB is also the excess of twice the rectangle AD, DB over twice the rectangle AC, CB,
while the squares on AD, DB exceed the squares on AC, CB by a rational area, for both are rational,
therefore twice the rectangle AD, DB also exceeds twice the rectangle AC, CB by a rational area:
which is impossible, for both are medial. [X. 26]

Therefore to a minor straight line only one straight line can be annexed which is incommensurable in square with the whole and which makes the squares on them added together rational, but twice the rectangle contained by them medial.

Q.E.D.

Proposition 83

To a straight line which produces with a rational area a medial whole only one straight line can be annexed which is incommensurable in square with the whole straight line and which with the whole straight line makes the sum of the squares on them medial, but twice the rectangle contained by them rational.

Let AB be the straight line which produces with a rational area a medial whole,
and let BC be an annex to AB;
therefore AC, CB are straight lines incommensurable in square which fulfil the given conditions. [X. 77]

I say that no other straight line can be annexed to AB which fulfils the same conditions.

For, if possible, let BD be so annexed;
therefore AD, DB are also straight lines incommensurable in square which fulfil the given conditions. [X. 77]

Since then, as in the preceding cases,
the excess of the squares on AD, DB over the squares on AC, CB is also the excess of twice the rectangle AD, DB over twice the rectangle AC, CB,
while twice the rectangle AD, DB exceeds twice the rectangle AC, CB by a rational area,
for both are rational,
therefore the squares on AD, DB also exceed the squares on AC, CB by a rational area:
which is impossible, for both are medial. [X. 26]

Therefore no other straight line can be annexed to AB which is incommensurable

in square with the whole and which with the whole fulfils the aforesaid conditions;

therefore only one straight line can be so annexed.

Q.E.D.

Proposition 84

To a straight line which produces with a medial area a medial whole only one straight line can be annexed which is incommensurable in square with the whole straight line and which with the whole straight line makes the sum of the squares on them medial and twice the rectangle contained by them both medial and also incommensurable with the sum of the squares on them.

Let *AB* be the straight line which produces with a medial area a medial whole, and *BC* an annex to it;
therefore *AC*, *CB* are straight lines incommensurable in square which fulfil the aforesaid conditions. [X. 78]

I say that no other straight line can be annexed to *AB* which fulfils the aforesaid conditions.

For, if possible, let *BD* be so annexed,
so that *AD*, *DB* are also straight lines incommensurable in square which make the squares on *AD*, *DB* added together medial, twice the rectangle *AD*, *DB* medial, and also the squares on *AD*, *DB* incommensurable with twice the rectangle *AD*, *DB*. [X. 78]

Let a rational straight line *EF* be set out,
let *EG* equal to the squares on *AC*, *CB* be applied to *EF*, producing *EM* as breadth,
and let *HG* equal to twice the rectangle *AC*, *CB* be applied to *EF*, producing *HM* as breadth;
therefore the remainder, the square on *AB* [II. 7], is equal to *EL*;
therefore *AB* is the "side" of *EL*.

Again, let *EI* equal to the squares on *AD*, *DB* be applied to *EF*, producing *EN* as breadth.

But the square on *AB* is also equal to *EL*;
therefore the remainder, twice the rectangle *AD*, *DB* [II. 7], is equal to *HI*.

Now, since the sum of the squares on *AC*, *CB* is medial and is equal to *EG*,
therefore *EG* is also medial.

And it is applied to the rational straight line *EF*, producing *EM* as breadth;

therefore *EM* is rational and incommensurable in length with *EF.* [x. 22]

Again, since twice the rectangle *AC, CB* is medial and is equal to *HG,*
therefore *HG* is also medial.

And it is applied to the rational straight line *EF,* producing *HM* as breadth;
therefore *HM* is rational and incommensurable in length with *EF.* [x. 22]

And, since the squares on *AC, CB* are incommensurable with twice the rectangle *AC, CB,*

EG is also incommensurable with *HG*;
therefore *EM* is also incommensurable in length with *MH*. [vi. 1, x. 11]

And both are rational;
therefore *EM, MH* are rational straight lines commensurable in square only;
therefore *EH* is an apotome, and *HM* an annex to it. [x. 73]

Similarly we can prove that *EH* is again an apotome and *HN* an annex to it.

Therefore to an apotome different rational straight lines are annexed which are commensurable with the wholes in square only:
which was proved impossible, [x. 79]

Therefore no other straight line can be so annexed to *AB.*

Therefore to *AB* only one straight line can be annexed which is incommensurable in square with the whole and which with the whole makes the squares on them added together medial, twice the rectangle contained by them medial, and also the squares on them incommensurable with twice the rectangle contained by them.

Q.E.D.

Definitions III

1. Given a rational straight line and an apotome, if the square on the whole be greater than the square on the annex by the square on a straight line commensurable in length with the whole, and the whole be commensurable in length with the rational straight line set out, let the apotome be called a *first apotome.*

2. But if the annex be commensurable in length with the rational straight line set out, and the square on the whole be greater than that on the annex by the square on a straight line commensurable with the whole, let the apotome be called a *second apotome.*

3. But if neither be commensurable in length with the rational straight line set out, and the square on the whole be greater than the square on the annex by the square on a straight line commensurable with the whole, let the apotome be called a *third apotome.*

4. Again, if the square on the whole he greater than the square on the annex by the square on a straight line incommensurable with the whole, then, if the whole be commensurable in length with the rational straight line set out, let the apotome be called a *fourth apotome;*

5. if the annex be so commensurable, a *fifth;*

6. and, if neither, a *sixth.*

Proposition 85

To find the first apotome.

Let a rational straight line *A* be set out,
and let *BG* be commensurable in length with *A*;
<div style="text-align:center">therefore BG is also rational.</div>

Let two square numbers *DE, EF* be set out, and let their difference *FD* not be square;
therefore neither has *ED* to *DF* the ratio which a square number has to a square number.

Let it be contrived that,
as *ED* is to *DF,* so is the square on *BG* to the square on *GC*; **[X. 6, Por.]**
therefore the square on *BG* is commensurable with the square on *GC*. **[X. 6]**

But the square on *BG* is rational;
<div style="text-align:center">therefore the square on GC is also rational;
therefore GC is also rational.</div>

And, since *ED* has not to *DF* the ratio which a square number has to a square number,
therefore neither has the square on *BG* to the square on *GC* the ratio which a square number has to a square number;
<div style="text-align:center">therefore BG is incommensurable in length with GC. [X. 9]</div>

And both are rational;
<div style="text-align:center">therefore BG, GC are rational straight lines commensurable in square only;
therefore BC is an apotome. [X. 73]</div>

I say next that it is also a first apotome.

For let the square on *H* be that by which the square on *BG* is greater than the square on *GC*.

Now since, as *ED* is to *FD,* so is the square on *BG* to the square on *GC*,
therefore also, *convertendo,* **[V. 19, Por.]**
<div style="text-align:center">as DE is to EF, so is the square on GB to the square on H.</div>

But *DE* has to *EF* the ratio which a square number has to a square number,
for each is square;
therefore the square on *GB* also has to the square on *H* the ratio which a square number has to a square number;
<div style="text-align:center">therefore BG is commensurable in length with H. [X. 9]</div>

And the square on *BG* is greater than the square on *GC* by the square on *H*;
therefore the square on *BG* is greater than the square on *GC* by the square on a straight line commensurable in length with *BG*.

And the whole *BG* is commensurable in length with the rational straight line *A* set out.

Therefore BC is a first apotome. [X. Deff. III. 1]

Therefore the first apotome BC has been found.

(Being) that which it was required to find.

Proposition 86

To find the second apotome.

Let a rational straight line A be set out, and GC commensurable in length with A;
 therefore GC is rational.

Let two square numbers DE, EF be set out, and let their difference DF not be square.

Now let it be contrived that, as FD is to DE, so is the square on CG to the square on GB. [X. 6, Por.]

Therefore the square on CG is commensurable with the square on GB. [X. 6]

But the square on CG is rational;
 therefore the square on GB is also rational;
 therefore BG is rational.

And, since the square on GC has not to the square on GB the ratio which a square number has to a square number,
 CG is incommensurable in length with GB. [X. 9]

And both are rational;
therefore CG, GB are rational straight lines commensurable in square only;
 therefore BC is an apotome. [X. 73]

I say next that it is also a second apotome.

For let the square on H be that by which the square on BG is greater than the square on GC.

Since then, as the square on BG is to the square on GC, so is the number ED to the number DF,
therefore, *convertendo*,
 as the square on BG is to the square on H, so is DE to EF. [V. 19, Por.]

And each of the numbers DE, EF is square;
therefore the square on BG has to the square on H the ratio which a square number has to a square number;
 therefore BG is commensurable in length with H. [X. 9]

And the square on BG is greater than the square on GC by the square on H;
therefore the square on BG is greater than the square on GC by the square on a straight line commensurable in length with BG.

And CG, the annex, is commensurable with the rational straight line A set out.

Therefore BC is a second apotome. [X. Deff. III. 2]

Therefore the second apotome BC has been found.

Q.E.D.

Proposition 87

To find the third apotome.

Let a rational straight line *A* be set out,
let three numbers *E*, *BC*, *CD* be set out which have not to one another the ratio which a square number has to a square number, but let *CB* have to *BD* the ratio which a square number has to a square number.

Let it be contrived that, as *E* is to *BC*, so is the square on *A* to the square on *FG*,
and, as *BC* is to *CD*, so is the square on *FG* to the square on *GH*. [X. 6, Por.]

Since then, as *E* is to *BC*, so is the square on *A* to the square on *FG*,
therefore the square on *A* is commensurable with the square on *FG*. [X. 6]

But the square on *A* is rational;
therefore the square on *FG* is also rational;
therefore *FG* is rational.

And, since *E* has not to *BC* the ratio which a square number has to a square number,
therefore neither has the square on *A* to the square on *FG* the ratio which a square number has to a square number;
therefore *A* is incommensurable in length with *FG*. [X. 9]

Again, since, as *BC* is to *CD*, so is the square on *FG* to the square on *GH*,
therefore the square on *FG* is commensurable with the square on *GH*. [X. 6]

But the square on *FG* is rational;
therefore the square on *GH* is also rational;
therefore *GH* is rational.

And, since *BC* has not to *CD* the ratio which a square number has to a square number,
therefore neither has the square on *FG* to the square on *GH* the ratio which a square number has to a square number;
therefore *FG* is incommensurable in length with *GH*. [X. 9]

And both are rational;
therefore *FG*, *GH* are rational straight lines commensurable in square only;
therefore *FH* is an apotome. [X. 73]

I say next that it is also a third apotome.

For since, as *E* is to *BC*, so is the square on *A* to the square on *FG*,
and, as *BC* is to *CD*, so is the square on *FG* to the square on *HG*,
therefore, *ex aequali*, as *E* is to *CD*, so is the square on *A* to the square on *HG*. [V. 22]

But *E* has not to *CD* the ratio which a square number has to a square number;
therefore neither has the square on *A* to the square on *GH* the ratio which a square number has to a square number;
therefore *A* is incommensurable in length with *GH*. [X. 9]

Therefore neither of the straight lines *FG, GH* is commensurable in length with the rational straight line *A* set out.

Now let the square on *K* be that by which the square on *FG* is greater than the square on *GH*.

Since then, as *BC* is to *CD*, so is the square on *FG* to the square on *GH*,
 therefore, *convertendo*, as *BC* is to *BD*, so is the square on *FG* to the square on *K*.
 [V. 19, Por.]

But *BC* has to *BD* the ratio which a square number has to a square number; therefore the square on *FG* also has to the square on *K* the ratio which a square number has to a square number.

Therefore *FG* is commensurable in length with *K*, [X. 9]
and the square on *FG* is greater than the square on *GH* by the square on a straight line commensurable with *FG*.

And neither of the straight lines *FG, GH* is commensurable in length with the rational straight line *A* set out;
 therefore *FH* is a third apotome. [X. Deff. III. 3]

Therefore the third apotome *FH* has been found.

 Q.E.D.

Proposition 88

To find the fourth apotome.

Let a rational straight line *A* be set out, and *BG* commensurable in length with it;
 therefore *BG* is also rational.

Let two numbers *DF, FE* be set out such that the whole *DE* has not to either of the numbers *DF, EF* the ratio which a square number has to a square number.

Let it be contrived that, as *DE* is to *EF,* so is the square on *BG* to the square on *GC;*
 [X. 6, Por.]
therefore the square on *BG* is commensurable with the square on *GC*. [X. 6]

But the square on *BG* is rational;
 therefore the square on *GC* is also rational;
 therefore *GC* is rational.

Now, since *DB* has not to *EF* the ratio which a square number has to a square number,
therefore neither has the square on *BG* to the square on *GC* the ratio which a square number has to a square number;
 therefore *BG* is incommensurable in length with *GC*. [X. 9]

And both are rational;
 therefore *BG, GC* are rational straight lines commensurable in square only;

therefore *BC* is an apotome. [X. 73]

Now let the square on *H* be that by which the square on *BG* is greater than the square on *GC*.

Since then, as *DE* is to *EF*, so is the square on *BG* to the square on *GC*, therefore also, *convertendo*,

as *ED* is to *DF*, so is the square on *GB* to the square on *H*. [V. 19, Por.]

But *ED* has not to *DF* the ratio which a square number has to a square number; therefore neither has the square on *GB* to the square on *H* the ratio which a square number has to a square number;

therefore *BG* is incommensurable in length with *H*. [X. 9]

And the square on *BG* is greater than the square on *GC* by the square on *H*; therefore the square on *BG* is greater than the square on *GC* by the square on a straight line incommensurable with *BG*.

And the whole *BG* is commensurable in length with the rational straight line *A* set out.

Therefore *BC* is a fourth apotome. [X. Deff. III. 4]

Therefore the fourth apotome has been found.

Q.E.D.

Proposition 89

To find the fifth apotome.

Let a rational straight line *A* be set out, and let *CG* be commensurable in length with *A*;

therefore *CG* is rational.

Let two numbers *DF*, *FE* be set out such that *DE* again has not to either of the numbers *DF*, *FE* the ratio which a square number has to a square number;

and let it be contrived that, as *FE* is to *ED*, so is the square on *CG* to the square on *GB*.

Therefore the square on *GB* is also rational; [X. 6]

therefore *BG* is also rational.

Now since, as *DE* is to *EF*, so is the square on *BG* to the square on *GC*, while *DE* has not to *EF* the ratio which a square number has to a square number, therefore neither has the square on *BG* to the square on *GC* the ratio which a square number has to a square number;

therefore *BG* is incommensurable in length with *GC*. [X. 9]

And both are rational;

therefore *BG*, *GC* are rational straight lines commensurable in square only;

<div align="center">therefore BC is an apotome. [X. 73]</div>

I say next that it is also a fifth apotome.

For let the square on H be that by which the square on BG is greater than the square on GC.

Since then, as the square on BG is to the square on GC, so is DE to EF, therefore, convertendo,

<div align="center">as ED is to DF, so is -the square on BG to the square on H. [V. 19, Por.]</div>

But ED has not to DF the ratio which a square number has to a square number; therefore neither has the square on BG to the square on H the ratio which a square number has to a square number;

<div align="center">therefore BG is incommensurable in length with H. [X. 9]</div>

And the square on BG is greater than the square on GC by the square on H; therefore the square on GB is greater than the square on GC by the square on a straight line incommensurable in length with GB.

And the annex CG is commensurable in length with the rational straight line A set out;

<div align="center">therefore BC is a fifth apotome. [X. Deff. III. 5]</div>

Therefore the fifth apotome BC has been found.

<div align="right">Q.E.D.</div>

Proposition 90

To find the sixth apotome.

Let a rational straight line A be set out, and three numbers E, BC, CD not having to one another the ratio which a square number has to a square number;
and further let CB also not have to BD the ratio which a square number has to a square number.

Let it be contrived that, as E is to BC, so is the square on A to the square on FG, and, as BC is to CD, so is the square on FG to the square on GH. [X. 6, Por.]

Now since, as E is to BC, so is the square on A to the square on FG,

<div align="center">therefore the square on A is commensurable with the square on FG. [X. 6]</div>

But the square on A is rational;

<div align="center">therefore the square on FG is also rational;</div>
<div align="center">therefore FG is also rational.</div>

And, since E has not to BC the ratio which a square number has to a square number,
therefore neither has the square on A to the square on FG the ratio which a square number has to a square number;

<div align="center">therefore A is incommensurable in length with FG. [X. 9]</div>

Again, since, as BC is to CD, so is the square on FG to the square on GH,

<div align="center">therefore the square on FG is commensurable with the square on GH. [X. 6]</div>

<div align="right">329</div>

But the square on *FG* is rational;
therefore the square on *GH* is also rational;
therefore *GH* is also rational.

And, since *BC* has not to *CD* the ratio which a square number has to a square number,
therefore neither has the square on *FG* to the square on *GH* the ratio which a square number has to a square number;
therefore *FG* is incommensurable in length with *GH*.

[X. 9]

And both are rational;
therefore *FG*, *GH* are rational straight lines commensurable in square only;
therefore *FH* is an apotome. [X. 73]

I say next that it is also a sixth apotome.

For since, as *E* is to *BC*, so is the square on *A* to the square on *FG*,
and, as *BC* is to *CD*, so is the square on *FG* to the square on *GH*,
therefore, *ex aequali*,
as *E* is to *CD*, so is the square on *A* to the square on *GH*. [V. 22]

But *E* has not to *CD* the ratio which a square number has to a square number;
therefore neither has the square on *A* to the square on *GH* the ratio which a square number has to a square number;
therefore *A* is incommensurable in length with *GH*; [X. 9]
therefore neither of the straight lines *FG*, *GH* is commensurable in length with the rational straight line *A*.

Now let the square on *K* be that by which the square on *FG* is greater than the square on *GH*.

Since then, as *BC* is to *CD*, so is the square on *FG* to the square on *GH*,
therefore, *convertendo*,
as *CB* is to *BD*, so is the square on *FG* to the square on *K*. [V. 19, Por.]

But *CB* has not to *BD* the ratio which a square number has to a square number;
therefore neither has the square on *FG* to the square on *K* the ratio which a square number has to a square number;
therefore *FG* is incommensurable in length with *K*. [X. 9]

And the square on *FG* is greater than the square on *GH* by the square on *K*;
therefore the square on *FG* is greater than the square on *GH* by the square on a straight line incommensurable in length with *FG*.

And neither of the straight lines *FG*, *GH* is commensurable with the rational straight line *A* set out.

Therefore *FH* is a sixth apotome. [X. Deff. III. 6]

Therefore the sixth apotome *FH* has been found.

Q.E.D.

Proposition 91

If an area be contained by a rational straight line and a first apotome, the "side" of the area is an apotome.

For let the area AB be contained by the rational straight line AC and the first apotome AD;
I say that the "side" of the area AB is an apotome.

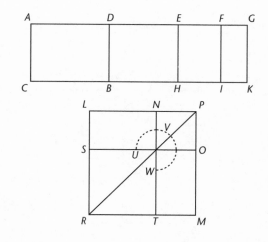

For, since AD is a first apotome, let DG be its annex;
therefore AG, GD are rational straight lines commensurable in square only. [X. 73]

And the whole AG is commensurable with the rational straight line AC set out, and the square on AG is greater than the square on GD by the square on a straight line commensurable in length with AG; [X. Deff. III. 1]
if therefore there be applied to AG a parallelogram equal to the fourth part of the square on DG and deficient by a square figure, it divides it into commensurable parts. [X. 17]

Let DG be bisected at E,
let there be applied to AG a parallelogram equal to the square on EG and deficient by a square figure,
and let it be the rectangle AF, FG;
 therefore AF is commensurable with FG.

And through the points E, F, G let EH, FI, GK be drawn parallel to AC.

Now, since AF is commensurable in length with FG,
therefore AG is also commensurable in length with each of the straight lines AF, FG. [X. 15]

But AG is commensurable with AC;
therefore each of the straight lines AF, FG is commensurable in length with AC. [X. 12]

And AC is rational;
 therefore each of the straight lines AF, FG is also rational,
 so that each of the rectangles AI, FK is also rational. [X. 19]

Now, since DE is commensurable in length with EG,
therefore DG is also commensurable in length with each of the straight lines DE, EG. [X. 15]

But DG is rational and incommensurable in length with AC;
therefore each of the straight lines DE, EG is also rational and incommensurable in length with AC; [X. 13]
 therefore each of the rectangles DH, EK is medial. [X. 21]

Now let the square LM be made equal to AI, and let there be subtracted the square NO having a common angle with it, the angle LPM, and equal to FK;
 therefore the squares LM, NO are about the same diameter. [VI. 26]

Let PR be their diameter, and let the figure be drawn.

Since then the rectangle contained by AF, FG is equal to the square on EG,
 therefore, as AF is to EG, so is EG to FG. [VI. 17]

But, as AF is to EG, so is AI to EK,
and, as EG is to FG, so is EK to KF; [VI. 1]

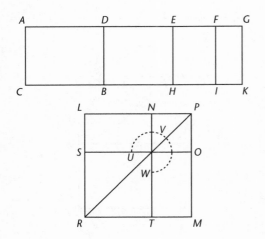

therefore *EK* is a mean proportional between *AI, KF*. [V. 11]

But *MN* is also a mean proportional between *LM, NO*, as was before proved, [Lemma after X. 53]

and *AI* is equal to the square *LM*, and *KF* to *NO*;

therefore *MN* is also equal to *EK*.

But *EK* is equal to *DH*, and *MN* to *LO*; therefore *DK* is equal to the gnomon *UVW* and *NO*.

But *AK* is also equal to the squares *LM, NO*; therefore the remainder *AB* is equal to *ST*.

But *ST* is the square on *LN*; therefore the square on *LN* is equal to *AB*; therefore *LN* is the "side" of *AB*.

I say next that *LN* is an apotome.

For, since each of the rectangles *AI, FK* is rational, and they are equal to *LM, NO*, therefore each of the squares *LM, NO*, that is, the squares on *LP, PN* respectively, is also rational; therefore each of the straight lines *LP, PN* is also rational.

Again, since *DH* is medial and is equal to *LO*, therefore *LO* is also medial.

Since, then, *LO* is medial, while *NO* is rational, therefore *LO* is incommensurable with *NO*.

But, as *LO* is to *NO*, so is *LP* to *PN*; [VI. 1] therefore *LP* is incommensurable in length with *PN*. [X. 11]

And both are rational; therefore *LP, PN* are rational straight lines commensurable in square only; therefore *LN* is an apotome. [X. 73]

And it is the "side" of the area *AB*; therefore the "side" of the area *AB* is an apotome.

Therefore etc.

Q.E.D.

Proposition 92

If an area be contained by a rational straight line and a second apotome, the "side" of the area is a first apotome of a medial straight line.

For let the area *AB* be contained by the rational straight line *AC* and the second apotome *AD*;
I say that the "side" of the area *AB* is a first apotome of a medial straight line.

For let *DG* be the annex to *AD*;
therefore *AG, GD* are rational straight lines
commensurable in square only, [X. 73]
and the annex *DG* is commensurable with
the rational straight line *AC* set out,
while the square on the whole *AG* is
greater than the square on the annex *GD*
by the square on a straight line commen-
surable in length with *AG*. [X. Deff. III. 2]

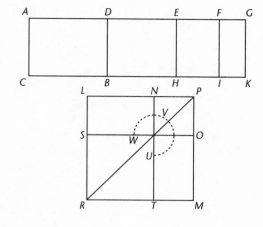

Since, then, the square on *AG* is greater
than the square on *GD* by the square on a
straight line commensurable with *AG*,
therefore, if there be applied to *AG* a
parallelogram equal to the fourth part of
the square on *GD* and deficient by a square
figure, it divides it into commensurable
parts. [X. 17]

Let then *DG* be bisected at *E*,
let there be applied to *AG* a parallelogram equal to the square on *EG* and deficient
by a square figure,
and let it be the rectangle *AF, FG*;

therefore *AF* is commensurable in length with *FG*.

Therefore *AG* is also commensurable in length with each of the straight lines
AF, FG. [X. 15]

But *AG* is rational and incommensurable in length with *AC*;
therefore each of the straight lines *AF, FG* is also rational and incommensurable
in length with *AC*; [X. 13]

therefore each of the rectangles *AI, FK* is medial. [X. 21]

Again, since *DE* is commensurable with *EG*,

therefore *DG* is also commensurable with each of the straight lines *DE, EG*.
 [X. 15]

But *DG* is commensurable in length with *AC*.

Therefore each of the rectangles *DH, EK* is rational. [X. 19]

Let then the square *LM* be constructed equal to *AI*,
and let there be subtracted *NO* equal to *FK* and being about the same angle with
LM, namely the angle *LPM*;

therefore the squares *LM, NO* are about the same diameter. [VI. 26]

Let *PR* be their diameter, and let the figure be drawn.

Since then *AI, FK* are medial and are equal to the squares on *LP, PN*,

the squares on *LP, PN* are also medial;

therefore *LP, PN* are also medial straight lines commensurable in square only.

And, since the rectangle *AF, FG* is equal to the square on *EG*,

therefore, as *AF* is to *EG*, so is *EG* to *FG*, [VI. 17]
while, as *AF* is to *EG*, so is *AI* to *EK*,
and, as *EG* is to *FG*, so is *EK* to *FK*; [VI. 1]

therefore *EK* is a mean proportional between *AI, FK*. [V. 11]

But *MN* is also a mean proportional between the squares *LM, NO*,

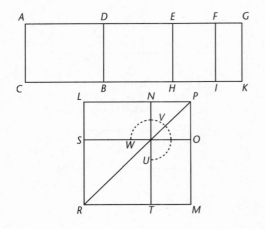

and *AI* is equal to *LM*, and *FK* to *NO*;
therefore *MN* is also equal to *EK*.

But *DH* is equal to *EK*,
and *LO* equal to *MN*;
therefore the whole *DK* is equal to the gnomon *UVW* and *NO*.

Since, then, the whole *AK* is equal to *LM*, *NO*,
and, in these, *DK* is equal to the gnomon *UVW* and *NO*,
therefore the remainder *AB* is equal to *TS*.

But *TS* is the square on *LN*;
therefore the square on *LN* is equal to the area *AB*;
therefore *LN* is the "side" of the area *AB*.

I say that *LN* is a first apotome of a medial straight line.

For, since *EK* is rational and is equal to *LO*,
therefore *LO*, that is, the rectangle *LP*, *PN*, is rational.

But *NO* was proved medial;
therefore *LO* is incommensurable with *NO*.

But, as *LO* is to *NO*, so is *LP* to *PN*; [VI. 1]
therefore *LP*, *PN* are incommensurable in length. [X. 11]

Therefore *LP*, *PN* are medial straight lines commensurable in square only, which contain a rational rectangle;
therefore *LN* is a first apotome of a medial straight line. [X. 74]

And it is the "side" of the area *AB*.

Therefore the "side" of the area *AB* is a first apotome of a medial straight line.

Q.E.D.

Proposition 93

If an area be contained by a rational straight line and a third apotome, the "side" of the area is a second apotome of a medial straight line.

For let the area *AB* be contained by the rational straight line *AC* and the third apotome *AD*;
I say that the "side" of the area *AB* is a second apotome of a medial straight line.

For let *DG* be the annex to *AD*;
therefore *AG*, *GD* are rational straight lines commensurable in square only,
and neither of the straight lines *AG*, *GD* is commensurable in length with the rational straight line *AC* set out,
while the square on the whole *AG* is greater than the square on the annex *DG* by the square on a straight line commensurable with *AG*. [X. Deff. III. 3]

Since, then, the square on *AG* is greater than the square on *GD* by the square on a straight line commensurable with *AG*,

therefore, if there be applied to *AG* a parallelogram equal to the fourth part of the square on *DG* and deficient by a square figure, it will divide it into commensurable parts. [X. 17]

Let then *DG* be bisected at *E*,
let there be applied to *AG* a parallelogram equal to the square on *EG* and deficient by a square figure,
and let it be the rectangle *AF, FG*.

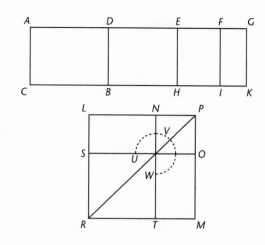

Let *EH, FI, GK* be drawn through the points *E, F, G* parallel to *AC*.

Therefore *AF, FG* are commensurable; therefore *AI* is also commensurable with *FK*. [VI. 1, X. 11]

And, since *AF, FG* are commensurable in length, therefore *AG* is also commensurable in length with each of the straight lines *AF, FG*. [X. 15]

But *AG* is rational and incommensurable in length with *AC*;
 so that *AF, FG* are so also. [X. 13]

 Therefore each of the rectangles *AI, FK* is medial. [X. 21]

Again, since *DE* is commensurable in length with *EG*, therefore *DG* is also commensurable in length with each of the straight lines *DE, EG*. [X. 15]

But *GD* is rational and incommensurable in length with *AC*; therefore each of the straight lines *DE, EG* is also rational and incommensurable in length with *AC*; [X. 13]
 therefore each of the rectangles *DH, EK* is medial. [X. 21]

And, since *AG, GD* are commensurable in square only,
 therefore *AG* is incommensurable in length with *GD*.

But *AG* is commensurable in length with *AF*, and *DG* with *EG*;
 therefore *AF* is incommensurable in length with *EG*. [X. 13]

But, as *AF* is to *EG*, so is *AI* to *EK*; [VI. 1]
 therefore *AI* is incommensurable with *EK*. [X. 11]

Now let the square *LM* be constructed equal to *AI*, and let there be subtracted *NO* equal to *FK* and being about the same angle with *LM*;
 therefore *LM, NO* are about the same diameter. [VI. 26]

Let *PR* be their diameter, and let the figure be drawn.

Now, since the rectangle *AF, FG* is equal to the square on *EG*,
 therefore, as *AF* is to *EG*, so is *EG* to *FG*. [VI. 17]

But, as *AF* is to *EG*, so is *AI* to *EK*,
and, as *EG* is to *FG*, so is *EK* to *FK*; [VI. 1]
 therefore also, as *AI* is to *EK*, so is *EK* to *FK*; [V. 11]
 therefore *EK* is a mean proportional between *AI, FK*.

But *MN* is also a mean proportional between the squares *LM, NO*,

335

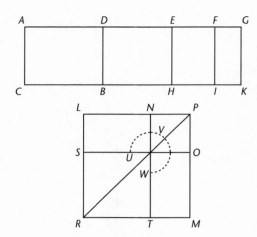

and *AI* is equal to *LM*, and *FK* to *NO*;
therefore *EK* is also equal to *MN*.

But *MN* is equal to *LO*,
and *EK* equal to *DH*;
therefore the whole *DK* is also equal to the gnomon *UVW* and *NO*.

But *AK* is also equal to *LM*, *NO*;
therefore the remainder *AB* is equal to *ST*, that is, to the square on *LN*;
therefore *LN* is the "side" of the area *AB*.

I say that *LN* is a second apotome of a medial straight line.

For, since *AI*, *FK* were proved medial, and are equal to the squares on *LP*, *PN*,
therefore each of the squares on *LP*, *PN* is also medial;
therefore each of the straight lines *LP*, *PN* is medial.

And, since *AI* is commensurable with *FK*, [VI. 1, X. 11]
therefore the square on *LP* is also commensurable with the square on *PN*.

Again, since *AI* was proved incommensurable with *EK*,
therefore *LM* is also incommensurable with *MN*,
that is, the square on *LP* with the rectangle *LP*, *PN*;
so that *LP* is also incommensurable in length with *PN*; [VI. 1, X. 11]
therefore *LP*, *PN* are medial straight lines commensurable in square only.

I say next that they also contain a medial rectangle.

For, since *EK* was proved medial, and is equal to the rectangle *LP*, *PN*,
therefore the rectangle *LP*, *PN* is also medial,
so that *LP*, *PN* are medial straight lines commensurable in square only which contain a medial rectangle.

Therefore *LN* is a second apotome of a medial straight line; [X. 75]
and it is the "side" of the area *AB*.

Therefore the "side" of the area *AB* is a second apotome of a medial straight line.

Q.E.D.

Proposition 94

If an area be contained by a rational straight line and a fourth apotome, the "side" of the area is minor.

For let the area *AB* be contained by the rational straight line *AC* and the fourth apotome *AD*;
I say that the "side" of the area *AB* is minor.

For let *DG* be the annex to *AD*;
therefore *AG*, *GD* are rational straight lines commensurable in square only,
AG is commensurable in length with the rational straight line *AC* set out,
and the square on the whole *AG* is greater than the square on the annex *DG* by the square on a straight line incommensurable in length with *AG*, [X. Deff. III. 4]

Since, then, the square on *AG* is greater than the square on *GD* by the square on a straight line incommensurable in length with *AG*,

therefore, if there be applied to *AG* a parallelogram equal to the fourth part of the square on *DG* and deficient by a square figure, it will divide it into incommensurable parts. [X. 18]

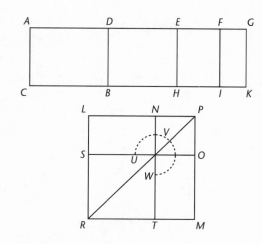

Let then *DG* be bisected at *E*,

let there be applied to *AG* a parallelogram equal to the square on *EG* and deficient by a square figure,

and let it be the rectangle *AF, FG*;

therefore *AF* is incommensurable in length with *FG*.

Let *EH, FI, GK* be drawn through *E, F, G* parallel to *AC, BD*.

Since, then, *AG* is rational and commensurable in length with *AC*,
 therefore the whole *AK* is rational. [X. 19]

Again, since *DG* is incommensurable in length with *AC*, and both are rational,
 therefore *DK* is medial. [X. 21]

Again, since *AF* is incommensurable in length with *FG*,
 therefore *AI* is also incommensurable with *FK*. [VI. 1, X. 11]

Now let the square *LM* be constructed equal to *AI*,
and let there be subtracted *NO* equal to *FK* and about the same angle, the angle *LPM*.

 Therefore the squares *LM, NO* are about the same diameter. [VI. 26]

Let *PR* be their diameter, and let the figure be drawn.

Since, then, the rectangle *AF, FG* is equal to the square on *EG*,
 therefore, proportionally, as *AF* is to *EG*, so is *EG* to *FG*. [VI. 17]

But, as *AF* is to *EG*, so is *AI* to *EK*,
and, as *EG* is to *FG*, so is *EK* to *FK*; [VI. 1]
 therefore *EK* is a mean proportional between *AI, FK*. [V. 11]

But *MN* is also a mean proportional between the squares *LM, NO*,
and *AI* is equal to *LM*, and *FK* to *NO*;
 therefore *EK* is also equal to *MN*.

But *DH* is equal to *EK*, and *LO* is equal to *MN*;
 therefore the whole *DK* is equal to the gnomon *UVW* and *NO*.

Since, then, the whole *AK* is equal to the squares *LM, NO*,
and, in these, *DK* is equal to the gnomon *UVW* and the square *NO*,
 therefore the remainder *AB* is equal to *ST*, that is, to the square on *LN*;
 therefore *LN* is the "side" of the area *AB*.

I say that *LN* is the irrational straight line called minor.

For, since *AK* is rational and is equal to the squares on *LP, PN*,
 therefore the sum of the squares on *LP, PN* is rational.

Again, since *DK* is medial,

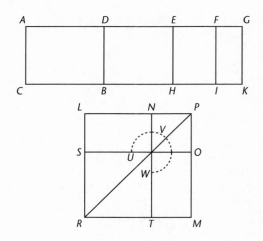

and *DK* is equal to twice the rectangle *LP, PN,*
therefore twice the rectangle *LP, PN* is medial.

And, since *AI* was proved incommensurable with *FK,*
therefore the square on *LP* is also incommensurable with the square on *PN.*

Therefore *LP, PN* are straight lines incommensurable in square which make the sum of the squares on them rational, but twice the rectangle contained by them medial.

Therefore *LN* is the irrational straight line called minor; [X. 76]

and it is the "side" of the area *AB.*

Therefore the "side" of the area *AB* is minor.

Q.E.D.

Proposition 95

If an area be contained by a rational straight line and a fifth apotome, the "side" of the area is a straight line which produces with a rational area a medial whole.

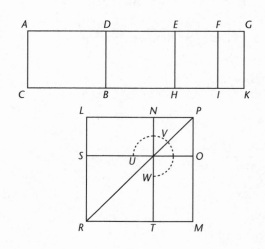

For let the area *AB* be contained by the rational straight line *AC* and the fifth apotome *AD;*
I say that the "side" of the area *AB* is a straight line which produces with a rational area a medial whole.

For let *DG* be the annex to *AD;*
therefore *AG, GD* are rational straight lines commensurable in square only,
the annex *GD* is commensurable in length with the rational straight line *AC* set out,
and the square on the whole *AG* is greater than the square on the annex *DG* by the square on a straight line incommensurable with *AG.* [X. Deff. III. 5]

Therefore, if there be applied to *AG* a parallelogram equal to the fourth part of the square on *DG* and deficient by a square figure, it will divide it into incommensurable parts. [X. 18]

Let then *DG* be bisected at the point *E,*
let there be applied to *AG* a parallelogram equal to the square on *EG* and deficient by a square figure, and let it be the rectangle *AF, FG;*
 therefore *AF* is incommensurable in length with *FG.*

Now, since AG is incommensurable in length with CA, and both are rational,
therefore AK is medial. [X. 21]

Again, since DG is rational and commensurable in length with AC,
DK is rational. [X. 19]

Now let the square LM be constructed equal to AI, and let the square NO equal to FK and about the same angle, the angle LPM, be subtracted;
therefore the squares LM, NO are about the same diameter. [VI. 26]

Let PR be their diameter, and let the figure be drawn.

Similarly then we can prove that LN is the "side" of the area AB.

I say that LN is the straight line which produces with a rational area a medial whole.

For, since AK was proved medial and is equal to the squares on LP, PN,
therefore the sum of the squares on LP, PN is medial.

Again, since DK is rational and is equal to twice the rectangle LP, PN,
the latter is itself also rational.

And, since AI is incommensurable with FK,
therefore the square on LP is also incommensurable with the square on PN;
therefore LP, PN are straight lines incommensurable in square whieh make the sum of the squares on them medial but twice the rectangle contained by them rational.

Therefore the remainder LN is the irrational straight line called that which produces with a rational area a medial whole; [X. 77]
and it is the "side" of the area AB.

Therefore the "side" of the area AB is a straight line which produces with a rational area a medial whole.

Q.E.D.

Proposition 96

If an area be contained by a rational straight line and a sixth apotome, the "side" of the area is a straight line which produces with a medial area a medial whole.

For let the area AB be contained by the rational straight line AC and the sixth apotome AD;
I say that the "side" of the area AB is a straight line which produces with a medial area a medial whole.

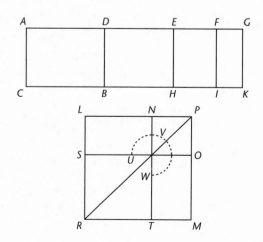

For let DG be the annex to AD;
therefore AG, GD are rational straight lines commensurable in square only,
neither of them is commensurable in length with the rational straight line AC set out,
and the square on the whole AG is greater than the square on the annex DG by the

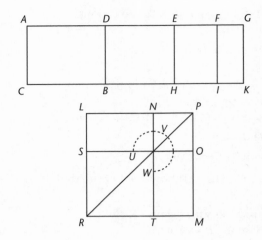

square on a straight line incommensurable in length with AG.　　[X. Deff. III. 6]

Since, then, the square on AG is greater than the square on GD by the square on a straight line incommensurable in length with AG,

therefore, if there be applied to AG a parallelogram equal to the fourth part of the square on DG and deficient by a square figure, it will divide it into incommensurable parts.　　[X. 18]

Let then DG be bisected at E,

let there be applied to AG a parallelogram equal to the square on EG and deficient by a square figure,

and let it be the rectangle AF, FG;

therefore AF is incommensurable in length with FG.

But, as AF is to FG, so is AI to FK;　　[VI. 1]

therefore AI is incommensurable with FK.　　[X. 11]

And, since AG, AC are rational straight lines commensurable in square only,

AK is medial.　　[X. 21]

Again, since AC, DG are rational straight lines and incommensurable in length,

DK is also medial.　　[X. 21]

Now, since AG, GD are commensurable in square only,

therefore AG is incommensurable in length with GD.

But, as AG is to GD, so is AK to KD;　　[VI. 1]

therefore AK is incommensurable with KD.　　[X. 11]

Now let the square LM be constructed equal to AI,

and let NO equal to FK, and about the same angle, be subtracted;

therefore the squares LM, NO are about the same diameter.　　[VI. 26]

Let PR be their diameter, and let the figure be drawn.

Then in manner similar to the above we can prove that LN is the "side" of the area AB.

I say that LN is a straight line which produces with a medial area a medial whole.

For, since AK was proved medial and is equal to the squares on LP, PN,

therefore the sum of the squares on LP, PN is medial.

Again, since DK was proved medial and is equal to twice the rectangle LP, PN,

twice the rectangle LP, PN is also medial.

And, since AK was proved incommensurable with DK,

the squares on LP, PN are also incommensurable with twice the rectangle LP, PN.

And, since AI is incommensurable with FK,

therefore the square on LP is also incommensurable with the square on PN;

therefore LP, PN are straight lines incommensurable in square which make the sum of the squares on them medial, twice the rectangle contained by them medial, and further, the squares on them incommensurable with twice the rectangle contained by them.

Therefore LN is the irrational straight line called that which produces with a medial area a medial whole; [X. 78]
and it is the "side" of the area AB.

Therefore the "side" of the area is a straight line which produces with a medial area a medial whole.

Q.E.D.

Proposition 97

The square on an apotome applied to a rational straight line produces as breadth a first apotome.

Let AB be an apotome, and CD rational,
and to CD let there be applied CE equal to the square on AB and producing CF as breadth;
I say that CF is a first apotome.

For let BG be the annex to AB;
therefore AG, GB are rational straight lines commensurable in square only. [X. 73]

To CD let there be applied CH equal to the square on AG, and KL equal to the square on BG.

Therefore the whole CL is equal to the squares on AG, GB, and, in these, CE is equal to the square on AB;
therefore the remainder FL is equal to twice the rectangle AG, GB. [II. 7]

Let FM be bisected at the point N,
and let NO be drawn through N parallel to CD;
therefore each of the rectangles FO, LN is equal to the rectangle AG, GB.

Now, since the squares on AG, GB are rational,
and DM is equal to the squares on AG, GB,
therefore DM is rational.

And it has been applied to the rational straight line CD, producing CM as breadth;
therefore CM is rational and commensurable in length with CD. [X. 20]

Again, since twice the rectangle AG, GB is medial, and FL is equal to twice the rectangle AG, GB,
therefore FL is medial.

And it is applied to the rational straight line CD, producing FM as breadth;
therefore FM is rational and incommensurable in length with CD. [X. 22]

And, since the squares on AG, GB are rational,

while twice the rectangle *AG, GB* is medial,
therefore the squares on *AG, GB* are incommensurable with twice the rectangle *AG, GB*.

And *CL* is equal to the squares on *AG, GB*,
and *FL* to twice the rectangle *AG, GB*,
therefore *DM* is incommensurable with *FL*.

But, as *DM* is to *FL*, so is *CM* to *FM*; [VI. 1]
therefore *CM* is incommensurable in length with *FM*. [X. 11]

And both are rational;
therefore *CM, MF* are rational straight lines commensurable in square only;
therefore *CF* is an apotome. [X. 73]

I say next that it is also a first apotome.

For, since the rectangle *AG, GB* is a mean proportional between the squares on *AG, GB*,
and *CH* is equal to the square on *AG*,
KL equal to the square on *BG*,
and *NL* equal to the rectangle *AG, GB*,
therefore *NL* is also a mean proportional between *CH, KL*;
therefore, as *CH* is to *NL*, so is *NL* to *KL*.

But, as *CH* is to *NL*, so is *CK* to *NM*,
and, as *NL* is to *KL*, so is *NM* to *KM*; [VI. 1]
therefore the rectangle *CK, KM* is equal to the square on *NM* [VI. 17], that is, to the fourth part of the square on *FM*.

And, since the square on *AG* is commensurable with the square on *GB*,
CH is also commensurable with *KL*.

But, as *CH* is to *KL*, so is *CK* to *KM*; [VI. 1]
therefore *CK* is commensurable with *KM*. [X. 11]

Since, then, *CM, MF* are two unequal straight lines,
and to *CM* there has been applied the rectangle *CK, KM* equal to the fourth part of the square on *FM* and deficient by a square figure,
while *CK* is commensurable with *KM*,
therefore the square on *CM* is greater than the square on *MF* by the square on a straight line commensurable in length with *CM*. [X. 17]

And *CM* is commensurable in length with the rational straight line *CD* set out;
therefore *CF* is a first apotome. [X. Deff. III. 1]

Therefore etc.

Q.E.D.

Proposition 98

The square on a first apotome of a medial straight line applied to a rational straight line produces as breadth a second apotome.

Let *AB* be a first apotome of a medial straight line and *CD* a rational straight line, and to *CD* let there be applied *CE* equal to the square on *AB*, producing *CF* as breadth;

I say that *CF* is a second apotome.

For let *BG* be the annex to *AB*;
therefore *AG*, *GB* are medial straight lines commensurable in square only which contain a rational rectangle. [X. 74]

To *CD* let there be applied *CH* equal to the square on *AG*, producing *CK* as breadth, and *KL* equal to the square on *GB*, producing *KM* as breadth;
therefore the whole *CL* is equal to the squares on *AG*, *GB*;
therefore *CL* is also medial. [X. 15 and 23, Por.]

And it is applied to the rational straight line *CD*, producing *CM* as breadth;
therefore *CM* is rational and incommensurable in length with *CD*. [X. 22]

Now, since *CL* is equal to the squares on *AG*, *GB*,
and, in these, the square on *AB* is equal to *CE*,
therefore the remainder, twice the rectangle *AG*, *GB*, is equal to *FL*. [II. 7]

But twice the rectangle *AG*, *GB* is rational;
therefore *FL* is rational.

And it is applied to the rational straight line *FE*, producing *FM* as breadth;
therefore *FM* is also rational and commensurable in length with *CD*. [X. 20]

Now, since the sum of the squares on *AG*, *GB*, that is, *CL*, is medial, while twice the rectangle *AG*, *GB*, that is, *FL*, is rational,
therefore *CL* is incommensurable with *FL*.

But, as *CL* is to *FL*, so is *CM* to *FM*; [VI. 1]
therefore *CM* is incommensurable in length with *FM*. [X. 11]

And both are rational;
therefore *CM*, *MF* are rational straight lines commensurable in square only;
therefore *CF* is an apotome. [X. 73]

I say next that it is also a second apotome.

For let *FM* be bisected at *N*,
and let *NO* be drawn through *N* parallel to *CD*;
therefore each of the rectangles *FO*, *NL* is equal to the rectangle *AG*, *GB*.

Now, since the rectangle *AG*, *GB* is a mean proportional between the squares on *AG*, *GB*,

and the square on *AG* is equal to *CH*,
the rectangle *AG, GB* to *NL*,
and the square on *BG* to *KL*,
 therefore *NL* is also a mean proportional between *CH, KL*;
 therefore, as *CH* is to *NL*, so is *NL* to *KL*.

But, as *CH* is to *NL*, so is *CK* to *NM*,
and, as *NL* is to *KL*, so is *NM* to *MK*; **[VI. 1]**
 therefore, as *CK* is to *NM*, so is *NM* to *KM*; **[V. 11]**
therefore the rectangle *CK, KM* is equal to the square on *NM* **[VI. 17]**, that is, to the
fourth part of the square on *FM*.

Since, then, *CM, MF* are two unequal straight lines, and the rectangle *CK, KM*
equal to the fourth part of the square on *MF* and deficient by a square figure has
been applied to the greater, *CM*, and divides it into commensurable parts,
therefore the square on *CM* is greater than the square on *MF* by the square on a
straight line commensurable in length with *CM*. **[X. 17]**

And the annex *FM* is commensurable in length with the rational straight line *CD*
set out;
 therefore *CF* is a second apotome. **[X. Deff. III. 2]**

Therefore etc.

 Q.E.D.

Proposition 99

*The square on a second apotome of a medial straight line applied to a rational
straight line produces as breadth a third apotome.*

Let *AB* be a second apotome of a medial straight line, and *CD* rational,
and to *CD* let there be applied *CE* equal to the square on *AB*, producing *CF* as
breadth;
I say that *CF* is a third apotome.

For let *BG* be the annex to *AB*;

therefore *AG, GB* are medial straight lines commensurable in square only which contain a medial rectangle. [X. 75]

Let *CH* equal to the square on *AG* be applied to *CD,* producing *CK* as breadth, and let *KL* equal to the square on *BG* be applied to *KH,* producing *KM* as breadth;

therefore the whole *CL* is equal to the squares on *AG, GB*;

therefore *CL* is also medial. [X. 15 and 23, Por.]

And it is applied to the rational straight line *CD,* producing *CM* as breadth;

therefore *CM* is rational and incommensurable in length with *CD.* [X. 22]

Now, since the whole *CL* is equal to the squares on *AG, GB,*
and, in these, *CE* is equal to the square on *AB,*

therefore the remainder *LF* is equal to twice the rectangle *AG, GB.* [II. 7]

Let then *FM* be bisected at the point *N,*
and let *NO* be drawn parallel to *CD;*

therefore each of the rectangles *FO, NL* is equal to the rectangle *AG, GB.*

But the rectangle *AG, GB* is medial;

therefore *FL* is also medial.

And it is applied to the rational straight line *EF,* producing *FM* as breadth;

therefore *FM* is also rational and incommensurable in length with *CD.* [X. 22]

And, since *AG, GB* are commensurable in square only,

therefore *AG* is incommensurable in length with *GB*;

therefore the square on *AG* is also incommensurable with the rectangle *AG, GB.* [VI. 1, X. 11]

But the squares on *AG, GB* are commensurable with the square on *AG,*
and twice the rectangle *AG, GB* with the rectangle *AG, GB*;
therefore the squares on *AG, GB* are incommensurable with twice the rectangle *AG, GB.* [X. 13]

But *CL* is equal to the squares on *AG, GB,*
and *FL* is equal to twice the rectangle *AG, GB*;

therefore *CL* is also incommensurable with *FL.*

But, as *CL* is to *FL,* so is *CM* to *FM*; [VI. 1]

therefore *CM* is incommensurable in length with *FM.* [X. 11]

And both are rational;

therefore *CM, MF* are rational straight lines commensurable in square only;

therefore *CF* is an apotome. [X. 73]

I say next that it is also a third apotome.

For, since the square on *AG* is commensurable with the square on *GB,*

therefore *CH* is also commensurable with *KL,*

so that *CK* is also commensurable with *KM.* [VI. 1, X. 11]

And, since the rectangle *AG, GB* is a mean proportional between the squares on *AG, GB,*
and *CH* is equal to the square on *AG,*
KL equal to the square on *GB,*
and *NL* equal to the rectangle *AG, GB*;

therefore *NL* is also a mean proportional between *CH, KL*;

therefore, as *CH* is to *NL,* so is *NL* to *KL.*

But, as *CH* is to *NL*, so is *CK* to *NM*,
and, as *NL* is to *KL*, so is *NM* to *KM*; [VI. 1]
 therefore, as *CK* is to *MN*, so is *MN* to *KM*; [V. 11]
therefore the rectangle *CK*, *KM* is equal to ⟨the square on *MN*, that is, to⟩ the fourth part of the square on *FM*.

Since, then, *CM*, *MF* are two unequal straight lines, and a parallelogram equal to the fourth part of the square on *FM* and deficient by a square figure has been applied to *CM*, and divides it into commensurable parts,
therefore the square on *CM* is greater than the square on *MF* by the square on a straight line commensurable with *CM*. [X. 17]

And neither of the straight lines *CM*, *MF* is commensurable in length with the rational straight line *CD* set out;
 therefore *CF* is a third apotome. [X. Deff. III. 3]
Therefore etc.

 Q.E.D.

Proposition 100

The square on a minor straight line applied to a rational straight line produces as breadth a fourth apotome.

Let *AB* be a minor and *CD* a rational straight line, and to the rational straight line *CD* let *CE* be applied equal to the square on *AB* and producing *CF* as breadth;
 I say that *CF* is a fourth apotome.

For let *BG* be the annex to *AB*;
therefore *AG*, *GB* are straight lines incommensurable in square which make the sum of the squares on *AG*, *GB* rational, but twice the rectangle *AG*, *GB* medial. [X. 76]

To *CD* let there be applied *CH* equal to the square on *AG* and producing *CK* as breadth,
and *KL* equal to the square on *BG*, producing *KM* as breadth;
 therefore the whole *CL* is equal to the squares on *AG*, *GB*.

And the sum of the squares on *AG, GB* is rational;
> therefore *CL* is also rational.

And it is applied to the rational straight line *CD*, pruducing *CM* as breadth;
> therefore *CM* is also rational and commensurable in length with *CD*. [X. 20]

And, since the whole *CL* is equal to the squares on *AG, GB*, and, in these, *CE* is equal to the square on *AB*,
> therefore the remainder *FL* is equal to twice the rectangle *AG, GB*. [II. 7]

Let then *FM* be bisected at the point *N*,
and let *NO* be drawn through *N* parallel to either of the straight lines *CD, ML*;
> therefore each of the rectangles *FO, NL* is equal to the rectangle *AG, GB*.

And, since twice the rectangle *AG, GB* is medial and is equal to *FL*,
> therefore *FL* is also medial.

And it is applied to the rational straight line *FE*, producing *FM* as breadth;
> therefore *FM* is rational and incommensurable in length with *CD*. [X. 22]

And, since the sum of the squares on *AG, GB* is rational,
while twice the rectangle *AG, GB* is medial,
> the squares on *AG, GB* are incommensurable with twice the rectangle *AG, GB*.

But *CL* is equal to the squares on *AG, GB*,
and *FL* equal to twice the rectangle *AG, GB*;
> therefore *CL* is incommensurable with *FL*.

But, as *CL* is to *FL*, so is *CM* to *MF*; [VI. 1]
> therefore *CM* is incommensurable in length with *MF*. [X. 11]

And both are rational;
> therefore *CM, MF* are rational straight lines commensurable in square only;
> therefore *CF* is an apotome. [X. 73]

I say that it is also a fourth apotome.

For, since *AG, GB* are incommensurable in square,
> therefore the square on *AG* is also incommensurable with the square on *GB*.

And *CH* is equal to the square on *AG*,
and *KL* equal to the square on *GB*;
> therefore *CH* is incommensurable with *KL*.

But, as *CH* is to *KL*, so is *CK* to *KM*; [VI. 1]
> therefore *CK* is incommensurable in length with *KM*. [X. 11]

And, since the rectangle *AG, GB* is a mean proportional between the squares on *AG, GB*,
and the square on *AG* is equal to *CH*,
the square on *GB* to *KL*,
and the rectangle *AG, GB* to *NL*,
> therefore *NL* is a mean proportional between *CH, KL*;
> therefore, as *CH* is to *NL*, so is *NL* to *KL*.

But, as *CH* is to *NL*, so is *CK* to *NM*,
and, as *NL* is to *KL*, so is *NM* to *KM*; [VI. 1]
> therefore, as *CK* is to *MN*, so is *MN* to *KM*; [V. 11]
therefore the rectangle *CK, KM* is equal to the square on *MN* [VI. 17], that is, to the fourth part of the square on *FM*.

Since then *CM, MF* are two unequal straight lines, and the rectangle *CK, KM*

equal to the fourth part of the square on *MF* and deficient by a square figure has been applied to *CM* and divides it into incommensurable parts,
therefore the square on *CM* is greater than the square on *MF* by the square on a straight line incommensurable with *CM*. [X. 18]

And the whole *CM* is commensurable in length with the rational straight line *CD* set out;

therefore *CF* is a fourth apotome. [X. Deff. III. 4]

Therefore etc.

Q.E.D.

Proposition 101

The square on the straight line which produces with a rational area a medial whole, if applied to a rational straight line, produces as breadth a fifth apotome.

Let *AB* be the straight line which produces with a rational area a medial whole, and *CD* a rational straight line, and to *CD* let *CE* be applied equal to the square on *AB* and producing *CF* as breadth;
I say that *CF* is a fifth apotome.

For let *BG* be the annex to *AB*;

therefore *AG*, *GB* are straight lines incommensurable in square which make the sum of the squares on them medial but twice the rectangle contained by them rational. [X. 77]

To *CD* let there be applied *CH* equal to the square on *AG*, and *KL* equal to the square on *GB*;

therefore the whole *CL* is equal to the squares on *AG*, *GB*.

But the sum of the squares on *AG*, *GB* together is medial;

therefore *CL* is medial.

And it is applied to the rational straight line *CD*, producing *CM* as breadth;

therefore *CM* is rational and incommensurable with *CD*. [X. 22]

And, since the whole *CL* is equal to the squares on *AG*, *GB*,
and, in these, *CE* is equal to the square on *AB*,

therefore the remainder *FL* is equal to twice the rectangle *AG*, *GB*. [II. 7]

Let then *FM* be bisected at *N*,
and through *N* let *NO* be drawn parallel to either of the straight lines *CD, ML*;
therefore each of the rectangles *FO, NL* is equal to the rectangle *AG, GB*.

And, since twice the rectangle *AG, GB* is rational and equal to *FL*,
therefore *FL* is rational.

And it is applied to the rational straight line *EF*, producing *FM* as breadth;
therefore *FM* is rational and commensurable in length with *CD*. [X. 20]

Now, since *CL* is medial, and *FL* rational,
therefore *CL* is incommensurable with *FL*.

But, as *CL* is to *FL*, so is *CM* to *MF*; [VI. 1]
therefore *CM* is incommensurable in length with *MF*. [X. 11]

And both are rational;
therefore *CM, MF* are rational straight. lines commensurable in square only;
therefore *CF* is an apotome. [X. 73]

I say next that it is also a fifth apotome.

For we can prove similarly that the rectangle *CK, KM* is equal to the square
on *NM*, that is, to the fourth part of the square on *FM*.

And, since the square on *AG* is incommensurable with the square on *GB*,
while the square on *AG* is equal to *CH*,
and the square on *GB* to *KL*,
therefore *CH* is incommensurable with *KL*.

But, as *CH* is to *KL*, so is *CK* to *KM*; [VI. 1]
therefore *CK* is incommensurable in length with *KM*. [X. 11]

Since then *CM, MF* are two unequal straight lines,
and a parallelogram equal to the fourth part of the square on *FM* and deficient by a
square figure has been applied to *CM*, and divides it into incommensurable parts,
therefore the square on *CM* is greater than the square on *MF* by the square on a
straight line incommensurable with *CM*. [X. 18]

And the annex *FM* is commensurable with the rational straight line *CD* set out;
therefore *CF* is a fifth apotome. [X. Deff III. 5]

Q.E.D.

Proposition 102

The square on the straight line which produces with a medial area a medial whole, if applied to a rational straight line, produces as breadth a sixth apotome.

Let *AB* be the straight line which produces with a medial area a medial whole,
and *CD* a rational straight line,

and to CD let CE be applied equal to the square on AB and producing CF as breadth;
I say that CF is a sixth apotome.

For let BG be the annex to AB;
therefore AG, GB are straight lines incommensurable in square which make the sum of the squares on them medial, twice the rectangle AG, GB medial, and the squares on AG, GB incommensurable with twice the rectangle AG, GB. [X. 78]

Now to CD let there be applied CH equal to the square on AG and producing CK as breadth,
and KL equal to the square on BG;
 therefore the whole CL is equal to the squares on AG, GB;
 therefore CL is also medial.

And it is applied to the rational straight line CD, producing CM as breadth;
 therefore CM is rational and incommensurable in length with CD. [X. 22]

Since now CL is equal to the squares on AG, GB,
and, in these, CE is equal to the square on AB,
 therefore the remainder FL is equal to twice the rectangle AG, GB. [II. 7]

And twice the rectangle AG, GB is medial;
 therefore FL is also medial.

And it is applied to the rational straight line FE, producing FM as breadth;
therefore FM is rational and incommensurable in length with CD. [X. 22]

And, since the squares on AG, GB are incommensurable with twice the rectangle AG, GB,
and CL is equal to the squares on AG, GB,
and FL equal to twice the rectangle AG, GB,
 therefore CL is incommensurable with FL.

But, as CL is to FL, so is CM to MF; [VI. 1]
 therefore CM is incommensurable in length with MF. [X. 11]

And both are rational.

Therefore CM, MF are rational straight lines commensurable in square only;
 therefore CF is an apotome. [X. 73]

I say next that it is also a sixth apotome.

For, since FL is equal to twice the rectangle AG, GB,
let FM be bisected at N,
and let NO be drawn through N parallel to CD;
 therefore each of the rectangles FO, NL is equal to the rectangle AG, GB.

And, since AG, GB are incommensurable in square,
 therefore the square on AG is incommensurable with the square on GB.

But CH is equal to the square on AG,

and *KL* is equal to the square on *GB*;

therefore *CH* is incommensurable with *KL*.

But, as *CH* is to *KL*, so is *CK* to *KM*; [VI. 1]

therefore *CK* is incommensurable with *KM*. [X. 11]

And, since the rectangle *AG, GB* is a mean proportional between the squares on *AG, GB*,

and *CH* is equal to the square on *AG*,

KL equal to the square on *GB*,

and *NL* equal to the rectangle *AG, GB*,

therefore *NL* is also a mean proportional between *CH, KL*;

therefore, as *CH* is to *NL*, so is *NL* to *KL*.

And for the same reason as before the square on *CM* is greater than the square on *MF* by the square on a straight line incommensurable with *CM*. [X. 18]

And neither of them is commensurable with the rational straight line *CD* set out;

therefore *CH* is a sixth apotome. [X. Deff. III. 6]

Q.E.D.

Proposition 103

A straight line commensurable in length with an apotome is an apotome and the same in order.

Let *AB* be an apotome,

and let *CD* be commensurable in length with *AB*;

I say that *CD* is also an apotome and the same in order with *AB*.

For, since *AB* is an apotome, let *BE* be the annex to it;

therefore *AE, EB* are rational straight lines commensurable in square only

[X. 73]

Let it be contrived that the ratio of *BE* to *DF* is the same as the ratio of *AB* to *CD*;

[VI. 12]

therefore also, as one is to one, so are all to all; [V. 12]

therefore also, as the whole *AE* is to the whole *CF*, so is *AB* to *CD*.

But *AB* is commensurable in length with *CD*.

Therefore *AE* is also commensurable with *CF*, and *BE* with *DF*. [X. 11]

And *AE, EB* are rational straight lines commensurable in square only;

therefore *CF, FD* are also rational straight lines commensurable in square only.

[X. 13]

Now since, as *AE* is to *CF*, so is *BE* to *DF*,

alternately therefore, as *AE* is to *EB*, so is *CF* to *FD*. [V. 16]

And the square on *AE* is greater than the square on *EB* either by the square on a straight line commensurable with *AE* or by the square on a straight line incommensurable with it.

If then the square on *AE* is greater than the square on *EB* by the square on a straight line commensurable with *AE*, the square on *CF* will also be greater than the square on *FD* by the square on a straight line commensurable with *CF*. [X. 14]

And, if *AE* is commensurable in length with the rational straight line set out, *CF* is so also, [X. 12]

if *BE*, then *DF* also, [X. 12]

and, if neither of the straight lines *AE, EB*, then neither of the straight lines *CF, FD*. [X. 13]

But, if the square on *AE* is greater than the square on *EB* by the square on a straight line incommensurable with *AE*,

the square on *CF* will also be greater than the square on *FD* by the square on a straight line incommensurable with *CF*. [X. 14]

And, if *AE* is commensurable in length with the rational straight line set out, *CF* is so also,

if *BE*, then *DF* also, [X. 12]

and, if neither of the straight lines *AE, EB*, then neither of the straight lines *CF, FD*. [X. 13]

Therefore *CD* is an apotome and the same in order with *AB*.

Q.E.D.

Proposition 104

A straight line commensurable with an apotome of a medial straight line is an apotome of a medial straight line and the same in order.

Let *AB* be an apotome of a medial straight line, and let *CD* be commensurable in length with *AB*; I say that *CD* is also an apotome of a medial straight line and the same in order with *AB*.

For, since *AB* is an apotome of a medial straight line, let *EB* be the annex to it.

Therefore *AE, EB* are medial straight lines commensurable in square only. [X. 74, 75]

Let it be contrived that, as *AB* is to *CD*, so is *BE* to *DF*; [VI. 12]

therefore *AE* is also commensurable with *CF*, and *BE* with *DF*. [V. 12, X. 11]

But *AE, EB* are medial straight lines commensurable in square only;

therefore *CF, FD* are also medial straight lines [X. 23] commensurable in square only; [X. 13]

therefore *CD* is an apotome of a medial straight line. [X. 74, 75]

I say next that it is also the same in order with *AB*.

Since, as *AE* is to *EB*, so is *CF* to *FD*,

therefore also, as the square on *AE* is to the rectangle *AE, EB*, so is the square on *CF* to the rectangle *CF, FD*.

But the square on *AE* is commensurable with the square on *CF*;

therefore the rectangle *AE, EB* is also commensurable with the rectangle *CF, FD*. [V. 16, X. 11]

Therefore, if the rectangle *AE, EB* is rational, the rectangle *CF, FD* will also be rational, [X. Def. 4]

and if the rectangle *AE, EB* is medial, the rectangle *CF, FD* is also medial. [X. 23, Por.]

Therefore *CD* is an apotome of a medial straight line and the same in order with *AB*. [X. 74, 75]

Q.E.D.

Proposition 105

A straight line commensurable with a minor straight line is minor.

Let *AB* be a minor straight line, and *CD* commensurable with *AB*;
I say that *CD* is also minor.

Let the same construction be made as before;
then, since *AE*, *EB* are incommensurable in square, [X. 76]
therefore *CF*, *FD* are also incommensurable in square. [X. 13]

Now since, as *AE* is to *EB*, so is *CF* to *FD*, [V. 12, V. 16]
therefore also, as the square on *AE* is to the square on *EB*, so is the square on *CF* to the square on *FD*. [VI. 22]

Therefore, *componendo*, as the squares on *AE*, *EB* are to the square on *EB*, so are the squares on *CF*, *FD* to the square on *FD*. [V. 18]

But the square on *BE* is commensurable with the square on *DF*;
therefore the sum of the squares on *AE*, *EB* is also commensurable with the sum of the squares on *CF*, *FD*. [V. 16, X. 11]

But the sum of the squares on *AE*, *EB* is rational; [X. 76]
therefore the sum of the squares on *CF*, *FD* is also rational. [X. Def. 4]

Again, since, as the square on *AE* is to the rectangle *AE*, *EB*, so is the square on *CF* to the rectangle *CF*, *FD*,
while the square on *AE* is commensurable with the square on *CF*,
therefore the rectangle *AE*, *EB* is also commensurable with the rectangle *CF*, *FD*.

But the rectangle *AE*, *EB* is medial; [X. 76]
therefore the rectangle *CF*, *FD* is also medial; [X. 23, Por.]
therefore *CF*, *FD* are straight lines incommensurable in square which make the sum of the squares on them rational, but the rectangle contained by them medial.

Therefore *CD* is minor. [X. 76]

Q.E.D.

Proposition 106

A straight line commensurable with that which produces with a rational area a medial whole is a straight line which produces with a rational area a medial whole.

Let *AB* be a straight line which produces with a rational area a medial whole, and *CD* commensurable with *AB*;

I say that *CD* is also a straight line which produces with a rational area a medial whole.

For let *BE* be the annex to *AB*;
therefore *AE, EB* are straight lines incommensurable in square which make the sum of the squares on *AE, EB* medial, but the rectangle contained by them rational. [X. 77]

Let the same construction be made.

Then we can prove, in manner similar to the foregoing, that *CF, FD* are in the same ratio as *AE, EB*,
the sum of the squares on *AE, EB* is commensurable with the sum of the squares on *CF, FD*,
and the rectangle *AE, EB* with the rectangle *CF, FD*;
so that *CF, FD* are also straight lines incommensurable in square which make the sum of the squares on *CF, FD* medial, but the rectangle contained by them rational.

Therefore *CD* is a straight line which produces with a rational area a medial whole. [X. 77]

Q.E.D.

Proposition 107

A straight line commensurable with that which produces with a medial area a medial whole is itself also a straight line which produces with a medial area a medial whole.

Let *AB* be a straight line which produces with a medial area a medial whole, and let *CD* be commensurable with *AB*;
I say that *CD* is also a straight line which produces with a medial area a medial whole.

For let *BE* be the annex to *AB*,
and let the same construction be made;
therefore *AE, EB* are straight lines incommensurable in square which make the sum of the squares on them medial, the rectangle contained by them medial, and further, the sum of the squares on them incommensurable with the rectangle contained by them. [X. 78]

Now, as was proved, *AE, EB* are commensurable with *CF, FD*,

the sum of the squares on *AE, EB* with the sum of the squares on *CF, FD,* and the rectangle *AE, EB* with the rectangle *CF, FD*;

therefore *CF, FD* are also straight lines incommensurable in square which make the sum of the squares on them medial, the rectangle contained by them medial, and further the sum of the squares on them incommensurable with the rectangle contained by them.

Therefore *CD* is a straight line which produces with a medial area a medial whole. [X. 78]

Q.E.D.

Proposition 108

If from a rational area a medial area be subtracted, the "side" of the remaining area becomes one of two irrational straight lines, either an apotome or a minor straight line.

For from the rational area *BC* let the medial area *BD* be subtracted;
I say that the "side" of the remainder *EC* becomes one of two irrational straight lines, either an apotome or a minor straight line.

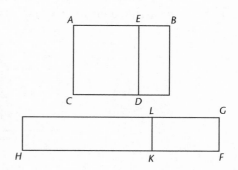

For let a rational straight line *FG* be set out, to *FG* let there be applied the rectangular parallelogram *GH* equal to *BC,* and let *GK* equal to *DB* be subtracted;
 therefore the remainder *EC* is equal to *LH.*

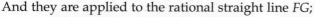

Since, then, *BC* is rational, and *BD* medial, while *BC* is equal to *GH,* and *BD* to *GK,*
 therefore *GH* is rational, and *GK* medial.

And they are applied to the rational straight line *FG*;
 therefore *FH* is rational and commensurable in length with *FG,* [X. 20]
while *FK* is rational and incommensurable in length with *FG*; [X. 22]
 therefore *FH* is incommensurable in length with *FK.* [X. 13]

Therefore *FH, FK* are rational straight lines commensurable in square only;
 therefore *KH* is an apotome [X. 73], and *KF* the annex to it.

Now the square on *HF* is greater than the square on *FK* by the square on a straight line either commensurable with *HF* or not commensurable.

First, let the square on it be greater by the square on a straight line commensurable with it.

Now the whole *HF* is commensurable in length with the rational straight line *FG* set out;
 therefore *KH* is a first apotome. [X. Deff. III. 1]

But the "side" of the rectangle contained by a rational straight line and a first apotome is an apotome. [X. 91]

 Therefore the "side" of *LH,* that is, of *EC,* is an apotome.

355

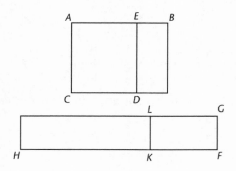

But, if the square on *HF* is greater than the square on *FK* by the square on a straight line incommensurable with *HF*,
while the whole *FH* is commensurable in length with the rational straight line *FG* set out,

> *KH* is a fourth apotome. [X. Deff. III. 4]

But the "side" of the rectangle contained by a rational straight line and a fourth apotome is minor. [X. 94]

Q.E.D.

Proposition 109

If from a medial area a rational area be subtracted, there arise two other irrational straight lines, either a first apotome of a medial straight line or a straight line which produces with a rational area a medial whole.

For from the medial area *BC* let the rational area *BD* be subtracted.

I say that the "side" of the remainder *EC* becomes one of two irrational straight lines, either a first apotome of a medial straight line or a straight line which produces with a rational area a medial whole.

For let a rational straight line *FG* be set out,
and let the areas be similarly applied.

It follows then that *FH* is rational and incommensurable in length with *FG*,
while *KF* is rational and commensurable in length with *FG*;
 therefore *FH*, *FK* are rational straight lines commensurable in square only;
 [X. 13]
 therefore *KH* is an apotome, and *FK* the annex to it. [X. 73]

Now the square on *HF* is greater than the square on *FK* either by the square on a straight line commensurable with *HF* or by the square on a straight line incommensurable with it.

If then the square on *HF* is greater than the square on *FK* by the square on a straight line commensurable with *HF*,
while the annex *FK* is commensurable in length with the rational straight line *FG* set out,

> *KH* is a second apotome. [X. Deff. III. 2]

But *FG* is rational;
so that the "side" of *LH*, that is, of *EC*, is a first apotome of a medial straight line.

<div align="right">[X. 92]</div>

But, if the square on *HF* is greater than the square on *FK* by the square on a straight line incommensurable with *HF*,
while the annex *FK* is commensurable in length with the rational straight line *FG* set out,

<div align="center">*KH* is a fifth apotome;</div> <div align="right">[X. Deff. III. 5]</div>

so that the "side" of *EC* is a straight line which produces with a rational area a medial whole.

<div align="right">[X. 95]</div>

<div align="right">Q.E.D.</div>

Proposition 110

If from a medial area there be subtracted a medial area incommensurable with the whole, the two remaining irrational straight lines arise, either a second apotome of a medial straight line or a straight line which produces with a medial area a medial whole.

For, as in the foregoing figures, let there be subtracted from the medial area *BC* the medial area *BD* incommensurable with the whole;
I say that the "side" of *EC* is one of two irrational straight lines, either a second apotome of a medial straight line or a straight line which produces with a medial area a medial whole.

For, since each of the rectangles *BC*, *BD* is medial,
and *BC* is incommensurable with *BD*,
it follows that each of the straight lines *FH*, *FK* will be rational and incommensurable in length with *FG*.

<div align="right">[X. 22]</div>

And, since *BC* is incommensurable with *BD*,
that is, *GH* with *GK*,

<div align="center">*HF* is also incommensurable with *FK*;</div> <div align="right">[VI. 1, X. 11]</div>

<div align="center">therefore *FH*, *FK* are rational straight lines commensurable in square only;</div>

<div align="center">therefore *KH* is an apotome.</div> <div align="right">[X. 73]</div>

If then the square on *FH* is greater than the square on *FK* by the square on a straight line commensurable with *FH*,
while neither of the straight lines *FH*, *FK* is commensurable in length with the rational straight line *FG* set out,

<div align="center">*KH* is a third apotome.</div> <div align="right">[X. Deff. III. 3]</div>

<div align="right">357</div>

But *KL* is rational,
and the rectangle contained by a rational straight line and a third apotome is irrational,
and the "side" of it is irrational, and is called a second apotome of a medial straight line; [X. 93]
so that the "side" of *LH*, that is, of *EC*, is a second apotome of a medial straight line.

But, if the square on *FH* is greater than the square on *FK* by the square on a straight line incommensurable with *FH*,
while neither of the straight lines *HF, FK* is commensurable in length with *FG*,
KH is a sixth apotome. [X. Deff. III. 6]

But the "side" of the rectangle contained by a rational straight line and a sixth apotome is a straight line which produces with a medial area a medial whole. [X. 96]

Therefore the "side" of *LH*, that is, of *EC*, is a straight line which produces with a medial area a medial whole.

Q.E.D.

Proposition 111

The apotome is not the same with the binomial straight line.

Let *AB* be an apotome;
I say that *AB* is not the same with the binomial straight line.

For, if possible, let it be so;
let a rational straight line *DC* be set out, and to *CD* let there be applied the rectangle *CE* equal to the square on *AB* and producing *DE* as breadth.

Then, since *AB* is an apotome,
DE is a first apotome. [X. 97]

Let *EF* be the annex to it;
therefore *DF, FE* are rational straight lines commensurable in square only,
the square on *DF* is greater than the square on *FE* by the square on a straight line commensurable with *DF*,
and *DF* is commensurable in length with the rational straight line *DC* set out. [X. Deff. III. 1]

Again, since *AB* is binomial,

therefore DE is a first binomial straight line. [X. 60]

Let it be divided into its terms at G, and let DG be the greater term;
therefore DG, GE are rational straight lines commensurable in square only,
the square on DG is greater than the square on GE by the square on a straight line commensurable with DG,
and the greater term DG is commensurable in length with the rational straight line DC set out. [X. Deff. II. 1]

Therefore DF is also commensurable in length with DG; [X. 12]
 therefore the remainder GF is also commensurable in length with DF. [X. 15]

But DF is incommensurable in length with EF;
 therefore FG is also incommensurable in length with EF. [X. 13]

Therefore GF, FE are rational straight lines commensurable in square only;
 therefore EG is an apotome. [X. 73]

But it is also rational:
which is impossible.

Therefore the apotome is not the same with the binomial straight line.

 Q.E.D.

The apotome and the irrational straight lines following it are neither the same with the medial straight line nor with one another.

For the square on a medial straight line, if applied to a rational straight line, produces as breadth a straight line rational and incommensurable in length with that to which it is applied, [X. 22]
while the square on an apotome, if applied to a rational straight line, produces as breadth a first apotome, [X. 97]
the square on a first apotome of a medial straight line, if applied to a rational straight line, produces as breadth a second apotome, [X. 98]
the square on a second apotome of a medial straight line, if applied to a rational straight line, produces as breadth a third apotome, [X. 99]
the square on a minor straight line, if applied to a rational straight line, produces as breadth a fourth apotome, [X. 100]
the square on the straight line which produces with a rational area a medial whole, if applied to a rational straight line, produces as breadth a fifth apotome, [X. 101]
and the square on the straight line which produces with a medial area a medial whole, if applied to a rational straight line, produces as breadth a sixth apotome. [X. 102]

Since then the said breadths differ from the first and from one another, from the first because it is rational, and from one another since they are not the same in order,
it is clear that the irrational straight lines themselves also differ from one another.

And, since the apotome has been proved not to be the same as the binomial straight line, [X. 111]
but, if applied to a rational straight line, the straight lines following the apotome produce, as breadths, each according to its own order, apotomes, and those

following the binomial straight line themselves also, according to their order, produce the binomials as breadths,

therefore those following the apotome are different, and those following the binomial straight line are different, so that there are, in order, thirteen irrational straight lines in all,

 Medial,
 Binomial,
 First bimedial,
 Second bimedial,
 Major,
 "Side" of a rational plus a medial area,
 "Side" of the sum of two medial areas,
 Apotome,
 First apotome of a medial straight line,
 Second apotome of a medial straight line,
 Minor,
 Producing with a rational area a medial whole,
 Producing with a medial area a medial whole.

Proposition 112

The square on a rational straight line applied to the binomial straight line produces as breadth an apotome the terms of which are commensurable with the terms of the binomial and moreover in the same ratio; and further, the apotome so arising will have the same order as the binomial straight line.

Let *A* be a rational straight line,

let *BC* be a binomial, and let *DC* be its greater term;

let the rectangle *BC, EF* be equal to the square on *A*;

I say that *EF* is an apotome the terms of which are commensurable with *CD, DB*, and in the same ratio, and further, *EF* will have the same order as *BC*.

For again let the rectangle *BD, G* be equal to the square on *A*.

Since, then, the rectangle *BC, EF* is equal to the rectangle *BD, G*,

 therefore, as *CB* is to *BD*, so is *G* to *EF.* [VI. 16]

But *CB* is greater than *BD*;

 therefore *G* is also greater than *EF.* [V. 16, V. 14]

Let *EH* be equal to *G*;

 therefore, as *CB* is to *BD*, so is *HE* to *EF*;

 therefore, *separando*, as *CD* is to *BD*, so is *HF* to *FE.* [V. 17]

Let it be contrived that, as *HF* is to *FE*, so is *FK* to *KE*;

therefore also the whole *HK* is to the whole *KF* as *FK* is to *KE*;

for, as one of the antecedents is to one of the consequents, so are all the

antecedents to all the consequents. [V. 12]

But, as FK is to KE, so is CD to DB; [V. 11]
 therefore also, as HK is to KF, so is CD to DB. [V. 11]

But the square on CD is commensurable with the square on DB; [X. 36]
 therefore the square on HK is also commensurable with the square on KF.
 [VI. 22, X. 11]

And, as the square on HK is to the square on KF, so is HK to KE, since the three
straight lines HK, KF, KE are proportional. [V. Def. 9]

Therefore HK is commensurable in length with KE,
 so that HE is also commensurable in length with EK. [X. 15]

Now, since the square on A is equal to the rectangle EH, BD,
while the square on A is rational,
 therefore the rectangle EH, BD is also rational.

And it is applied to the rational straight line BD;
 therefore EH is rational and commensurable in length with BD; [X. 20]
so that EK, being commensurable with it, is also rational and commensurable in
length with BD.

Since, then, as CD is to DB, so is FK to KE,
while CD, DB are straight lines commensurable in square only,
 therefore FK, KE are also commensurable in square only. [X. 11]

But KE is rational;
 therefore FK is also rational.

Therefore FK, KE are rational straight lines commensurable in square only;
 therefore EF is an apotome. [X. 73]

Now the square on CD is greater than the square on DB either by the square on
a straight line commensurable with CD or by the square on a straight line
incommensurable with it.

If then the square on CD is greater than the square on DB by the square on a
straight line commensurable with CD, the square on FK is also greater than the
square on KE by the square on a straight line commensurable with FK. [X. 14]

And, if CD is commensurable in length with the rational straight line set out,
 so also is FK; [X. 11, 12]
if BD is so commensurable,
 so also is KE; [X. 12]
but, if neither of the straight lines CD, DB is so commensurable,
 neither of the straight lines FK, KE is so.

But, if the square on CD is greater than the square on DB by the square on a
straight line incommensurable with CD,
the square on FK is also greater than the square on KE by the square on a straight
line incommensurable with FK. [X. 14]

And, if CD is commensurable with the rational straight line set out,
 so also is FK;
if BD is so commensurable,
 so also is KE;
but, if neither of the straight lines CD, DB is so commensurable,
 neither of the straight lines FK, KE is so;

so that *FE* is an apotome, the terms of which, *FK*, *KE* are commensurable with the terms *CD*, *DB* of the binomial straight line and in the same ratio, and it has the same order as *BC*.

Q.E.D.

Proposition 113

The square on a rational straight line, if applied to an apotome, produces as breadth the binomial straight line the terms of which are commensurable with the terms of the apotome and in the same ratio; and further, the binomial so arising has the same order as the apotome.

Let *A* be a rational straight line and *BD* an apotome, and let the rectangle *BD*, *KH* be equal to the square on *A*, so that the square on the rational straight line *A* when applied to the apotome *BD* produces *KH* as breadth;

I say that *KH* is a binomial straight line the terms of which are commensurable with the terms of *BD* and in the same ratio; and further, *KH* has the same order as *BD*.

For let *DC* be the annex to *BD*;
therefore *BC*, *CD* are rational straight lines commensurable in square only. **[X. 73]**

Let the rectangle *BC*, *G* be also equal to the square on *A*.

But the square on *A* is rational;
therefore the rectangle *BC*, *G* is also rational.

And it has been applied to the rational straight line *BC*;
therefore *G* is rational and commensurable in length with *BC*. **[X. 20]**

Since now the rectangle *BC*, *G* is equal to the rectangle *BD*, *KH*,
therefore, proportionally, as *CB* is to *BD*, so is *KH* to *G*. **[VI. 16]**

But *BC* is greater than *BD*;
therefore *KH* is also greater than *G*. **[V. 16, V. 14]**

Let *KE* be made equal to *G*;
therefore *KE* is commensurable in length with *BC*.

And since, as *CB* is to *BD*, so is *HK* to *KE*,
therefore, *convertendo*, as *BC* is to *CD*, so is *KH* to *HE*. **[V. 19, Por.]**

Let it be contrived that, as *KH* is to *HE*, so is *HF* to *FE*;
therefore also the remainder *KF* is to *FH* as *KH* is to *HE*,
that is, as *BC* is to *CD*. **[V. 19]**

362

But *BC, CD* are commensurable in square only;

therefore *KF, FH* are also commensurable in square only. [X. 11]

And since, as *KH* is to *HE*, so is *KF* to *FH*,
while, as *KH* is to *HE*, so is *HF* to *FE*,

therefore also, as *KF* is to *FH*, so is *HF* to *FE*, [V. 11]

so that also, as the first is to the third, so is the square on the first to the square
on the second; [V. Def. 9]

therefore also, as *KF* is to *FE*, so is the square on *KF* to the square on *FH*.

But the square on *KF* is commensurable with the square on *FH*,
for *KF, FH* are commensurable in square;

therefore *KF* is also commensurable in length with *FE*, [X. 11]

so that *KF* is also commensurable in length with *KE*. [X. 15]

But *KE* is rational and commensurable in length with *BC*;
therefore *KF* is also rational and commensurable in length with *BC*. [X. 12]

And, since, as *BC* is to *CD*, so is *KF* to *FH*,

alternately, as *BC* is to *KF*, so is *DC* to *FH*. [V. 16]

But *BC* is commensurable with *KF*;

therefore *FH* is also commensurable in length with *CD*. [X. 11]

But *BC, CD* are rational straight lines commensurable in square only;
therefore *KF, FH* are also rational straight lines [X. Def. 3] commensurable in
square only;

therefore *KH* is binomial. [X. 36]

If now the square on *BC* is greater than the square on *CD* by the square on a
straight line commensurable with *BC*,
the square on *KF* will also be greater than the square on *FH* by the square on a
straight line commensurable with *KF*. [X. 14]

And, if *BC* is commensurable in length with the rational straight line set out,

so also is *KF*;

if *CD* is commensurable in length with the rational straight line set out,

so also is *FH*,

but, if neither of the straight lines *BC, CD*,

then neither of the straight lines *KF, FH*.

But, if the square on *BC* is greater than the square on *CD* by the square on a
straight line incommensurable with *BC*,
the square on *KF* is also greater than the square on *FH* by the square on a
straight line incommensurable with *KF*. [X. 14]

And, if *BC* is commensurable with the rational straight line set out,

so also is *KF*;

if *CD* is so commensurable,

so also is *FH*;

but, if neither of the straight lines *BC, CD*,

then neither of the straight lines *KF, FH*.

Therefore *KH* is a binomial straight line, the terms of which *KF, FH* are com-
mensurable with the terms *BC, CD* of the apotome and in the same ratio,
and further, *KH* has the same order as *BD*.

Q.E.D.

Proposition 114

If an area be contained by an apotome and the binomial straight line the terms of which are commensurable with the terms of the apotome and in the same ratio, the "side" of the area is rational.

For let an area, the rectangle *AB*, *CD*, be contained by the apotome *AB* and the binomial straight line *CD*,
and let *CE* be the greater term of the latter;
let the terms *CE*, *ED* of the binomial straight line be commensurable with the terms *AF*, *FB* of the apotome and in the same ratio;
and let the "side" of the rectangle *AB*, *CD* be *G*;
I say that *G* is rational.

For let a rational straight line *H* be set out,
and to *CD* let there be applied a rectangle equal to the square on *H* and producing *KL* as breadth.

Therefore *KL* is an apotome.

Let its terms be *KM*, *ML* commensurable with the terms *CE*, *ED* of the binomial straight line and in the same ratio. [X. 112]

But *CE*, *ED* are also commensurable with *AF*, *FB* and in the same ratio;
therefore, as *AF* is to *FB*, so is *KM* to *ML*.

Therefore, alternately, as *AF* is to *KM*, so is *BF* to *LM*;
therefore also the remainder *AB* is to the remainder *KL* as *AF* is to *KM*.

[V. 19]

But *AF* is commensurable with *KM*; [X. 12]
therefore *AB* is also commensurable with *KL*. [X. 11]

And, as *AB* is to *KL*, so is the rectangle *CD*, *AB* to the rectangle *CD*, *KL*; [VI. 1]
therefore the rectangle *CD*, *AB* is also commensurable with the rectangle *CD*, *KL*.

[X. 11]

But the rectangle *CD*, *KL* is equal to the square on *H*;
therefore the rectangle *CD*, *AB* is commensurable with the square on *H*.

But the square on *G* is equal to the rectangle *CD*, *AB*;
therefore the square on *G* is commensurable with the square on *H*.

But the square on *H* is rational;
therefore the square on *G* is also rational;
therefore *G* is rational.

And it is the "side" of the rectangle *CD*, *AB*.

Therefore etc.

Q.E.D.

PORISM. And it is made manifest to us by this also that it is possible for a rational area to be contained by irrational straight lines.

Proposition 115

From a medial straight line there arise irrational straight lines infinite in number, and none of them is the same as any of the preceding.

```
A ————————
B ————————————————————
C ————————————————
D ——————————————————
```

Let *A* be a medial straight line;
I say that from *A* there arise irrational straight lines infinite in number, and none of them is the same as any of the preceding.

Let a rational straight line *B* be set out, and let the square on *C* be equal to the rectangle *B*, *A*;

therefore *C* is irrational; **[X. Def. 4]**
for that which is contained by an irrational and a rational straight line is irrational.
 [deduction from X. 20]

And it is not the same with any of the preceding;
for the square on none of the preceding, if applied to a rational straight line produces as breadth a medial straight line.

Again, let the square on *D* be equal to the rectangle *B*, *C*;

therefore the square on *D* is irrational. **[deduction from X. 20]**

Therefore *D* is irrational; **[X. Def. 4]**
and it is not the same with any of the preceding, for the square on none of the preceding, if applied to a rational straight line, produces *C* as breadth.

Similarly, if this arrangement proceeds *ad infinitum*, it is manifest that from the medial straight line there arise irrational straight lines infinite in number, and none is the same with any of the preceding.

Q.E.D.

Book XI

Definitions

1. A *solid* is that which has length, breadth, and depth.

2. An extremity of a solid is a surface.

3. A *straight line* is at *right angles to a plane,* when it makes right angles with all the straight lines which meet it and are in the plane.

4. A *plane* is *at right angles to a plane* when the straight lines drawn, in one of the planes, at right angles to the common section of the planes, are at right angles to the remaining plane.

5. The *inclination of a straight line to a plane* is, assuming a perpendicular drawn from the extremity of the straight line which is elevated above the plane to the plane, and a straight line joined from the point thus arising to the extremity of the straight line which is in the plane, the angle contained by the straight line so drawn and the straight line standing up.

6. The *inclination of a plane to a plane* is the acute angle contained by the straight lines drawn at right angles to the common section at the same point, one in each of the planes.

7. A plane is said to be *similarly inclined* to a plane as another is to another when the said angles of the inclinations are equal to one another.

8. *Parallel planes* are those which do not meet.

9. *Similar solid figures* are those contained by similar planes equal in multitude.

10. *Equal and similar solid figures* are those contained by similar planes equal in multitude and in magnitude.

11. A *solid angle* is the inclination constituted by more than two lines which meet one another and are not in the same surface, towards all the lines.

 Otherwise: A *solid angle* is that which is contained by more than two plane angles which are not in the same plane and are constructed to one point.

12. A *pyramid* is a solid figure, contained by planes, which is constructed from one plane to one point.

13. A *prism* is a solid figure contained by planes two of which, namely those which are opposite, are equal, similar and parallel, while the rest are parallelograms.

14. When, the diameter of a semicircle remaining fixed, the semicircle is carried round and restored again to the same position from which it began to be moved, the figure so comprehended is a *sphere*.

15. The *axis of the sphere* is the straight line which remains fixed and about which the semicircle is turned.

16. The *centre of the sphere* is the same as that of the semicircle.

17. A *diameter of the sphere* is any straight line drawn through the centre and terminated in both directions by the surface of the sphere.

18. When, one side of those about the right angle in a right-angled triangle remaining fixed, the triangle is carried round and restored again to the same position from which it began to be moved, the figure so comprehended is a *cone*.

 And, if the straight line which remains fixed be equal to the remaining side about the right angle which is carried round, the cone will be *right-angled*; if less, *obtuse-angled*; and if greater, *acute-angled*.

19. The *axis of the cone* is the straight line which remains fixed and about which the triangle is turned.

20. And the *base* is the circle described by the straight line which is carried round.

21. When, one side of those about the right angle in a rectangular parallelogram remaining fixed, the parallelogram is carried round and restored again to the same position from which it began to be moved, the figure so comprehended is a *cylinder*.

22. The *axis of the cylinder* is the straight line which remains fixed and about which the parallelogram is turned.

23. And the *bases* are the circles described by the two sides opposite to one another which are carried round.

24. *Similar cones and cylinders* are those in which the axes and the diameters of the bases are proportional.

25. A *cube* is a solid figure contained by six equal squares.

26. An *octahedron* is a solid figure contained by eight equal and equilateral triangles.

27. An *icosahedron* is a solid figure contained by twenty equal and equilateral triangles.

28. A *dodecahedron* is a solid figure contained by twelve equal, equilateral, and equiangular pentagons.

Proposition 1

A part of a straight line cannot be in the plane of reference and a part in a plane more elevated.

For, if possible, let a part *AB* of the straight line *ABC* be in the plane of reference, and a part *BC* in a plane more elevated.

There will then be in the plane of reference some straight line continuous with *AB* in a straight line.

Let it be *BD*;
therefore *AB* is a common segment of the two straight lines *ABC*, *ABD*:
which is impossible, inasmuch as, if we describe a circle with centre *B* and distance *AB*, the diameters will cut off unequal circumferences of the circle.

Therefore a part of a straight line cannot be in the plane of reference, and a part in a plane more elevated.

<div align="right">Q.E.D.</div>

Proposition 2

If two straight lines cut one another, they are in one plane, and every triangle is in one plane.

For let the two straight lines *AB*, *CD* cut one another at the point *E*;
I say that *AB*, *CD* are in one plane, and every triangle is in one plane.

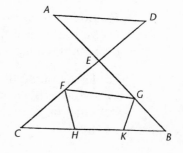

For let points *F*, *G* be taken at random on *EC*, *EB*,
let *CB*, *FG* be joined,
and let *FH*, *GK* be drawn across;
I say first that the triangle *ECB* is in one plane.

For, if part of the triangle *ECB*, either *FHC* or *GBK*, is in the plane of reference, and the rest in another,
a part also of one of the straight lines *EC*, *EB* will be in the plane of reference, and a part in another.

But, if the part *FCBG* of the triangle *ECB* be in the plane of reference, and the rest in another,
a part also of both the straight lines *EC*, *EB* will be in the plane of reference and a part in another:
which was proved absurd. [XI. 1]

<div align="center">Therefore the triangle ECB is in one plane.</div>

But, in whatever plane the triangle *ECB* is, in that plane also is each of the straight lines *EC*, *EB*,
and, in whatever plane each of the straight lines *EC*, *EB* is, in that plane are *AB*, *CD* also. [XI. 1]

Therefore the straight lines *AB*, *CD* are in one plane,
<div align="center">and every triangle is in one plane.</div>

<div align="right">Q.E.D.</div>

Proposition 3

If two planes cut one another, their common section is a straight line.

For let the two planes *AB, BC* cut one another, and let the line *DB* be their common section;
I say that the line *DB* is a straight line.

For, if not, from *D* to *B* let the straight line *DEB* be joined in the plane *AB,*
and in the plane *BC* the straight line *DFB.*

Then the two straight lines *DEB, DFB* will have the same extremities, and will clearly enclose an area:
which is absurd.

Therefore *DEB, DFB* are not straight lines.

Similarly we can prove that neither will there be any other straight line joined from *D* to *B* except *DB* the common section of the planes *AB, BC.*

Therefore etc.

Q.E.D.

Proposition 4

If a straight line be set up at right angles to two straight lines which cut one another, at their common point of section, it will also be at right angles to the plane through them.

For let a straight line *EF* be set up at right angles to the two straight lines *AB, CD,* which cut one another at the point *E,* from *E;*
I say that *EF* is also at right angles to the plane through *AB, CD.*

For let *AE, EB, CE, ED* be cut off equal to one another,
and let any straight line *GEH* be drawn across through *E,* at random;
let *AD, CB* be joined,
and further let *FA, FG, FD, FC, FH, FB* be joined from the point *F* taken at random [on *EF*].

Now, since the two straight lines *AE, ED* are equal to the two straight lines *CE, EB,* and contain equal angles, [I. 15]
therefore the base *AD* is equal to the base *CB,*
and the triangle *AED* will be equal to the triangle *CEB* [I. 4]
so that the angle *DAE* is also equal to the angle *EBC.*

But the angle *AEG* is also equal to the angle *BEH;* [I.15]
therefore *AGE, BEH* are two triangles which have two angles equal to two angles respectively, and one side equal to one side, namely that adjacent to the equal angles, that is to say, *AE* to *EB;*
therefore they will also have the remaining sides equal to the remaining sides.
 [I.26]

Therefore *GE* is equal to *EH*, and *AG* to *BH*.

And, since *AE* is equal to *EB*,
while *FE* is common and at right angles,
therefore the base *FA* is equal to the base *FB*. [I. 4]

For the same reason
FC is also equal to *FD*.

And, since *AD* is equal to *CB*,
and *FA* is also equal to *FB*,
the two sides *FA, AD* are equal to the two sides *FB, BC* respectively;
and the base *FD* was proved equal to the base *FC*;
therefore the angle *FAD* is also equal to the angle *FBC*. [I. 8]

And since, again, *AG* was proved equal to *BH*,
and further, *FA* also equal to *FB*,
the two sides *FA, AG* are equal to the two sides *FB, BH*.

And the angle *FAG* was proved equal to the angle *FBH*;
therefore the base *FG* is equal to the base *FH*. [I. 4]

Now since, again, GE was proved equal to *EH*,
and *EF* is common,
the two sides *GE, EF* are equal to the two sides *HE, EF*;
and the base *FG* is equal to the base *FH*;
therefore the angle *GEF* is equal to the angle *HEF*. [I. 8]

Therefore each of the angles *GEF, HEF* is right.

Therefore *FE* is at right angles to *GH* drawn at random through *E*.

Similarly we can prove that *FE* will also make right angles with all the straight lines which meet it and are in the plane of reference.

But a straight line is at right angles to a plane when it makes right angles with all the straight lines which meet it and are in that same plane; [XI. Def. 3]
therefore *FE* is at right angles to the plane of reference.

But the plane of reference is the plane through the straight lines *AB, CD*.

Therefore *FE* is at right angles to the plane through *AB, CD*.

Therefore etc.

Q.E.D.

Proposition 5

If a straight line be set up at right angles to three straight lines which meet one another, at their common point of section, the three straight lines are in one plane.

For let a straight line *AB* be set up at right angles to the three straight lines *BC, BD, BE*, at their point of meeting at *B*;
I say that *BC, BD, BE* are in one plane.

For suppose they are not, but, if possible, let *BD, BE* be in the plane of reference and *BC* in one more elevated;
let the plane through *AB, BC* be produced;
it will thus make, as common section in the plane of reference, a straight line. [XI. 3]

Let it make *BF*.

Therefore the three straight lines *AB*, *BC*, *BF* are in one plane, namely that drawn through *AB*, *BC*.

Now, since *AB* is at right angles to each of the straight lines *BD*, *BE*,
therefore *AB* is also at right angles to the plane through *BD*, *BE*. **[XI. 4]**

But the plane through *BD*, *BE* is the plane of reference;
therefore *AB* is at right angles to the plane of reference.

Thus *AB* will also make right angles with all the straight lines which meet it and are in the plane of reference. **[XI. Def. 3]**

But *BF* which is in the plane of reference meets it;
therefore the angle *ABF* is right.

But, by hypothesis, the angle *ABC* is also right;
therefore the angle *ABF* is equal to the angle *ABC*.

And they are in one plane:
which is impossible.

Therefore the straight line *BC* is not in a more elevated plane;
therefore the three straight lines *BC*, *BD*, *BE* are in one plane.

Therefore, if a straight line be set up at right angles to three straight lines, at their point of meeting, the three straight lines are in one plane.

Q.E.D.

Proposition 6

If two straight lines be at right angles to the same plane, the straight lines will be parallel.

For let the two straight lines *AB*, *CD* be at right angles to the plane of reference;
I say that *AB* is parallel to *CD*.

For let them meet the plane of reference at the points *B*, *D*,
let the straight line *BD* be joined,
let *DE* be drawn, in the plane of reference, at right angles to *BD*,
let *DE* be made equal to *AB*,
and let *BE*, *AE*, *AD* be joined.

Now, since *AB* is at right angles to the plane of reference, it will also make right angles with all the straight lines which meet it and are in the plane of reference. **[XI. Def. 3]**

But each of the straight lines *BD*, *BE* is in the plane of reference and meets *AB*;
therefore each of the angles *ABD*, *ABE* is right.

For the same reason
each of the angles *CDB*, *CDE* is also right.

And, since *AB* is equal to *DE*,

and *BD* is common,

 the two sides *AB*, *ED* are equal to the two sides *ED*, *DB*;

and they include right angles;

 therefore the base *AD* is equal to the base *BE*. [I. 4]

And, since *AB* is equal to *DE*,

while *AD* is also equal to *BE*

 the two sides *AB*, *BE* are equal to the two sides *ED*, *DA*;

and *AE* is their common base;

 therefore the angle *ABE* is equal to the angle *EDA*. [I.8]

But the angle *ABE* is right;

 therefore the angle *EDA* is also right;

 therefore *ED* is at right angles to *DA*.

But it is also at right angles to each of the straight lines *BD*, *DC*;

therefore *ED* is set up at right angles to the three straight lines *BD*, *DA*, *DC* at their point of meeting;

 therefore the three straight lines *BD*, *DA*, *DC* are in one plane. [XI. 5]

But, in whatever plane *DB*, *DA* are, in that plane is *AB* also,

for every triangle is in one plane; [XI. 2]

 therefore the straight lines *AB*, *BD*, *DC* are in one plane.

And each of the angles *ABD*, *BDC* is right;

 therefore *AB* is parallel to *CD*. [I. 28]

Therefore etc.

 Q.E.D.

Proposition 7

If two straight lines be parallel and points be taken at random on each of them, the straight line joining the points is in the same plane with the parallel straight line.

Let *AB*, *CD* be two parallel straight lines,

and let points *E*, *F* be taken at random on them respectively;

I say that the straight line joining the points *E*, *F* is in the same plane with the parallel straight lines.

For suppose it is not, but, if possible, let it be in a more elevated plane as *EGF*,

and let a plane be drawn through *EGF*;

it will then make, as section in the plane of reference, a straight line. [XI. 3]

Let it make it, as *EF*;

 therefore the two straight lines *EGF*, *EF* will enclose an area:

which is impossible.

Therefore the straight line joined from *E* to *F* is not in a plane more elevated;

therefore the straight line joined from *E* to *F* is in the plane through the parallel straight lines *AB*, *CD*.

Therefore etc.

 Q.E.D.

Proposition 8

If two straight lines be parallel, and one of them be at right angles to any plane, the remaining one will also be at right angles to the same plane.

Let *AB, CD* be two parallel straight lines, and let one of them, *AB*, be at right angles to the plane of reference;
I say that the remaining one, *CD*, will also be at right angles to the same plane.

For let *AB, CD* meet the plane of reference at the points *B, D,*
and let *BD* be joined;
therefore *AB, CD, BD* are in one plane. [XI. 7]

Let *DE* be drawn, in the plane of reference, at right angles to *BD,*
let *DE* be made equal to *AB,*
and let *BE, AE, AD* be joined.

Now, since *AB* is at right angles to the plane of reference, therefore *AB* is also at right angles to all the straight lines which meet it and are in the plane of reference;
[XI. Def. 3]
therefore each of the angles *ABD, ABE* is right.

And, since the straight line *BD* has fallen on the parallels *AB, CD,*
therefore the angles *ABD, CDB* are equal to two right angles. [I. 29]

But the angle *ABD* is right;
therefore the angle *CDB* is also right;
therefore *CD* is at right angles to *BD.*

And, since *AB* is equal to *DE,*
and *BD* is common,
the two sides *AB, BD* are equal to the two sides *ED, DB;*
and the angle *ABD* is equal to the angle *EDB,*
for each is right;
therefore the base *AD* is equal to the base *BE.*

And, since *AB* is equal to *DE,*
and *BE* to *AD,*
the two sides *AB, BE* are equal to the two sides *ED, DA* respectively,
and *AE* is their common base;
therefore the angle *ABE* is equal to the angle *EDA.*

But the angle *ABE* is right;
therefore the angle *EDA* is also right;
therefore *ED* is at right angles to *AD.*

But it is also at right angles to *DB;*
therefore *ED* is also at right angles to the plane through *BD, DA.* [XI. 4]

Therefore *ED* will also make right angles with all the straight lines which meet it and are in the plane through *BD, DA.*

But *DC* is in the plane through *BD, DA,* inasmuch as *AB, BD* are in the plane through *BD, DA,*
[XI. 2]

and *DC* is also in the plane in which *AB, BD* are.

Therefore *ED* is at right angles to *DC*,

so that *CD* is also at right angles to *DE*.

But *CD* is also at right angles to *BD*.

Therefore *CD* is set up at right angles to the two straight lines *DE, DB* which cut one another, from the point of section at *D*;

so that *CD* is also at right angles to the plane through *DE, DB*. [XI. 4]

But the plane through *DE, DB* is the plane of reference;

therefore *CD* is at right angles to the plane of reference.

Therefore etc.

Q.E.D.

Proposition 9

Straight lines which are parallel to the same straight line and are not in the same plane with it are also parallel to one another.

For let each of the straight lines *AB, CD* be parallel to *EF*, not being in the same plane with it;
I say that *AB* is parallel to *CD*.

For let a point *G* be taken at random on *EF*,
and from it let there be drawn *GH*, in the plane through *EF, AB*, at right angles to *EF*, and *GK* in the plane through *FE, CD* again at right angles to *EF*.

Now, since *EF* is at right angles to each of the straight lines *GH, GK*,

therefore *EF* is also at right angles to the plane through *GH, GK*. [XI. 4]

And *EF* is parallel to *AB*;

therefore *AB* is also at right angles to the plane through *HG, GK*. [XI. 8]

For the same reason

CD is also at right angles to the plane through *HG, GK*;

therefore each of the straight lines *AB, CD* is at right angles to the plane through *HG, GK*.

But, if two straight lines be at right angles to the same plane, the straight lines are parallel; [XI. 6]

therefore *AB* is parallel to *CD*.

Q.E.D.

Proposition 10

If two straight lines meeting one another be parallel to two straight lines meeting one another not in the same plane, they will contain equal angles.

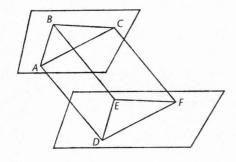

For let the two straight lines *AB, BC* meeting one another be parallel to the two straight lines *DE, EF* meeting one another, not in the same plane;
I say that the angle *ABC* is equal to the angle *DEF*.

For let *BA, BC, ED, EF* be cut off equal to one another, and let *AD, CF, BE, AC, DF* be joined.

Now, since *BA* is equal and parallel to *ED*, therefore *AD* is also equal and parallel to *BE*.
[I. 33]

For the same reason

CF is also equal and parallel to *BE*.

Therefore each of the straight lines *AD, CF* is equal and parallel to *BE*.

But straight lines which are parallel to the same straight line and are not in the same plane with it are parallel to one another; [XI. 9]
therefore *AD* is parallel and equal to *CF*.

And *AG, DF* join them;
therefore *AC* is also equal and parallel to *DF*. [I. 33]

Now, since the two sides *AB, BC* are equal to the two sides *DE, EF*, and the base *AC* is equal to the base *DF*,
therefore the angle *ABC* is equal to the angle *DEF*. [I. 8]

Therefore etc.

Q.E.D.

Proposition 11

From a given elevated point to draw a straight line perpendicular to a given plane.

Let *A* be the given elevated point, and the plane of reference the given plane;
thus it is required to draw from the point *A* a straight line perpendicular to the plane of reference.

Let any straight line *BC* be drawn, at random, in the plane of reference,
and let *AD* be drawn from the point *A* perpendicular to *BC*.
[I. 12]

If then *AD* is also perpendicular to the plane of reference, that which was enjoined will have been done.

But, if not, let *DE* be drawn from the point *D* at right angles to *BC* and in the plane of reference, [I. 11]
let *AF* be drawn from *A* perpendicular to *DE*, [I. 12]
and let *GH* be drawn through the point *F* parallel to *BC*. [I. 31]

Now, since BC is at right angles to each of the straight lines DA, DE,
therefore BC is also at right angles to the plane through ED, DA. [XI. 4]

And GH is parallel to it;
but, if two straight lines be parallel, and one of them be at right angles to any plane, the remaining one will also be at right angles to the same plane; [XI. 8]
therefore GH is also at right angles to the plane through ED, DA.

Therefore GH is also at right angles to all the straight lines which meet it and are in the plane through ED, DA. [XI. Def. 3]

But AF meets it and is in the plane through ED, DA;
therefore GH is at right angles to FA,
so that FA is also at right angles to GH.

But AF is also at right angles to DE;
therefore AF is at right angles to each of the straight lines GH, DE.

But, if a straight line be set up at right angles to two straight lines which cut one another, at the point of section, it will also be at right angles to the plane through them; [XI.4]
therefore FA is at right angles to the plane through ED, GH.

But the plane through ED, GH is the plane of reference;
therefore AF is at right angles to the plane of reference.

Therefore from the given elevated point A the straight line AF has been drawn perpendicular to the plane of reference.

Q.E.F.

Proposition 12

To set up a straight line at right angles to a given plane from a given point in it.

Let the plane of reference be the given plane,
and A the point in it;
thus it is required to set up from the point A a straight line at right angles to the plane of reference.

Let any elevated point B be conceived,
from B let BC be drawn perpendicular to the plane of reference, [XI. 11]
and through the point A let AD be drawn parallel to BC. [I. 31]

Then, since AD, CB are two parallel straight lines, while one of them, BC, is at right angles to the plane of reference,
therefore the remaining one, AD, is also at right angles to the plane of reference. [XI. 8]

Therefore AD has been set up at right angles to the given plane from the point A in it.

Q.E.F.

Proposition 13

From the same point two straight lines cannot be set up at right angles to the same plane on the same side.

For, if possible, from the same point A let the two straight lines AB, AC be set up at right angles to the plane of reference and on the same side,
and let a plane be drawn through BA, AC;
it will then make, as section through A in the plane of reference, a straight line. [XI. 3]

Let it make DAE;
therefore the straight lines AB, AC, DAE are in one plane.

And, since CA is at right angles to the plane of reference, it will also make right angles with all the straight lines which meet it and are in the plane of reference.
[XI. Def. 3]

But DAE meets it and is in the plane of reference;
therefore the angle CAE is right.

For the same reason
the angle BAE is also right;
therefore the angle CAE is equal to the angle BAE.
And they are in one plane:
which is impossible.

Therefore etc.

Q.E.D.

Proposition 14

Planes to which the same straight line is at right angles will be parallel.

For let any straight line AB be at right angles to each of the planes CD, EF;
I say that the planes are parallel.

For, if not, they will meet when produced.

Let them meet;
they will then make, as common section, a straight line. [XI. 3]

Let them make GH;
let a point K be taken at random on GH,
and let AK, BK be joined.

Now, since *AB* is at right angles to the plane *EF*,
therefore *AB* is also at right angles to *BK* which is a straight line in the plane *EF* produced; [XI. Def. 3]

therefore the angle *ABK* is right.

For the same reason

the angle *BAK* is also right.

Thus, in the triangle *ABK*, the two angles *ABK*, *BAK* are equal to two right angles:
which is impossible. [I. 17]

Therefore the planes *CD*, *EF* will not meet when produced;

therefore the planes *CD*, *EF* are parallel. [XI. Def. 8]

Therefore planes to which the same straight line is at right angles are parallel.

Q.E.D.

Proposition 15

If two straight lines meeting one another be parallel to two straight lines meeting one another, not being in the same plane, the planes through them are parallel.

For let the two straight lines *AB*, *BC* meeting one another be parallel to the two straight lines *DE*, *EF* meeting one another, not being in the same plane;
I say that the planes produced through *AB*, *BC* and *DE*, *EF* will not meet one another.

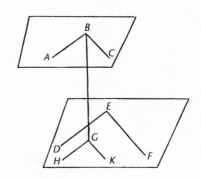

For let *BG* be drawn from the point *B* perpendicular to the plane through *DE*, *EF*, [XI. 11]
and let it meet the plane at the point G;
through G let *GH* be drawn parallel to *ED*, and *GK* parallel to *EF*. [I. 31]

Now, since *BG* is at right angles to the plane through *DE*, *EF*,
therefore it will also make right angles with all the straight lines which meet it and are in the plane through *DE*, *EF*. [XI. Def. 3]

But each of the straight lines *GH*, *GK* meets it and is in the plane through *DE*, *EF*;
therefore each of the angles *BGH*, *BGK* is right.

And, since *BA* is parallel to *GH*, [XI. 9]
therefore the angles *GBA*, *BGH* are equal to two right angles. [I. 29]

But the angle *BGH* is right;

therefore the angle *GBA* is also right;

therefore *GB* is at right angles to *BA*.

For the same reason

GB is also at right angles to *BC*.

Since then the straight line *GB* is set up at right angles to the two straight lines *BA*, *BC* which cut one another,

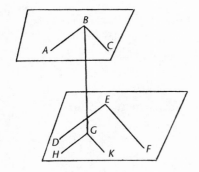

therefore *GB* is also at right angles to the plane through *BA, BC*. [XI. 4]

But planes to which the same straight line is at right angles are parallel; [XI. 14]
therefore the plane through *AB, BC* is parallel to the plane through *DE, EF*.

Therefore, if two straight lines meeting one another be parallel to two straight lines meeting one another, not in the same plane, the planes through them are parallel.

Q.E.D.

Proposition 16

If two parallel planes be cut by any plane, their common sections are parallel.

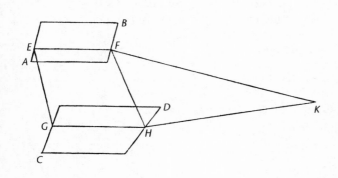

For let the two parallel planes *AB, CD* be cut by the plane *EFGH*,
and let *EF, GH* be their common sections;
I say that *EF* is parallel to *GH*.

For, if not, *EF, GH* will, when produced, meet either in the direction of *F, H* or of *E, G*.

Let them be produced, as in the direction of *F, H*, and let them, first, meet at *K*.

Now, since *EFK* is in the plane *AB*,
 therefore all the points on *EFK* are also in the plane *AB*. [XI. 1]

But *K* is one of the points on the straight line *EFK*;
 therefore *K* is in the plane *AB*.

For the same reason
 K is also in the plane *CD*;
therefore the planes *AB, CD* will meet when produced.

But they do not meet, because they are, by hypothesis, parallel;
therefore the straight lines *EF, GH* will not meet when produced in the direction of *F, H*.

Similarly we can prove that neither will the straight lines *EF, GH* meet when produced in the direction of *E, G*.

But straight lines which do not meet in either direction are parallel. [I. Def. 23]
 Therefore *EF* is parallel to *GH*.

Therefore etc.

Q.E.D.

Proposition 17

If two straight lines be cut by parallel planes, they will be cut in the same ratios.

For let the two straight lines *AB, CD* be cut
by the parallel planes *GH, KL, MN* at the
points *A, E, B* and *C, F, D*;
I say that, as the straight line *AE* is to *EB*, so
is *CF* to *FD*.

For let *AC, BD, AD* be joined,
let *AD* meet the plane *KL* at the point *O*,
and let *EO, OF* be joined.

Now, since the two parallel planes *KL, MN*
are cut by the plane *EBDO*,
their common sections *EO, BD* are parallel.
 [XI. 16]

For the same reason, since the two parallel
planes *GH, KL* are cut by the plane *AOFC*,
 their common sections *AC, OF* are parallel. **[XI. 16]**

And, since the straight line *EO* has been drawn parallel to *BD*, one of the sides
of the triangle *ABD*,
 therefore, proportionally, as *AE* is to *EB*, so is *AO* to *OD*. **[VI. 2]**

Again, since the straight line *OF* has been drawn parallel to *AC*, one of the sides
of the triangle *ADC*,
 proportionally, as *AO* is to *OD*, so is *CF* to *FD*. **[VI. 2]**

But it was also proved that, as *AO* is to *OD*, so is *AE* to *EB*;
 therefore also, as *AE* is to *EB*, so is *CF* to *FD*. **[V. 11]**

Therefore etc.

Q.E.D.

Proposition 18

*If a straight line be at right angles to any plane, all the planes through it will
also be at right angles to the same plane.*

For let any straight line *AB* be at right angles to the
plane of reference;
I say that all the planes through *AB* are also at right
angles to the plane of reference.

For let the plane *DE* be drawn through *AB*,
let *CE* be the common section of the plane *DE* and
the plane of reference,
let a point *F* be taken at random on *CE*,
and from *F* let *FG* be drawn in the plane *DE* at right angles to *CE*. **[I. 11]**

Now, since *AB* is at right angles to the plane of reference, *AB* is also at right
angles to all the straight lines which meet it and are in the plane of reference;
 [IX. Def. 3]

so that it is also at right angles to *CE*;
therefore the angle *ABF* is right.

But the angle *GFB* is also right;
therefore *AB* is parallel to *FG*. [I. 28]

But *AB* is at right angles to the plane of reference;
therefore *FG* is also at right angles to the plane of reference. [XI. 8]

Now a plane is at right angles to a plane, when the straight lines drawn, in one of the planes, at right angles to the common section of the planes are at right angles to the remaining plane. [XI. Def. 4]

And *FG*, drawn in one of the planes *DE* at right angles to *CE*, the common section of the planes, was proved to be at right angles to the plane of reference; therefore the plane *DE* is at right angles to the plane of reference.

Similarly also it can be proved that all the planes through *AB* are at right angles to the plane of reference.

Therefore etc.

Q.E.D.

Proposition 19

If two planes which cut one another be at right angles to any plane, their common section will also be at right angles to the same plane.

For let the two planes *AB*, *BC* be at right angles to the plane of reference,
and let *BD* be their common section;
I say that *BD* is at right angles to the plane of reference.

For suppose it is not, and from the point *D* let *DE* be drawn in the plane *AB* at right angles to the straight line *AD*,
and *DF* in the plane *BC* at right angles to *CD*.

Now, since the plane *AB* is at right angles to the plane of reference,
and *DE* has been drawn in the plane *AB* at right angles to *AD*, their common section,
therefore *DE* is at right angles to the plane of reference. [XI. Def. 4]

Similarly we can prove that
DF is also at right angles to the plane of reference.

Therefore from the same point *D* two straight lines have been set up at right angles to the plane of reference on the same side:
which is impossible. [XI. 13]

Therefore no straight line except the common section *DB* of the planes *AB*, *BC* can be set up from the point *D* at right angles to the plane of reference.

Therefore etc.

Q.E.D.

Proposition 20

If a solid angle be contained by three plane angles, any two, taken together in any manner, are greater than the remaining one.

For let the solid angle at *A* be contained by the three plane angles *BAC, CAD, DAB*; I say that any two of the angles *BAC, CAD, DAB*, taken together in any manner, are greater than the remaining one.

If now the angles *BAC, CAD, DAB* are equal to one another, it is manifest that any two are greater than the remaining one.

But, if not, let *BAC* be greater,
and on the straight line *AB*, and at the point *A* on it, let the angle *BAE* be constructed, in the plane through *BA, AC*, equal to the angle *DAB*;
let *AE* be made equal to *AD*,
and let *BEC*, drawn across through the point *E*, cut the straight lines *AB, AC* at the points *B, C*;
let *DB, DC* be joined.

Now, since *DA* is equal to *AE*,
and *AB* is common,

two sides are equal to two sides;
and the angle *DAB* is equal to the angle *BAE*;
therefore the base *DB* is equal to the base *BE*. [I. 4]

And, since the two sides *BD, DC* are greater than *BC*, [I. 20]
and of these *DB* was proved equal to *BE*,
therefore the remainder *DC* is greater than the remainder *EC*.

Now, since *DA* is equal to *AE*,
and *AC* is common,
and the base *DC* is greater than the base *EC*,
therefore the angle *DAC* is greater than the angle *EAC*. [I. 25]

But the angle *DAB* was made equal to the angle *BAE*;
therefore the angles *DAB, DAC* are greater than the angle *BAC*.

Similarly we can prove that the remaining angles also, taken together two and two, are greater than the remaining one.

Therefore etc.

Q.E.D.

Proposition 21

Any solid angle is contained by plane angles less than four right angles.

Let the angle at *A* be a solid angle contained by the plane angles *BAC, CAD, DAB;*
I say that the angles *BAC, CAD, DAB* are less than four right angles.

For let points *B, C, D* be taken at random on the straight lines *AB, AC, AD* respectively,
and let *BC, CD, DB* be joined.

Now, since the solid angle at *B* is contained by the three plane angles *CBA, ABD, CBD,*
 any two are greater than the remaining one;

[XI. 20]

therefore the angles *CBA, ABD* are greater than the angle *CBD.*

For the same reason
 the angles *BCA, ACD* are also greater than the angle *BCD,*
 and the angles *CDA, ADB* are greater than the angle *CDB;*
therefore the six angles *CBA, ABD, BCA, ACD, CDA, ADB* are greater than the three angles *CBD, BCD, CDB.*

But the three angles *CBD, BDC, BCD* are equal to two right angles; [I. 32]
therefore the six angles *CBA, ABD, BCA, ACD, CDA, ADB* are greater than two right angles.

And, since the three angles of each of the triangles *ABC, ACD, ADB* are equal to two right angles,
therefore the nine angles of the three triangles, the angles *CBA, ACB, BAC, ACD, CDA, CAD, ADB, DBA, BAD* are equal to six right angles;
and of them the six angles *ABC, BCA, ACD, CDA, ADB, DBA* are greater than two right angles;
therefore the remaining three angles *BAC, CAD, DAB* containing the solid angle are less than four right angles.

Therefore etc.

Q.E.D.

Proposition 22

If there be three plane angles of which two, taken together in any manner, are greater than the remaining one, and they are contained by equal straight lines, it is possible to construct a triangle out of the straight lines joining the extremities of the equal straight lines.

Let there be three plane angles *ABC, DEF, GHK,* of which two, taken together in any manner, are greater than the remaining one, namely

the angles *ABC, DEF* greater than the angle *GHK,*
the angles *DEF, GHK* greater than the angle *ABC,*
and, further, the angles *GHK, ABC* greater than the angle *DEF;*
let the straight lines *AB, BC, DE, EF, GH, HK* be equal,
and let *AC, DF, GK* be joined;
I say that it is possible to construct a triangle out of straight lines equal to *AC, DF, GK,* that is, that any two of the straight lines *AC, DF, GK* are greater than the remaining one.

Now, if the angles *ABC, DEF, GHK* are equal to one another, it is manifest that, *AC, DF, GK* being equal also, it is possible to construct a triangle out of straight lines equal to *AC, DF, GK.*

But, if not, let them be unequal,
and on the straight line *HK,* and at the point *H* on it, let the angle *KHL* be constructed equal to the angle *ABC;*
let *HL* be made equal to one of the straight lines *AB, BC, DE, EF, GH, HK,*
and let *KL, GL* be joined.

Now, since the two sides *AB, BC* are equal to the two sides *KH, HL,*
and the angle at *B* is equal to the angle *KHL,*
> therefore the base *AC* is equal to the base *KL.* [I. 4]

And, since the angles *ABC, GHK* are greater than the angle *DEF,*
while the angle *ABC* is equal to the angle *KHL,*
> therefore the angle *GHL* is greater than the angle *DEF.*

And, since the two sides *GH, HL* are equal to the two sides *DE, EF,*
and the angle *GHL* is greater than the angle *DEF,*
> therefore the base *GL* is greater than the base *DF.* [I. 24]

But *GK, KL* are greater than *GL.*

Therefore *GK, KL* are much greater than *DF.*

But *KL* is equal to *AC;*
> therefore *AC, GK* are greater than the remaining straight line *DF.*

Similarly we can prove that
> *AC, DF* are greater than *GK,*
> and further, *DF, GK* are greater than *AC.*

Therefore it is possible to construct a triangle out of straight lines equal to *AC, DF, GK.*

<div align="right">Q.E.D.</div>

Proposition 23

To construct a solid angle out of three plane angles two of which, taken together in any manner, are greater than the remaining one: thus the three angles must be less than four right angles.

Let the angles *ABC, DEF, GHK* be the three given plane angles, and let two of these, taken together in any manner, be greater than the remaining one, while, further, the three are less than four right angles;

thus it is required to construct a solid angle out of angles equal to the angles *ABC, DEF, GHK*.

Let *AB, BC, DE, EF, GH, HK* be cut off equal to one another,
and let *AC, DF, GK* be joined;
it is therefore possible to construct a triangle out of straight lines equal to *AC, DF, GK*. **[XI. 22]**

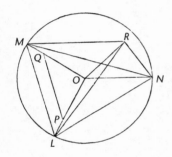

Let *LMN* be so constructed that *AC* is equal to *LM*, *DF* to *MN*, and further, *GK* to *NL*,
let the circle *LMN* be described about the triangle *LMN*,
let its centre be taken, and let it be *O*;
let *LO, MO, NO* be joined;
I say that *AB* is greater than *LO*.

For, if not, *AB* is either equal to *LO*, or less.

First, let it be equal.

Then, since *AB* is equal to *LO*,
while *AB* is equal to *BC*, and *OL* to *OM*,
the two sides *AB, BC* are equal to the two sides *LO, OM* respectively;
and, by hypothesis, the base *AC* is equal to the base *LM*;
therefore the angle *ABC* is equal to the angle *LOM*. **[I. 8]**

For the same reason
the angle *DEF* is also equal to the angle *MON*,
and further the angle *GHK* to the angle *NOL*;
therefore the three angles *ABC, DEF, GHK* are equal to the three angles *LOM, MON, NOL*.

But the three angles *LOM, MON, NOL* are equal to four right angles;
therefore the angles *ABC, DEF, GKH* are equal to four right angles.

But they are also, by hypothesis, less than four right angles:
which is absurd.

Therefore *AB* is not equal to *LO*.

I say next that neither is *AB* less than *LO*.

For, if possible, let it be so,
and let *OP* be made equal to *AB*, and *OQ* equal to *BC*,
and let *PQ* be joined.

Then, since *AB* is equal to *BC*,
OP is also equal to *OQ*,

so that the remainder *LP* is equal to *QM*.

Therefore *LM* is parallel to *PQ*, [VI. 2]

and *LMO* is equiangular with *PQO*; [I. 29]

therefore, as *OL* is to *LM*, so is *OP* to *PQ*; [VI. 4]

and alternately, as *LO* is to *OP*, so is *LM* to *PQ*. [V. 16]

But *LO* is greater than *OP*;

therefore *LM* is also greater than *PQ*.

But *LM* was made equal to *AC*;

therefore *AC* is also greater than *PQ*.

Since, then, the two sides *AB*, *BC* are equal to the two sides *PO*, *OQ*, and the base *AC* is greater than the base *PQ*,

therefore the angle *ABC* is greater than the angle *POQ*. [I. 25]

Similarly we can prove that

the angle *DEF* is also greater than the angle *MON*,

and the angle *GHK* greater than the angle *NOL*.

Therefore the three angles *ABC*, *DEF*, *GHK* are greater than the three angles *LOM*, *MON*, *NOL*.

But, by hypothesis, the angles *ABC*, *DEF*, *GHK* are less than four right angles;

therefore the angles *LOM*, *MON*, *NOL* are much less than four right angles.

But they are also equal to four right angles:

which is absurd.

Therefore *AB* is not less than *LO*.

And it was proved that neither is it equal;

therefore *AB* is greater than *LO*.

Let then *OR* be set up from the point *O* at right angles to the plane of the circle *LMN*, [XI. 12]

and let the square on *OR* be equal to that area by which the square on *AB* is greater than the square on *LO*; [Lemma]

let *RL*, *RM*, *RN* be joined.

Then, since *RO* is at right angles to the plane of the circle *LMN*,

therefore *RO* is also at right angles to each of the straight lines *LO*, *MO*, *NO*.

And, since *LO* is equal to *OM*,

while *OR* is common and at right angles,

therefore the base *RL* is equal to the base *RM*. [I. 4]

For the same reason

RN is also equal to each of the straight lines *RL*, *RM*;

therefore the three straight lines *RL*, *RM*, *RN* are equal to one another.

Next, since by hypothesis the square on *OR* is equal to that area by which the square on *AB* is greater than the square on *LO*,

therefore the square on *AB* is equal to the squares on *LO*, *OR*.

But the square on *LR* is equal to the squares on *LO*, *OR*, for the angle *LOR* is right;

[I.47]

therefore the square on *AB* is equal to the square on *RL*;

therefore *AB* is equal to *RL*.

But each of the straight lines *BC*, *DE*, *EF*, *GH*, *HK* is equal to *AB*,

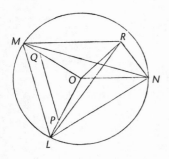

while each of the straight lines *RM, RN* is equal to *RL*; therefore each of the straight lines *AB, BC, DE, EF, GH, HK* is equal to each of the straight lines *RL, RM, RN*.

And, since the two sides *LR, RM* are equal to the two sides *AB, BC*,

and the base *LM* is by hypothesis equal to the base *AC*, therefore the angle *LRM* is equal to the angle *ABC*.

[I. 8]

For the same reason
the angle *MRN* is also equal to the angle *DEF*,
and the angle *LRN* to the angle *GHK*.

Therefore, out of the three plane angles *LRM, MRN, LRN*, which are equal to the three given angles *ABC, DEF, GHK*, the solid angle at *R* has been constructed, which is contained by the angles *LRM, MRN, LRN*.

Q.E.F.

LEMMA. But how it is possible to take the square on *OR* equal to that area by which the square on *AB* is greater than the square on *LO*, we can show as follows.

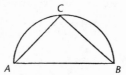

Let the straight lines *AB, LO* be set out,
and let *AB* be the greater;
let the semicircle *ABC* be described on *AB*,
and into the semicircle *ABC* let *AC* be fitted equal to the straight line *LO*, not being greater than the diameter *AB*; [IV. 1]
let *CB* be joined.

Since then the angle *ACB* is an angle in the semicircle *ACB*,
therefore the angle *ACB* is right. [III. 31]

Therefore the square on *AB* is equal to the squares on *AC, CB*. [I. 47]

Hence the square on *AB* is greater than the square on *AC* by the square on *CB*.

But *AC* is equal to *LO*.

Therefore the square on *AB* is greater than the square on *LO* by the square on *CB*.

If then we cut off *OR* equal to *BC*, the square on *AB* will be greater than the square on *LO* by the square on *OR*.

Q.E.F.

Proposition 24

If a solid be contained by parallel planes, the opposite planes in it are equal and parallelogrammic.

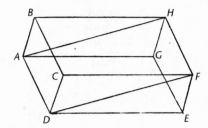

For let the solid *CDHG* be contained by the parallel planes *AC, GF, AH, DF, BF, AE*; I say that the opposite planes in it are equal and parallelogrammic.

For, since the two parallel planes *BG, CE* are cut by the plane *AC*,
their common sections are parallel. [XI. 16]

Therefore *AB* is parallel to *DC*.

Again, since the two parallel planes *BF, AE* are cut by the plane *AC*,
their common sections are parallel. [XI. 16]

Therefore *BC* is parallel to *AD*.

But *AB* was also proved parallel to *DC*;
therefore *AC* is a parallelogram.

Similarly we can prove that each of the planes *DF, FG, GB, BF, AE* is a parallelogram.

Let *AH, DF* be joined.

Then, since *AB* is parallel to *DC*, and *BH* to *CF*,
the two straight lines *AB, BH* which meet one another are parallel to the two straight lines *DC, CF* which meet one another, not in the same plane;
therefore they will contain equal angles; [XI. 10]
therefore the angle *ABH* is equal to the angle *DCF*.

And, since the two sides *AB, BH* are equal to the two sides *DC, CF*, [I. 34]
and the angle *ABH* is equal to the angle *DCF*,
therefore the base *AH* is equal to the base *DF*,
and the triangle *ABH* is equal to the triangle *DCF*. [I. 4]

And the parallelogram *BG* is double of the triangle *ABH*, and the parallelogram *CE* double of the triangle *DCF*; [I. 34]
therefore the parallelogram *BG* is equal to the parallelogram *CE*.

Similarly we can prove that
AC is also equal to *GF*,
and *AE* to *BF*.

Therefore etc.

Q.E.D.

Proposition 25

If a parallelepipedal solid be cut by a plane which is parallel to the opposite planes, then, as the base is to the base, so will the solid be to the solid.

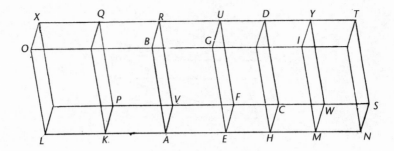

For let the parallelepipedal solid *ABCD* be cut by the plane *FG* which is parallel to the opposite planes *RA, DH*;
I say that, as the base *AEFV* is to the base *EHCF*, so is the solid *ABFU* to the solid *EGCD*.

For let *AH* be produced in each direction,
let any number of straight lines whatever, *AK, KL*, be made equal to *AE*,
and any number whatever, *HM, MN*, equal to *EH*;
and let the parallelograms *LP, KV, HW, MS* and the solids *LQ, KR, DM, MT* be completed.

Then, since the straight lines *LK, KA, AE* are equal to one another,
the parallelograms *LP, KV, AF* are also equal to one another,
KO, KB, AG are equal to one another,
and further, *LX, KQ, AR* are equal to one another, for they are opposite.

[XI. 24]

For the same reason
the parallelograms *EC, HW, MS* are also equal to one another,
HG, HI, IN are equal to one another,
and further, *DH, MY, NT* are equal to one another.

Therefore in the solids *LQ, KR, AU* three planes are equal to three planes.

But the three planes are equal to the three opposite;
therefore the three solids *LQ, KR, AU* are equal to one another.

[XI. Def. 10]

For the same reason
the three solids *ED, DM, MT* are also equal to one another.

Therefore, whatever multiple the base *LF* is of the base *AF*, the same multiple also is the solid *LU* of the solid *AU*.

For the same reason,
whatever multiple the base *NF* is of the base *FH*, the same multiple also is the solid *NU* of the solid *HU*.

And, if the base *LF* is equal to the base *NF*, the solid *LU* is also equal to the solid *NU*;
if the base *LF* exceeds the base *NF*, the solid *LU* also exceeds the solid *NU*;
and, if one falls short, the other falls short.

Therefore, there being four magnitudes, the two bases *AF, FH*, and the two solids *AU, UH*,
equimultiples have been taken of the base *AF* and the solid *AU*, namely the base *LF* and the solid *LU*,
and equimultiples of the base *HF* and the solid *HU*, namely the base *NF* and the solid *NU*,

and it has been proved that, if the base *LF* exceeds the base *FN*, the solid *LU* also exceeds the solid *NU*,
if the bases are equal, the solids are equal,
and if the base falls short, the solid falls short,

Therefore, as the base *AF* is to the base *FH*, so is the solid *AU* to the solid *UH*.

[V. Def. 5]

Q.E.D.

Proposition 26

On a given straight line, and at a given point on it, to construct a solid angle equal to a given solid angle.

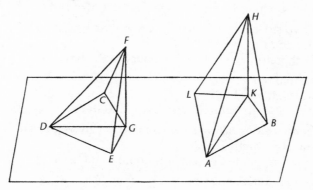

Let *AB* be the given straight line, *A* the given point on it, and the angle at *D*, contained by the angles *EDC, EDF, FDC*, the given solid angle;
thus it is required to construct on the straight line *AB*, and at the point *A* on it, a solid angle equal to the solid angle at *D*.

For let a point *F* be taken at random on *DF*,
let *FG* be drawn from *F* perpendicular to the plane through *ED, DC*, and let it meet the plane at *G*, [XI. 11]
let *DG* be joined,
let there be constructed on the straight line *AB* and at the point *A* on it the angle *BAL* equal to the angle *EDC*, and the angle *BAK* equal to the angle *EDG*, [I. 23]
let *AK* be made equal to *DG*,
let *KH* be set up from the point *K* at right angles to the plane through *BA, AL*,

[XI. 12]

let *KH* be made equal to *GF*,
and let *HA* be joined;
I say that the solid angle at *A*, contained by the angles *BAL, BAH, HAL* is equal to the solid angle at *D* contained by the angles *EDC, EDF, FDC*.

For let *AB, DE* be cut off equal to one another,
and let *HB, KB, FE, GE* be joined.

Then, since *FG* is at right angles to the plane of reference, it will also make right angles with all the straight lines which meet it and are in the plane of reference;

[XI. Def. 3]

therefore each of the angles *FGD, FGE* is right.

391

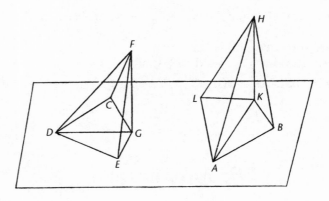

For the same reason

> each of the angles *HKA, HKB* is also right.

And, since the two sides *KA, AB* are equal to the two sides *GD, DE* respectively, and they contain equal angles,

> therefore the base *KB* is equal to the base *GE*. [I. 4]

But *KH* is also equal to *GF*,
and they contain right angles;

> therefore *HB* is also equal to *FE*. [I. 4]

Again, since the two sides *AK, KH* are equal to the two sides *DG, GF*,
and they contain right angles,

> therefore the base *AH* is equal to the base *FD*. [I. 4]

But *AB* is also equal to *DE*;

> therefore the two sides *HA, AB* are equal to the two sides *DF, DE*.

And the base *HB* is equal to the base *FE*;

> therefore the angle *BAH* is equal to the angle *EDF*. [I. 8]

For the same reason

> the angle *HAL* is also equal to the angle *FDC*.

And the angle *BAL* is also equal to the angle *EDC*.

Therefore on the straight line *AB*, and at the point *A* on it, a solid angle has been constructed equal to the given solid angle at *D*.

> Q.E.F.

Proposition 27

On a given straight line to describe a parallelepipedal solid similar and similarly situated to a given parallelepipedal solid.

Let *AB* be the given straight line and *CD* the given parallelepipedal solid; thus it is required to describe on the given straight line *AB* a parallelepipedal solid similar and similarly situated to the given parallelepipedal solid *CD*.

For on the straight line *AB* and at the point *A* on it let the solid angle, contained by the angles *BAH, HAK, KAB*, be constructed equal to the solid

angle at *C*, so that the angle *BAH* is equal to the angle *ECF*, the angle *BAK* equal to the angle *EGG*, and the angle *KAH* to the angle *GCF*;
and let it be contrived that,
as *EC* is to *CG*, so is *BA* to *AK*,
and, as *GC* is to *CF*, so is *KA* to *AH*. [VI. 12]

Therefore also, *ex aequali*,

<div align="center">as EC is to CF, so is BA to AH. [V. 22]</div>

Let the parallelogram *HB* and the solid *AL* be completed.

Now since, as *EC* is to *CG*, so is *BA* to *AK*,
and the sides about the equal angles *ECG*, *BAK* are thus proportional,
therefore the parallelogram *GE* is similar to the parallelogram *KB*.

For the same reason
the parallelogram *KH* is also similar to the parallelogram *GF*,
and further, *FE* to *HB*;
therefore three parallelograms of the solid *CD* are similar to three parallelograms of the solid *AL*.

But the former three are both equal and similar to the three opposite parallelograms,
and the latter three are both equal and similar to the three opposite parallelograms;
therefore the whole solid *CD* is similar to the whole solid *AL*. [XI. Def. 9]

Therefore on the given straight line *AB* there has been described *AL* similar and similarly situated to the given parallelepipedal solid *CD*.

<div align="right">Q.E.F.</div>

Proposition 28

If a parallelepipedal solid be cut by a plane through the diagonals of the opposite planes, the solid will be bisected by the plane.

For let the parallelepipedal solid *AB* be cut by the plane *CDEF* through the diagonals *CF, DE* of opposite planes;
I say that the solid *AB* will be bisected by the plane *CDEF*.

For, since the triangle *CGF* is equal to the triangle *CFB*,
<div align="center">[I. 34]</div>
and *ADE* to *DEH*,
while the parallelogram *CA* is also equal to the parallelogram *EB*, for they are opposite,
and *GE* to *CH*,
therefore the prism contained by the two triangles *CGF*,

ADE and the three parallelograms *GE, AC, CE* is also equal to the prism contained by the two triangles *CFB, DEH* and the three parallelograms *CH, BE, CE*;

for they are contained by planes equal both in multitude and in magnitude. **[XI. Def. 10]**

Hence the whole solid *AB* is bisected by the plane *CDEF*.

Q.E.D.

Proposition 29

Parallelepipedal solids which are on the same base and of the same height, and in which the extremities of the sides which stand up are on the same straight lines, are equal to one another.

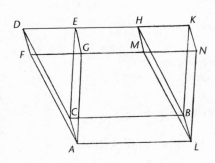

Let *CM, CN* be parallelepipedal solids on the same base *AB* and of the same height,

and let the extremities of their sides which stand up, namely *AG, AF, LM, LN, CD, CE, BH, BK,* be on the same straight lines *FN, DK*;

I say that the solid *CM* is equal to the solid *CN*.

For, since each of the figures *CH, CK* is a parallelogram, *CB* is equal to each of the straight lines *DH, EK*, **[I. 34]**

 hence *DH* is also equal to *EK*.

Let *EH* be subtracted from each;

 therefore the remainder *DE* is equal to the remainder *HK*.

Hence the triangle *DCE* is also equal to the triangle *HBK*, **[I. 8, 4]**

 and the parallelogram *DG* to the parallelogram *HN*. **[I. 36]**

For the same reason

 the triangle *AFG* is also equal to the triangle *MLN*.

But the parallelogram *CF* is equal to the parallelogram *BM*, and *CG* to *BN*, for they are opposite;

therefore the prism contained by the two triangles *AFG, DCE* and the three parallelograms *AD, DG, CG* is equal to the prism contained by the two triangles *MLN, HBK* and the three parallelograms *BM, HN, BN*.

Let there be added to each the solid of which the parallelogram *AB* is the base and *GEHM* its opposite;

therefore the whole parallelepipedal solid *CM* is equal to the whole parallelepipedal solid *CN*.

Therefore etc.

Q.E.D.

Proposition 30

Parallelepipedal solids which are on the same base and of the same height, and in which the extremities of the sides which stand up are not on the same straight lines, are equal to one another.

Let *CM*, *CN* be parallelepipedal solids on the same base *AB* and of the same height, and let the extremities of their sides which stand up, namely *AF*, *AG*, *LM*, *LN*, *CD*, *CE*, *BH*, *BK*, not be on the same straight lines; I say that the solid *CM* is equal to the solid *CN*.

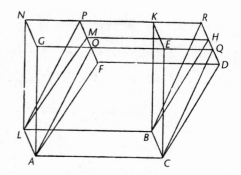

For let *NK*, *DH* be produced and meet one another at *R*, and further, let *FM*, *GE* be produced to *P*, *Q*; let *AO*, *LP*, *CQ*, *BR* be joined.

Then the solid *CM*, of which the parallelogram *ACBL* is the base, and *FDHM* its opposite, is equal to the solid *CP*, of which the parallelogram *ACBL* is the base, and *OQRP* its opposite; for they are on the same base *ACBL* and of the same height, and the extremities of their sides which stand up, namely *AF*, *AO*, *LM*, *LP*, *CD*, *CQ*, *BH*, *BR*, are on the same straight lines *FP*, *DR*. [XI. 29]

But the solid *CP*, of which the parallelogram *ACBL* is the base, and *OQRP* its opposite, is equal to the solid *CN*, of which the parallelogram *ACBL* is the base and *GEKN* its opposite; for they are again on the same base *ACBL* and of the same height, and the extremities of their sides which stand up, namely *AG*, *AO*, *CE*, *CQ*, *LN*, *LP*, *BK*, *BR*, are on the same straight lines *GQ*, *NR*.

Hence the solid *CM* is also equal to the solid *CN*.

Therefore etc.

Q.E.D.

Proposition 31

Parallelepipedal solids which are on equal bases and of the same height are equal to one another.

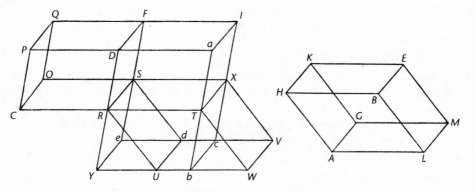

Let the parallelepipedal solids *AE*, *CF*, of the same height, be on equal bases *AB*, *CD*.

I say that the solid *AE* is equal to the solid *CF*.

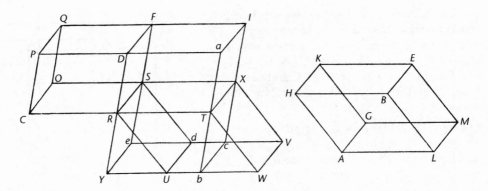

First, let the sides which stand up, *HK, BE, AG, LM, PQ, DF, CO, RS,* be at right angles to the bases *AB, CD;*
let the straight line *RT* be produced in a straight line with *CR;*
on the straight line *RT,* and at the point *R* on it, let the angle *TRU* be constructed equal to the angle *ALB,* [I. 23]
let *RT* be made equal to *AL,* and *RU* equal to *LB,*
and let the base *RW* and the solid *XU* be completed.

Now, since the two sides *TR, RU* are equal to the two sides *AL, LB,*
and they contain equal angles,
 therefore the parallelogram *RW* is equal and similar to the parallelogram *HL.*

Since again *AL* is equal to *RT,* and *LM* to *RS,*
and they contain right angles,
 therefore the parallelogram *RX* is equal and similar to the parallelogram *AM.*

For the same reason
 LE is also equal and similar to *SU;*
therefore three parallelograms of the solid *AE* are equal and similar to three parallelograms of the solid *XU.*

But the former three are equal and similar to the three opposite, and the latter three to the three opposite; [XI. 24]
therefore the whole parallelepipedal solid *AE* is equal to the whole parallelpipedal solid *XU.* [XI. Def. 10]

Let *DR, WU* be drawn through and meet one another at *Y,*
let *aTb* be drawn through *T* parallel to *DY,*
let *PD* be produced to *a,*
and let the solids *YX, RI* be completed.

Then the solid *XY,* of which the parallelogram *RX* is the base and *Yc* its opposite, is equal to the solid *XU* of which the parallelogram *RX* is the base and *UV* its opposite,
for they are on the same base *RX* and of the same height, and the extremities of their sides which stand up, namely *RY, RU, Tb, TW, Se, Sd, Xc, XV,* are on the same straight lines *YW, eV.* [XI. 29]

But the solid *XU* is equal to *AE;*
 therefore the solid *XY* is also equal to the solid *AE.*

And, since the parallelogram *RUWT* is equal to the parallelogram *YT,*
for they are on the same base *RT* and in the same parallels *RT, YW,* [I. 35]

while *RUWT* is equal to *CD*, since it is also equal to *AB*,
therefore the parallelogram *YT* is also equal to *CD*.

But *DT* is another parallelogram;
therefore, as the base *CD* is to *DT*, so is *YT* to *DT*. [V. 7]

And, since the parallelepipedal solid *CI* has been cut by the plane *RF* which is parallel to opposite planes,
as the base *CD* is to the base *DT*, so is the solid *CF* to the solid *RI*. [XI. 25]

For the same reason,
since the parallelepipedal solid *YI* has been cut by the plane *RX* which is parallel to opposite planes,
as the base *YT* is to the base *TD*, so is the solid *YX* to the solid *RI*. [XI. 25]

But, as the base *CD* is to *DT*, so is *YT* to *DT*;
therefore also, as the solid *CF* is to the solid *RI*, so is the solid *YX* to *RI*.
[V. 11]

Therefore each of the solids *CF*, *YX* has to *RI* the same ratio;
therefore the solid *CF* is equal to the solid *YX*. [V. 9]

But *YX* was proved equal to *AE*;
therefore *AE* is also equal to *CF*.

Next, let the sides standing up, *AG*, *HK*, *BE*, *LM*, *CN*, *PQ*, *DF*, *RS*, not be at right angles to the bases *AB*, *CD*;

I say again that the solid *AE* is equal to the solid *CF*.

For from the points *K*, *E*, *G*, *M*, *Q*, *F*, *N*, *S* let *KO*, *ET*, *GU*, *MV*, *QW*, *FX*, *NY*, *SI* be drawn perpendicular to the plane of reference, and let them meet the plane at the points *O*, *T*, *U*, *V*, *W*, *X*, *Y*, *I*,
and let *OT*, *OU*, *UV*, *TV*, *WX*, *WY*, *YI*, *IX* be joined.

Then the solid *KV* is equal to the solid *QI*,
for they are on the equal bases *KM*, *QS* and of the same height, and their sides which stand up are at right angles to their bases. [First part of this Prop.]

But the solid *KV* is equal to the solid *AE*,
and *QI* to *CF*;
for they are on the same base and of the same height, while the extremities of their sides which stand up are not on the same straight lines. [XI. 30]

Therefore the solid *AE* is also equal to the solid *CF*.

Therefore etc.

Q.E.D.

Proposition 32

Parallelepipedal solids which are of the same height are to one another as their bases.

Let *AB, CD* be parallelepipedal solids of the same height;
I say that the parallelepipedal solids *AB, CD* are to one another as their bases, that is, that, as the base *AE* is to the base *CF*, so is the solid *AB* to the solid *CD*.

For let *FH* equal to *AE* be applied to *FG*, **[I. 45]**
and, on *FH* as base, and with the same height as that of *CD*, let the parallelepipedal solid *GK* be completed.

Then the solid *AB* is equal to the solid *GK*;
for they are on equal bases *AE, FH* and of the same height. **[XI. 31]**

And, since the parallelepipedal solid *CK* is cut by the plane *DG* which is parallel to opposite planes,
 therefore, as the base *CF* is to the base *FH*, so is the solid *CD* to the solid *DH*.
 [XI. 25]

But the base *FH* is equal to the base *AE*,
and the solid *GK* to the solid *AB*;
therefore also, as the base *AE* is to the base *CF*, so is the solid *AB* to the solid *CD*.

Therefore etc.

 Q.E.D.

Proposition 33

Similar parallelepipedal solids are to one another in the triplicate ratio of their corresponding sides.

Let *AB, CD* be similar parallelepipedal solids,
and let *AE* be the side corresponding to *CF*;
I say that the solid *AB* has to the solid *CD* the ratio triplicate of that which *AE* has to *CF*.

For let *EK, EL, EM* be produced in a straight line with *AE, GE, HE*,
let *EK* be made equal to *CF*, *EL* equal to *FN*, and further, *EM* equal to *FR*,
and let the parallelogram *KL* and the solid *KP* be completed.

Now, since the two sides *KE, EL* are equal to the two sides *CF, FN*,
while the angle *KEL* is also equal to the angle *CFN*, inasmuch as the angle *AEG* is also equal to the angle *CFN* because of the similarity of the solids *AB, CD*,
 therefore the parallelogram *KL* is equal ⟨and similar⟩ to the parallelogram *CN*.

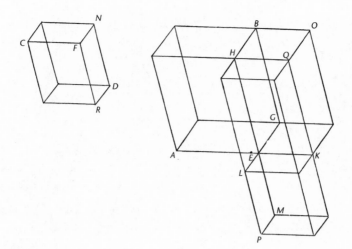

For the same reason

the parallelogram *KM* is also equal and similar to *CR*,

and further, *EP* to *DF*;

therefore three parallelograms of the solid *KP* are equal and similar to three parallelograms of the solid *CD*.

But the former three parallelograms are equal and similar to their opposites, and the latter three to their opposites; **[XI. 24]**

therefore the whole solid *KP* is equal and similar to the whole solid *CD*.

[XI. Def. 10]

Let the parallelogram *GK* be completed,

and on the parallelograms *GK*, *KL* as bases, and with the same height as that of *AB*, let the solids *EO*, *LQ* be completed.

Then since, owing to the similarity of the solids *AB*, *CD*,

as *AE* is to *CF*, so is *EG* to *FN*, and *EH* to *FR*,

while *CF* is equal to *EK*, *FN* to *EL*, and *FR* to *EM*,

therefore, as *AE* is to *EK*, so is *GE* to *EL*, and *HE* to *EM*.

But, as *AE* is to *EK*, so is *AG* to the parallelogram *GK*,

as *GE* is to *EL*, so is *GK* to *KL*,

and, as *HE* is to *EM*, so is *QE* to *KM*; **[VI. 1]**

therefore also, as the parallelogram *AG* is to *GK*, so is *GK* to *KL*, and *QE* to *KM*.

But, as *AG* is to *GK*, so is the solid *AB* to the solid *EO*,

as *GK* is to *KL*, so is the solid *OE* to the solid *QL*,

and, as *QE* is to *KM*, so is the solid *QL* to the solid *KP*; **[XI. 32]**

therefore also, as the solid *AB* is to *EO*, so is *EO* to *QL*, and *QL* to *KP*.

But, if four magnitudes be continuously proportional, the first has to the fourth the ratio triplicate of that which it has to the second; **[V. Def. 10]**

therefore the solid *AB* has to *KP* the ratio triplicate of that which *AB* has to *EO*.

But, as *AB* is to *EO*, so is the parallelogram *AG* to *GK*, and the straight line *AE* to *EK*; **[VI. 1]**

hence the solid *AB* has also to *KP* the ratio triplicate of that which *AE* has to *EK*.

But the solid *KP* is equal to the solid *CD*,

and the straight line *EK* to *CF*;

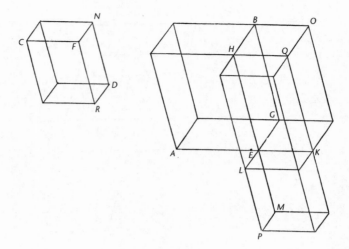

therefore the solid *AB* has also to the solid *CD* the ratio triplicate of that which the corresponding side of it, *AE*, has to the corresponding side *CF*.

Therefore etc.

<div align="right">Q.E.D.</div>

PORISM. From this it is manifest that, if four straight lines be [continuously] proportional, as the first is to the fourth, so will a parallelepipedal solid on the first be to the similar and similarly described parallelepipedal solid on the second, inasmuch as the first has to the fourth the ratio triplicate of that which it has to the second.

Proposition 34

In equal parallelepipedal solids the bases are reciprocally proportional to the heights; and those parallelepipedal solids in which the bases are reciprocally proportional to the heights are equal.

Let *AB*, *CD* be equal parallelepipedal solids;
I say that in the parallelepipedal solids *AB*, *CD* the bases are reciprocally proportional to the heights,

that is, as the base *EH* is to the base *NQ*, so is the height of the solid *CD* to the height of the solid *AB*.

First, let the sides which stand up, namely *AG, EF, LB, HK, CM, NO, PD, QR*, be at right angles to their bases;
I say that, as the base *EH* is to the base *NQ*, so is *CM* to *AG*.

If now the base *EH* is equal to the base *NQ*,
while the solid *AB* is also equal to the solid *CD*,
<div align="center">CM will also be equal to AG.</div>

For parallelepipedal solids of the same height are to one another as the bases;
<div align="right">[XI. 32]</div>
and, as the base *EH* is to *NQ*, so will *CM* be to *AG*,
and it is manifest that in the parallelepipedal solids *AB, CD* the bases are reciprocally proportional to the heights.

Next, let the base *EH* not be equal to the base *NQ*,
but let *EH* be greater.

Now the solid *AB* is equal to the solid *CD*;
<div align="center">therefore CM is also greater than AG.</div>
Let then *CT* be made equal to *AG*,
and let the parallelepipedal solid *VC* be completed on *NQ* as base and with *CT* as height.

Now, since the solid *AB* is equal to the solid *CD*,
and *CV* is outside them,
while equals have to the same the same ratio, <div align="right">[V. 7]</div>
therefore, as the solid *AB* is to the solid *CV*, so is the solid *CD* to the solid *CV*.

But, as the solid *AB* is to the solid *CV*, so is the base *EH* to the base *NQ*,
for the solids *AB, CV* are of equal height; <div align="right">[XI. 32]</div>
and, as the solid *CD* is to the solid *CV*, so is the base *MQ* to the base *TQ* [XI. 25]
and *CM* to *CT*; <div align="right">[VI. 1]</div>
<div align="center">therefore also, as the base EH is to the base NQ, so is MC to CT.</div>

But *CT* is equal to *AG*;
<div align="center">therefore also, as the base EH is to the base NQ, so is MC to AG.</div>

Therefore in the parallelepipedal solids *AB, CD* the bases are reciprocally proportional to the heights.

Again, in the parallelepipedal solids *AB, CD* let the bases be reciprocally proportional to the heights, that is, as the base *EH* is to the base *NQ*, so let the height of the solid *CD* be to the height of the solid *AB*;
I say that the solid *AB* is equal to the solid *CD*.

Let the sides which stand up be again at right angles to the bases.

Now, if the base *EH* is equal to the base *NQ*,
and, as the base *EH* is to the base *NQ*, so is the height of the solid *CD* to the height of the solid *AB*,
 therefore the height of the solid *CD* is also equal to the height of the solid *AB*.

But parallelepipedal solids on equal bases and of the same height are equal to one another; [XI. 31]
 therefore the solid *AB* is equal to the solid *CD*.

Next, let the base *EH* not be equal to the base *NQ*,
but let *EH* be greater;
therefore the height of the solid *CD* is also greater than the height of the solid *AB*,
 that is, *CM* is greater than *AG*.

Let *CT* be again made equal to *AG*,
and let the solid *CV* be similarly completed.

Since, as the base *EH* is to the base *NQ*, so is *MC* to *AG*,
while *AG* is equal to *CT*,
 therefore, as the base *EH* is to the base *NQ*, so is *CM* to *CT*.

But, as the base *EH* is to the base *NQ*, so is the solid *AB* to the solid *CV*,
for the solids *AB, CV* are of equal height; [XI. 32]
and, as *CM* is to *CT*, so is the base *MQ* to the base *QT* [VI. 1]
and the solid *CD* to the solid *CV*. [XI. 25]

Therefore also, as the solid *AB* is to the solid *CV*, so is the solid *CD* to the solid *CV*;
 therefore each of the solids *AB, CD* has to *CV* the same ratio.

Therefore the solid *AB* is equal to the solid *CD*. [V. 9]

Now let the sides which stand up, *FE, BL, GA, HK, ON, DP, MC, RQ*, not be at right angles to their bases;
let perpendiculars be drawn from the points *F, G, B, K, O, M, D, R* to the planes through *EH, NQ*,
and let them meet the planes at *S, T, U, V, W, X, Y, a,*

and let the solids *FV, Oa* be completed;

I say that, in this case too, if the solids *AB, CD* are equal, the bases are reciprocally proportional to the heights, that is, as the base *EH* is to the base *NQ*, so is the height of the solid *CD* to the height of the solid *AB*.

Since the solid *AB* is equal to the solid *CD*,
while *AB* is equal to *BT*,
for they are on the same base *FK* and of the same height; [XI. 29, 30]
and the solid *CD* is equal to *DX*,
for they are again on the same base *RO* and of the same height; [XI. 29, 30]
 therefore the solid *BT* is also equal to the solid *DX*.

Therefore, as the base *FK* is to the base *OR*, so is the height of the solid *DX* to the height of the solid *BT*. [Part I.]

But the base *FK* is equal to the base *EH*,
and the base *OR* to the base *NQ*;
therefore, as the base *EH* is to the base *NQ*, so is the height of the solid *DX* to the height of the solid *BT*.

But the solids *DX, BT* and the solids *DC, BA* have the same heights respectively;
therefore, as the base *EH* is to the base *NQ*, so is the height of the solid *DC* to the height of the solid *AB*.

Therefore in the parallelepipedal solids *AB, CD* the bases are reciprocally proportional to the heights.

Again, in the parallelepipedal solids *AB, CD* let the bases be reciprocally proportional to the heights,
that is, as the base *EH* is to the base *NQ*, so let the height of the solid *CD* be to the height of the solid *AB*;
I say that the solid *AB* is equal to the solid *CD*.

For, with the same construction,
since, as the base *EH* is to the base *NQ*, so is the height of the solid *CD* to the height of the solid *AB*,
while the base *EH* is equal to the base *FK*,
and *NQ* to *OR*,
therefore, as the base *FK* is to the base *OR*, so is the height of the solid *CD* to the height of the solid *AB*.

But the solids *AB, CD* and *BT, DX* have the same heights respectively;
therefore, as the base *FK* is to the base *OR*, so is the height of the solid *DX* to the height of the solid *BT*.

Therefore in the parallelepipedal solids *BT, DX* the bases are reciprocally proportional to the heights;
 therefore the solid *BT* is equal to the solid *DX*. [Part I.]

But *BT* is equal to *BA*,
for they are on the same base *FK* and of the same height; [XI. 29, 30]
and the solid *DX* is equal to the solid *DC* [XI. 29, 30]

Therefore the solid *AB* is also equal to the solid *CD*.

 Q.E.D.

Proposition 35

If there be two equal plane angles, and on their vertices there be set up elevated straight lines containing equal angles with the original straight lines respectively, if on the elevated straight lines points be taken at random and perpendiculars be drawn from them to the planes in which the original angles are, and if from the points so arising in the planes straight lines be joined to the vertices of the original angles, they will contain, with the elevated straight lines, equal angles.

Let the angles *BAC, EDF* be two equal rectilineal angles, and from the points *A, D* let the elevated straight lines *AG, DM* be set up containing, with the original straight lines, equal angles respectively, namely, the angle *MDE* to the angle *GAB* and the angle *MDF* to the angle *GAC*,

let points *G, M* be taken at random on *AG, DM*,

let *GL, MN* be drawn from the points *G, M* perpendicular to the planes through *BA, AC* and *ED, DF*, and let them meet the planes at *L, N*,

and let *LA, ND* be joined;

I say that the angle *GAL* is equal to the angle *MDN*.

Let *AH* be made equal to *DM*,

and let *HK* be drawn through the point *H* parallel to *GL*.

But *GL* is perpendicular to the plane through *BA, AC*;

 therefore *HK* is also perpendicular to the plane through *BA, AC*. **[XI. 8]**

From the points *K, N* let *KC, NF, KB, NE* be drawn perpendicular to the straight lines *AC, DF, AB, DE*,

and let *HC, CB, MF, FE* be joined.

Since the square on *HA* is equal to the squares on *HK, KA*,

and the squares on *KC, CA* are equal to the square on *KA*, **[I. 47]**

 therefore the square on *HA* is also equal to the squares on *HK, KC, CA*.

But the square on *HC* is equal to the squares on *HK, KC*; **[I. 47]**

 therefore the square on *HA* is equal to the squares on *HC, CA*.

 Therefore the angle *HCA* is right. **[I. 48]**

For the same reason

 the angle *DFM* is also right.

Therefore the angle *ACH* is equal to the angle *DFM*.

But the angle *HAC* is also equal to the angle *MDF*.

Therefore *MDF, HAC* are two triangles which have two angles equal to two angles respectively, and one side equal to one side, namely, that subtending one

of the equal angles, that is, HA equal to MD;
therefore they will also have the remaining sides equal to the remaining sides
respectively. [I. 26]

<div align="center">Therefore AC is equal to DF.</div>

Similarly we can prove that AB is also equal to DE.

Since then AC is equal to DF, and AB to DE,
<div align="center">the two sides CA, AB are equal to the two sides FD, DE.</div>

But the angle CAB is also equal to the angle FDE;
therefore the base BC is equal to the base EF, the triangle to the triangle, and the
remaining angles to the remaining angles; [I. 4]
<div align="center">therefore the angle ACB is equal to the angle DFE.</div>

But the right angle ACK is also equal to the right angle DFN;
 therefore the remaining angle BCK is also equal to the remaining angle EFN.

For the same reason
<div align="center">the angle CBK is also equal to the angle FEN.</div>

Therefore BCK, EFN are two triangles which have two angles equal to two
angles respectively, and one side equal to one side, namely, that adjacent to the
equal angles, that is, BC equal to EF;
 therefore they will also have the remaining sides equal to the remaining sides.
[I. 26]

<div align="center">Therefore CK is equal to FN.</div>

But AC is also equal to DF;
 therefore the two sides AC, CK are equal to the two sides DF, FN;
and they contain right angles.

<div align="center">Therefore the base AK is equal to the base DN. [I. 4]</div>

And, since AH is equal to DM,
<div align="center">the square on AH is also equal to the square on DM.</div>

But the squares on AK, KH are equal to the square on AH,
for the angle AKH is right; [I. 47]
and the squares on DN, NM are equal to the square on DM,
for the angle DNM is right; [I. 47]
 therefore the squares on AK, KH are equal to the squares on DN, NM;
and of these the square on AK is equal to the square on DN;
 therefore the remaining square on KH is equal to the square on NM;
<div align="center">therefore HK is equal to MN.</div>

And, since the two sides HA, AK are equal to the two sides MD, DN respectively,
and the base HK was proved equal to the base MN,
<div align="center">therefore the angle HAK is equal to the angle MDN. [I. 8]</div>

Therefore etc.

<div align="right">Q.E.D.</div>

PORISM. From this it is manifest that, if there be two equal plane angles, and if
there be set up on them elevated straight lines which are equal and contain
equal angles with the original straight lines respectively, the perpendiculars
drawn from their extremities to the planes in which are the original angles are
equal to one another.

Proposition 36

If three straight lines be proportional, the parallelepipedal solid formed out of the three is equal to the parallelepipedal solid on the mean which is equilateral, but equiangular with the aforesaid solid.

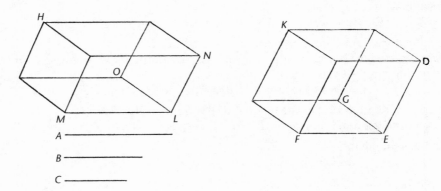

Let *A, B, C* be three straight lines in proportion, so that, as *A* is to *B*, so is *B* to *C*; I say that the solid formed out of *A, B, C* is equal to the solid on *B* which is equilateral, but equiangular with the aforesaid solid.

Let there be set out the solid angle at *E* contained by the angles *DEG, GEF, FED*, let each of the straight lines *DE, GE, EF* be made equal to *B*, and let the parallelepipedal solid *EK* be completed, let *LM* be made equal to *A*, and on the straight line *LM*, and at the point *L* on it, let there be constructed a solid angle equal to the solid angle at *E*, namely that contained by *NLO, OLM, MLN*; let *LO* be made equal to *B*, and *LN* equal to *C*.

Now, since, as *A* is to *B*, so is *B* to *C*, while *A* is equal to *LM*, *B* to each of the straight lines *LO, ED*, and *C* to *LN*, therefore, as *LM* is to *EF*, so is *DE* to *LN*.

Thus the sides about the equal angles *NLM, DEF* are reciprocally proportional; therefore the parallelogram *MN* is equal to the parallelogram *DF*. [VI. 14]

And, since the angles *DEF, NLM* are two plane rectilineal angles, and on them the elevated straight lines *LO, EG* are set up which are equal to one another and contain equal angles with the original straight lines respectively, therefore the perpendiculars drawn from the points *G, O* to the planes through *NL, LM* and *DE, EF* are equal to one another; [XI. 35, Por.] hence the solids *LH, EK* are of the same height.

But parallelepipedal solids on equal bases and of the same height are equal to one another; [XI. 31] therefore the solid *HL* is equal to the solid *EK*.

And *LH* is the solid formed out of *A, B, C*, and *EK* the solid on *B*; therefore the parallelepipedal solid formed out of *A, B, C* is equal to the solid on *B* which is equilateral, but equiangular with the aforesaid solid.

 Q.E.D.

Proposition 37

If four straight lines be proportional, the parallelepipedal solids on them which are similar and similarly described will also be proportional; and, if the parallelepipedal solids on them which are similar and similarly described be proportional, the straight lines will themselves also be proportional.

Let *AB*, *CD*, *EF*, *GH* be four straight lines in proportion, so that, as *AB* is to *CD*, so is *EF* to *GH*;
and let there be described on *AB*, *CD*, *EF*, *GH* the similar and similarly situated parallelepipedal solids *KA*, *LC*, *ME*, *NG*;
I say that, as *KA* is to *LC*, so is *ME* to *NG*.

For, since the parallelepipedal solid *KA* is similar to *LC*,
 therefore *KA* has to *LC* the ratio triplicate of that which *AB* has to *CD*.

<div align="right">[XI. 33]</div>

For the same reason
 ME also has to *NG* the ratio triplicate of that which *EF* has to *GH*. [XI. 33]

And, as *AB* is to *CD*, so is *EF* to *GH*.

 Therefore also, as *AK* is to *LC*, so is *ME* to *NG*.

Next, as the solid *AK* is to the solid *LC*, so let the solid *ME* be to the solid *NG*;
I say that, as the straight line *AB* is to *CD*, so is *EF* to *GH*.

For since, again, *KA* has to *LC* the ratio triplicate of that which *AB* has to *CD*,

<div align="right">[XI. 33]</div>

and *ME* also has to *NG* the ratio triplicate of that which *EF* has to *GH*, [XI. 33]
and, as *KA* is to *LC*, so is *ME* to *NG*,
 therefore also, as *AB* is to *CD*, so is *EF* to *GH*.

Therefore etc.

<div align="right">Q.E.D.</div>

Proposition 38

If the sides of the opposite planes of a cube be bisected, and planes be carried through the points of section, the common section of the planes and the diameter of the cube bisect one another.

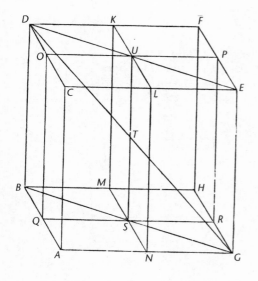

For let the sides of the opposite planes *CF, AH* of the cube *AF* be bisected at the points *K, L, M, N, O, Q, P, R*, and through the points of section let the planes *KN, OR* be carried;

let *US* be the common section of the planes, and *DG* the diameter of the cube *AF.*

I say that *UT* is equal to *TS*, and *DT* to *TG.*

For let *DU, UE, BS, SG* be joined.

Then, since *DO* is parallel to *PE*, the alternate angles *DOU, UPE* are equal to one another. [I. 29]

And, since *DO* is equal to *PE*, and *OU* to *UP*, and they contain equal angles, therefore the base *DU* is equal to the base *UE*, the triangle *DOU* is equal to the triangle *PUE*, and the remaining angles are equal to the remaining angles; [I. 4] therefore the angle *OUD* is equal to the angle *PUE.*

For this reason *DUE* is a straight line. [I. 14]

For the same reason, *BSG* is also a straight line, and *BS* is equal to *SG.*

Now, since *CA* is equal and parallel to *DB*, while *CA* is also equal and parallel to *EG*, therefore *DB* is also equal and parallel to *EG.* [XI. 9]

And the straight lines *DE, BG* join their extremities; therefore *DE* is parallel to *BG.* [I. 33]

Therefore the angle *EDT* is equal to the angle *BGT*, for they are alternate; [I. 29] and the angle *DTU* is equal to the angle *GTS.* [I. 15]

Therefore *DTU, GTS* are two triangles which have two angles equal to two angles, and one side equal to one side, namely that subtending one of the equal angles, that is, *DU* equal to *GS*, for they are the halves of *DE, BG*; therefore they will also have the remaining sides equal to the remaining sides. [I. 26]

Therefore *DT* is equal to *TG*, and *UT* to *TS.*

Therefore etc.

Q.E.D.

Proposition 39

If there be two prisms of equal height, and one have a parallelogram as base and the other a triangle, and if the parallelogram be double of the triangle, the prisms will be equal.

Let *ABCDEF, GHKLMN* be two prisms of equal height,
let one have the parallelogram *AF* as base, and the other the triangle *GHK*,
and let the parallelogram *AF* be double of the triangle *GHK*;
I say that the prism *ABCDEF* is equal to the prism *GHKLMN*.

For let the solids *AO, GP* be completed.

Since the parallelogram *AF* is double of the triangle *GHK*,
while the parallelogram *HK* is also double of the triangle *GHK*, [I. 34]
 therefore the parallelogram *AF* is equal to the parallelogram *HK*.

But parallelepipedal solids which are on equal bases and of the same height are
equal to one another; [XI. 31]
 therefore the solid *AO* is equal to the solid *GP.*

And the prism *ABCDEF* is half of the solid *AO,*
and the prism *GHKLMN* is half of the solid *GP*; [XI. 28]
 therefore the prism *ABCDEF* is equal to the prism *GHKLMN.*

Therefore etc.

Q.E.D.

Book XII

Proposition 1

Similar polygons inscribed in circles are to one another as the squares on the diameters.

 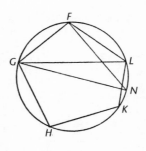

Let *ABC, FGH* be circles,
let *ABCDE, FGHKL* be similar polygons inscribed in them, and let *BM, GN* be diameters of the circles;
I say that, as the square on *BM* is to the square on *GN*, so is the polygon *ABCDE* to the polygon *FGHKL*.

For let *BE, AM, GL, FN* be joined.

Now, since the polygon *ABCDE* is similar to the polygon *FGHKL*,
 the angle *BAE* is equal to the angle *GFL*,
 and, as *BA* is to *AE*, so is *GF* to *FL*. **[VI. Def. 1]**

Thus *BAE, GFL* are two triangles which have one angle equal to one angle, namely the angle *BAE* to the angle *GFL*, and the sides about the equal angles proportional;
 therefore the triangle *ABE* is equiangular with the triangle *FGL*. **[VI. 6]**

Therefore the angle *AEB* is equal to the angle *FLG*.

But the angle *AEB* is equal to the angle *AMB*,
for they stand on the same circumference; **[III. 27]**
and the angle *FLG* to the angle *FNG*;
 therefore the angle *AMB* is also equal to the angle *FNG*.

But the right angle *BAM* is also equal to the right angle *GFN*; **[III. 31]**
 therefore the remaining angle is equal to the remaining angle. **[I. 32]**

Therefore the triangle *ABM* is equiangular with the triangle *FGN*.

 Therefore, proportionally, as *BM* is to *GN*, so is *BA* to *GF*. **[VI. 4]**

But the ratio of the square on *BM* to the square on *GN* is duplicate of the ratio of *BM* to *GN*,
and the ratio of the polygon *ABCDE* to the polygon *FGHKL* is duplicate of the ratio of *BA* to *GF*; **[VI. 20]**

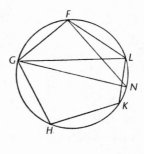

therefore also, as the square on *BM* is to the square on *GN*, so is the polygon *ABCDE* to the polygon *FGHKL*.

Therefore etc.

<div align="right">Q.E.D.</div>

Proposition 2

Circles are to one another as the squares on the diameters.

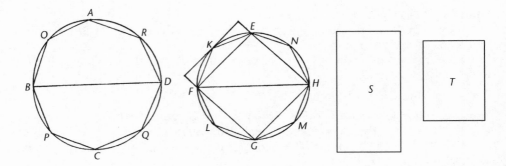

Let *ABCD, EFGH* be circles, and *BD, FH* their diameters;
I say that, as the circle *ABCD* is to the circle *EFGH*, so is the square on *BD* to the square on *FH*.

For, if the square on *BD* is not to the square on *FH* as the circle *ABCD* is to the circle *EFGH*,
then, as the square on *BD* is to the square on *FH*, so will the circle *ABCD* be either to some less area than the circle *EFGH*, or to a greater.

First, let it be in that ratio to a less area *S*.

Let the square *EFGH* be inscribed in the circle *EFGH*; then the inscribed square is greater than the half of the circle *EFGH*, inasmuch as, if through the points *E, F, G, H* we draw tangents to the circle, the square *EFGH* is half the square circumscribed about the circle, and the circle is less than the circumscribed square;
hence the inscribed square *EFGH* is greater than the half of the circle *EFGH*.

Let the circumferences *EF, FG, GH, HE* be bisected at the points *K, L, M, N,* and let *EK, KF, FL, LG, GM, MH, HN, NE* be joined;

therefore each of the triangles *EKF, FLG, GMH, HNE* is also greater than the half of the segment of the circle about it, inasmuch as, if through the points *K, L, M, N* we draw tangents to the circle and complete the parallelograms on the straight lines *EF, FG, GH, HE*, each of the triangles *EKF, FLG, GMH, HNE* will be half of the parallelogram about it,
while the segment about it is less than the parallelogram;
hence each of the triangles *EKF, FLG, GMH, HNE* is greater than the half of the segment of the circle about it.

Thus, by bisecting the remaining circumferences and joining straight lines, and by doing this continually, we shall leave some segments of the circle which will be less than the excess by which the circle *EFGH* exceeds the area *S*.

For it was proved in the first theorem of the tenth book that, if two unequal magnitudes be set out, and if from the greater there be subtracted a magnitude greater than the half, and from that which is left a greater than the half, and if this be done continually, there will be left some magnitude which will be less than the lesser magnitude set out.

Let segments be left such as described, and let the segments of the circle *EFGH* on *EK, KF, FL, LG, GM, MH, HN, NE* be less than the excess by which the circle *EFGH* exceeds the area *S*.

Therefore the remainder, the polygon *EKFLGMHN*, is greater than the area *S*.

Let there be inscribed, also, in the circle *ABCD* the polygon *AOBPCQDR* similar to the polygon *EKFLGMHN*;
therefore, as the square on *RD* is to the square on *FH*, so is the polygon *AOBPCQDR* to the polygon *EKFLGMHN*. [XII. 1]

But, as the square on *BD* is to the square on *FH*, so also is the circle *ABCD* to the area *S*;
therefore also, as the circle *ABCD* is to the area S, so is the polygon *AOBPCQDR* to the polygon *EKFLGMHN*; [V. 11]
therefore, alternately, as the circle *ABCD* is to the polygon inscribed in it, so is the area S to the polygon *EKFLGMHN*. [V. 16]

But the circle *ABCD* is greater than the polygon inscribed in it;
therefore the area S is also greater than the polygon *EKFLGMHN*.
But it is also less:
which is impossible.

Therefore, as the square on *BD* is to the square on *FH*, so is not the circle *ABCD* to any area less than the circle *EFGH*.

Similarly we can prove that neither is the circle *EFGH* to any area less than the circle *ABCD* as the square on *FH* is to the square on *BD*.

I say next that neither is the circle *ABCD* to any area greater than the circle *EFGH* as the square on *BD* is to the square on *FH*.

For, if possible, let it be in that ratio to a greater area *S*.

Therefore, inversely, as the square on *FH* is to the square on *DB*, so is the area S to the circle *ABCD*.

But, as the area S is to the circle *ABCD*, so is the circle *EFGH* to some area less than the circle *ABCD*;
therefore also, as the square on *FH* is to the square on *BD*, so is the circle *EFGH* to some area less than the circle *ABCD*: [V. 11]

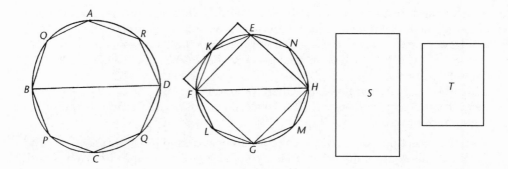

which was proved impossible.

Therefore, as the square on *BD* is to the square on *FH*, so is not the circle *ABCD* to any area greater than the circle *EFGH*.

And it was proved that neither is it in that ratio to any area less than the circle *EFGH*;

therefore, as the square on *BD* is to the square on *FH*, so is the circle *ABCD* to the circle *EFGH*.

Therefore etc.

<div align="right">Q.E.D.</div>

LEMMA. I say that, the area *S* being greater than the circle *EFGH*, as the area *S* is to the circle *ABCD*, so is the circle *EFGH* to some area less than the circle *ABCD*.

For let it be contrived that, as the area *S* is to the circle *ABCD*, so is the circle *EFGH* to the area *T*.

I say that the area *T* is less than the circle *ABCD*.

For since, as the area *S* is to the circle *ABCD*, so is the circle *EFGH* to the area *T*, therefore, alternately, as the area S is to the circle *EFGH*, so is the circle *ABCD* to the area *T*. [V. 16]

But the area S is greater than the circle *EFGH*;
therefore the circle *ABCD* is also greater than the area *T*.

Hence, as the area S is to the circle *ABCD*, so is the circle *EFGH* to some area less than the circle *ABCD*.

<div align="right">Q.E.D.</div>

Proposition 3

Any pyramid which has a triangular base is divided into two pyramids equal and similar to one another, similar to the whole and having triangular bases, and into two equal prisms; and the two prisms are greater than the half of the whole pyramid.

Let there be a pyramid of which the triangle *ABC* is the base and the point *D* the vertex;

I say that the pyramid *ABCD* is divided into two pyramids equal to one another, having triangular bases and similar to the whole pyramid, and into two equal prisms; and the two prisms are greater than the half of the whole pyramid.

For let *AB, BC, CA, AD, DB, DC* be bisected at the points *E, F, G, H, K, L,* and let *HE, EG, GH, HK, KL, LH, KF, FG* be joined.

Since *AE* is equal to *EB*, and *AH* to *DH*,
therefore *EH* is parallel to *DB*. [VI. 2]

For the same reason
 HK is also parallel to *AB*.

Therefore *HEBK* is a parallelogram;
 therefore *HK* is equal to *EB*. [I. 34]

But *EB* is equal to EA;
 therefore *AE* is also equal to *HK*.

But *AH* is also equal to *HD*;
 therefore the two sides *EA, AH* are equal to the two sides *KH, HD* respectively,
and the angle *EAH* is equal to the angle *KHD*;
 therefore the base *EH* is equal to the base *KD*. [I. 4]

Therefore the triangle *AEH* is equal and similar to the triangle *HKD*.

For the same reason
 the triangle *AHG* is also equal and similar to the triangle *HLD*.

Now, since two straight lines *EH, HG* meeting one another are parallel to two straight lines *KD, DL* meeting one another, and are not in the same plane,
 they will contain equal angles. [XI. 10]

Therefore the angle *EHG* is equal to the angle *KDL*.

And, since the two straight lines *EH, HG* are equal to the two *KD, DL* respectively, and the angle *EHG* is equal to the angle *KDL*,
 therefore the base *EG* is equal to the base *KL*; [I. 4]
 therefore the triangle *EHG* is equal and similar to the triangle *KDL*.

For the same reason
 the triangle *AEG* is also equal and similar to the triangle *HKL*.

Therefore the pyramid of which the triangle *AEG* is the base and the point *H* the vertex is equal and similar to the pyramid of which the triangle *HKL* is the base and the point *D* the vertex. [XI. Def. 10]

And, since *HK* has been drawn parallel to *AB*, one of the sides of the triangle *ADB*,
 the triangle *ADB* is equiangular to the triangle *DHK*, [1. 29]
and they have their sides proportional;
 therefore the triangle *ADB* is similar to the triangle *DHK*. [VI. Def. 1]

For the same reason
 the triangle *DBC* is also similar to the triangle *DKL*,
 and the triangle *ADC* to the triangle *DLH*.

Now, since the two straight lines *BA, AC* meeting one another are parallel to the two straight lines *KH, HL* meeting one another, not in the same plane,

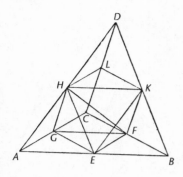

they will contain equal angles. [XI. 10]

Therefore the angle *BAC* is equal to the angle *KHL*.

And, as *BA* is to *AC*, so is *KH* to *HL*;
therefore the triangle *ABC* is similar to the triangle *HKL*.

Therefore also the pyramid of which the triangle *ABC* is the base and the point *D* the vertex is similar to the pyramid of which the triangle *HKL* is the base and the point *D* the vertex.

But the pyramid of which the triangle *HKL* is the base and the point *D* the vertex was proved similar to the pyramid of which the triangle *AEG* is the base and the point *H* the vertex.

Therefore each of the pyramids *AEGH*, *HKLD* is similar to the whole pyramid *ABCD*.

Next, since *BF* is equal to *FC*,
 the parallelogram *EBFG* is double of the triangle *GFC*.

And since, if there be two prisms of equal height, and one have a parallelogram as base, and the other a triangle, and if the parallelogram be double of the triangle, the prisms are equal, [XI. 39]
therefore the prism contained by the two triangles *BKF*, *EHG*, and the three parallelograms *EBFG*, *EBKH*, *HKFG* is equal to the prism contained by the two triangles *GFC*, *HKL* and the three parallelograms *KFCL*, *LCGH*, *HKFG*.

And it is manifest that each of the prisms, namely that in which the parallelogram *EBFG* is the base and the straight line *HK* is its opposite, and that in which the triangle *GFC* is the base and the triangle *HKL* its opposite, is greater than each of the pyramids of which the triangles *AEG*, *HKL* are the bases and the points *H*, *D* the vertices,
inasmuch as, if we join the straight lines *EF*, *EK*, the prism in which the parallelogram *EBFG* is the base and the straight line *HK* its opposite is greater than the pyramid of which the triangle *EBF* is the base and the point *K* the vertex.

But the pyramid of which the triangle *EBF* is the base and the point *K* the vertex is equal to the pyramid of which the triangle *AEG* is the base and the point *H* the vertex;
for they are contained by equal and similar planes.

Hence also the prism in which the parallelogram *EBFG* is the base and the straight line *HK* its opposite is greater than the pyramid of which the triangle *AEG* is the base and the point *H* the vertex.

But the prism in which the parallelogram *EBFG* is the base and the straight line *HK* its opposite is equal to the prism in which the triangle *GFC* is the base and the triangle *HKL* its opposite,
and the pyramid of which the triangle *AEG* is the base and the point *H* the vertex is equal to the pyramid of which the triangle *HKL* is the base and the point *D* the vertex.

Therefore the said two prisms are greater than the said two pyramids of which the triangles *AEG*, *HKL* are the bases and the points *H*, *D* the vertices.

Therefore the whole pyramid, of which the triangle *ABC* is the base and the point *D* the vertex, has been divided into two pyramids equal to one another and into two equal prisms, and the two prisms are greater than the half of the whole pyramid.

Q.E.D.

Proposition 4

If there be two pyramids of the same height which have triangular bases, and each of them be divided into two pyramids equal to one another and similar to the whole, and into two equal prisms, then, as the base of the one pyramid is to the base of the other pyramid, so will all the prisms in the one pyramid be to all the prisms, being equal in multitude, in the other pyramid.

 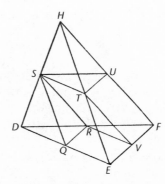

Let there be two pyramids of the same height which have the triangular bases *ABC*, *DEF*, and vertices the points *G*, *H*,
and let each of them be divided into two pyramids equal to one another and similar to the whole and into two equal prisms; [XII. 3]
I say that, as the base *ABC* is to the base *DEF*, so are all the prisms in the pyramid *ABCG* to all the prisms, being equal in multitude, in the pyramid *DEFH*,

For, since *BO* is equal to *OC*, and *AL* to *LC*.
therefore *LO* is parallel to *AB*,
and the triangle *ABC* is similar to the triangle *LOC*.

For the same reason
the triangle *DEF* is also similar to the triangle *RVF*.

And, since *BC* is double of *CO*, and *EF* of *FV*,
therefore, as *BC* is to *CO*, so is *EF* to *FV*.

And on *BC*, *CO* are described the similar and similarly situated rectilineal figures *ABC*, *LOC*, and on *EF*, *FV* the similar and similarly situated figures *DEF*, *RVF*;
therefore, as the triangle *ABC* is to the triangle *LOC*, so is the triangle *DEF* to the triangle *RVF*; [VI. 22]
therefore, alternately, as the triangle *ABC* is to the triangle *DEF*, so is the triangle *LOC* to the triangle *RVF*. [V. 16]

But, as the triangle *LOC* is to the triangle *RVF*, so is the prism in which the

triangle *LOC* is the base and *PMN* its opposite, to the prism in which the triangle *RVF* is the base and *STU* its opposite; **[Lemma following]**

therefore also, as the triangle *ABC* is to the triangle *DEF*, so is the prism in which the triangle *LOC* is the base and *PMN* its opposite, to the prism in which the triangle *RVF* is the base and *STU* its opposite.

But, as the said prisms are to one another, so is the prism in which the parallelogram *KBOL* is the base and the straight line *PM* its opposite, to the prism in which the parallelogram *QEVR* is the base and the straight line *ST* its opposite.

[XI. 39; cf. XII. 3]

Therefore also the two prisms, that in which the parallelogram *KBOL* is the base and *PM* its opposite, and that in which the triangle *LOC* is the base and *PMN* its opposite, are to the prisms in which *QEVR* is the base and the straight line *ST* its opposite and in which the triangle *RVF* is the base and *STU* its opposite in the same ratio. **[V. 12]**

Therefore also, as the base *ABC* is to the base *DEF*, so are the said two prisms to the said two prisms.

And similarly, if the pyramids *PMNG*, *STUH* be divided into two prisms and two pyramids,

as the base *PMN* is to the base *STU*, so will the two prisms in the pyramid *PMNG* be to the two prisms in the pyramid *STUH*.

But, as the base *PMN* is to the base *STU*, so is the base *ABC* to the base *DEF*;
for the triangles *PMN*, *STU* are equal to the triangles *LOC*, *RVF* respectively.

Therefore also, as the base *ABC* is to the base *DEF*, so are the four prisms to the four prisms.

And similarly also, if we divide the remaining pyramids into two pyramids and into two prisms, then, as the base *ABC* is to the base *DEF*, so will all the prisms in the pyramid *ABCG* be to all the prisms, being equal in multitude, in the pyramid *DEFH*.

Q.E.D.

LEMMA. But that, as the triangle *LOC* is to the triangle *RVF*, so is the prism in which the triangle *LOC* is the base and *PMN* its opposite, to the prism in which the triangle *RVF* is the base and *STU* its opposite, we must prove as follows.

For in the same figure let perpendiculars be conceived drawn from *G*, *H* to the planes *ABC*, *DEF*; these are of course equal because, by hypothesis, the pyramids are of equal height.

Now, since the two straight lines *GC* and the perpendicular from *G* are cut by the parallel planes *ABC*, *PMN*;

they will be cut in the same ratios. [XI. 17]

And *GC* is bisected by the plane *PMN* at *N*;
therefore the perpendicular from *G* to the plane *ABC* will also be bisected by the plane *PMN*.

For the same reason,
the perpendicular from *H* to the plane *DEF* will also be bisected by the plane *STU*.

And the perpendiculars from *G*, *H* to the planes *ABC*, *DEF* are equal;
therefore the perpendiculars from the triangles *PMN*, *STU* to the planes *ABC*, *DEF* are also equal.

Therefore the prisms in which the triangles *LOC*, *RVF* are bases, and *PMN*, *STU* their opposites, are of equal height.

Hence also the parallelepipedal solids described from the said prisms are of equal height and are to one another as their bases; [XI. 32]
therefore their halves, namely the said prisms, are to one another as the base *LOC* is to the base *RVF*.

Q.E.D.

Proposition 5

Pyramids which are of the same height and have triangular bases are to one another as the bases.

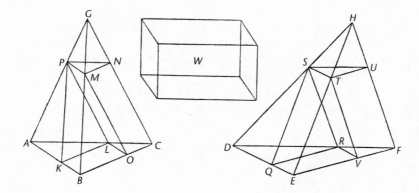

Let there by pyramids of the same height, of which the triangles *ABC*, *DEF* are the bases and the points *G*, *H* the vertices;
I say that, as the base *ABC* is to the base *DEF*, so is the pyramid *ABCG* to the pyramid *DEFH*.

For, if the pyramid *ABCG* is not to the pyramid *DEFH* as the base *ABC* is to the base *DEF*, then, as the base *ABC* is to the base *DEF*, so will the pyramid *ABCG* be either to some solid less than the pyramid *DEFH* or to a greater.

Let it, first, be in that ratio to a less solid *W*, and let the pyramid *DEFH* be divided into two pyramids equal to one another and similar to the whole and into two equal prisms;
then the two prisms are greater than the half of the whole pyramid. [XII. 3]

Again, let the pyramids arising from the division be similarly divided,

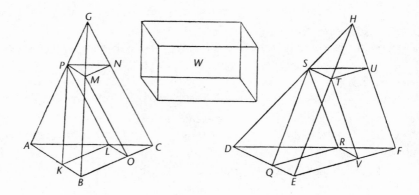

and let this be done continually until there are left over from the pyramid *DEFH* some pyramids which are less than the excess by which the pyramid *DEFH* exceeds the solid *W*. [X. 1]

Let such be left, and let them be, for the sake of argument, *DQRS, STUH*;
therefore the remainders, the prisms in the pyramid *DEFH*, are greater than the solid *W*.

Let the pyramid *ABCG* also be divided similarly, and a similar number of times, with the pyramid *DEFH*;
therefore, as the base *ABC* is to the base *DEF*, so are the prisms in the pyramid *ABCG* to the prisms in the pyramid *DEFH*. [XII. 4]

But, as the base *ABC* is to the base *DEF*, so also is the pyramid *ABCG* to the solid *W*;
therefore also, as the pyramid *ABCG* is to the solid *W*, so are the prisms in the pyramid *ABCG* to the prisms in the pyramid *DEFH*; [V. 11]
therefore, alternately, as the pyramid *ABCG* is to the prisms in it, so is the solid *W* to the prisms in the pyramid *DEFH*. [V. 16]

But the pyramid *ABCG* is greater than the prisms in it;
 therefore the solid *W* is also greater than the prisms in the pyramid *DEFH*.

But it is also less:
which is impossible.

Therefore the prism *ABCG* is not to any solid less than the pyramid *DEFH* as the base *ABC* is to the base *DEF*.

Similarly it can be proved that neither is the pyramid *DEFH* to any solid less than the pyramid *ABCG* as the base *DEF* is to the base *ABC*.

I say next that neither is the pyramid *ABCG* to any solid greater than the pyramid *DEFH* as the base *ABC* is to the base *DEF*.

For, if possible, let it be in that ratio to a greater solid *W*;
therefore, inversely, as the base *DEF* is to the base *ABC*, so is the solid *W* to the pyramid *ABCG*.

But, as the solid *W* is to the solid *ABCG*, so is the pyramid *DEFH* to some solid less than the pyramid *ABCG*, as was before proved; [XII. 2, Lemma]
therefore also, as the base *DEF* is to the base *ABC*, so is the pyramid *DEFH* to some solid less than the pyramid *ABCG*: [V. 11]
which was proved absurd.

Therefore the pyramid *ABCG* is not to any solid greater than the pyramid *DEFH* as the base *ABC* is to the base *DEF*.

But it was proved that neither is it in that ratio to a less solid.

Therefore, as the base *ABC* is to the base *DEF*, so is the pyramid *ABCG* to the pyramid *DEFH*.

<div align="right">Q.E.D.</div>

Proposition 6

Pyramids which are of the same height and have polygonal bases are to one another as the bases.

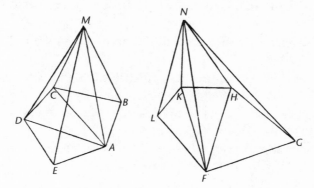

Let there be pyramids of the same height of which the polygons *ABCDE*, *FGHKL* are the bases and the points *M*, *N* the vertices;
I say that, as the base *ABCDE* is to the base *FGHKL*, so is the pyramid *ABCDEM* to the pyramid *FGHKLN*.

For let *AC*, *AD*, *FH*, *FK* be joined.

Since then *ABCM*, *ACDM* are two pyramids which have triangular bases and equal height,

<div align="center">they are to one another as the bases; [XII. 5]</div>
therefore, as the base *ABC* is to the base *ACD*, so is the pyramid *ABCM* to the pyramid *ACDM*.

And, *componendo*, as the base *ABCD* is to the base *ACD*, so is the pyramid *ABCDM* to the pyramid *ACDM*. [V. 18]

But also, as the base *ACD* is to the base *ADE*, so is the pyramid *ACDM* to the pyramid *ADEM*. [XII. 5]

Therefore, *ex aequali*, as the base *ABCD* is to the base *ADE*, so is the pyramid *ABCDM* to the pyramid *ADEM*. [V. 22]

And again, *componendo*, as the base *ABCDE* is to the base *ADE*, so is the pyramid *ABCDEM* to the pyramid *ADEM*. [V. 18]

Similarly also it can be proved that, as the base *FGHKL* is to the base *FGH*, so is the pyramid *FGHKLN* to the pyramid *FGHN*.

And, since *ADEM*, *FGHN* are two pyramids which have triangular bases and equal height,

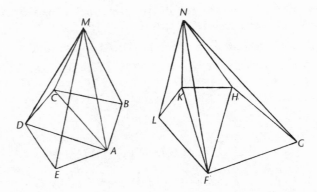

therefore, as the base *ADE* is to the base *FGH*, so is the pyramid *ADEM* to the pyramid *FGHN*. [XII. 5]

But, as the base *ADE* is to the base *ABCDE*, so was the pyramid *ADEM* to the pyramid *ABCDEM*.

Therefore also, *ex aequali*, as the base *ABCDE* is to the base *FGH*, so is the pyramid *ABCDEM* to the pyramid *FGHN*. [V. 22]

But further, as the base *FGH* is to the base *FGHKL*, so also was the pyramid *FGHN* to the pyramid *FGHKLN*.

Therefore also, *ex aequali*, as the base *ABCDE* is to the base *FGHKL*, so is the pyramid *ABCDEM* to the pyramid *FGHKLN*. [V. 22]

Q.E.D.

Proposition 7

Any prism which has a triangular base is divided into three pyramids equal to one another which have triangular bases.

Let there be a prism in which the triangle *ABC* is the base and *DEF* its opposite;
I say that the prism *ABCDEF* is divided into three pyramids equal to one another, which have triangular bases.

For let *BD, EC, CD* be joined.

Since *ABED* is a parallelogram, and *BD* is its diameter,
therefore the triangle *ABD* is equal to the triangle *EBD*; [I. 34]

therefore also the pyramid of which the triangle *ABD* is the base and the point *C* the vertex is equal to the pyramid of which the triangle *DEB* is the base and the point *C* the vertex. [XII. 5]

But the pyramid of which the triangle *DEB* is the base and the point *C* the vertex is the same with the pyramid of which the triangle *EBC* is the base and the point *D* the vertex;
for they are contained by the same planes.

Therefore the pyramid of which the triangle *ABD* is the base and the point *C* the vertex is also equal to the pyramid of which the triangle *EBC* is the base and the point *D* the vertex.

Again, since *FCBE* is a parallelogram,
and *CE* is its diameter,

<div align="center">the triangle CEF is equal to the triangle CBE. [I. 34]</div>

Therefore also the pyramid of which the triangle *BCE* is the base and the point *D* the vertex is equal to the pyramid of which the triangle *BCE* is the base and the point *D* the vertex.

<div align="right">[XII. 5]</div>

But the pyramid of which the triangle *BCE* is the base and the point *D* the vertex was proved equal to the pyramid of which the triangle *ABD* is the base and the point *C* the vertex;
therefore also the pyramid of which the triangle *CEF* is the base and the point *D* the vertex is equal to the pyramid of which the triangle *ABD* is the base and the point *C* the vertex;
therefore the prism *ABCDEF* has been divided into three pyramids equal to one another which have triangular bases.

And, since the pyramid of which the triangle *ABD* is the base and the point *C* the vertex is the same with the pyramid of which the triangle *CAB* is the base and the point *D* the vertex,
for they are contained by the same planes,
while the pyramid of which the triangle *ABD* is the base and the point *C* the vertex was proved to be a third of the prism in which the triangle *ABC* is the base and *DEF* its opposite, therefore also the pyramid of which the triangle *ABC* is the base and the point *D* the vertex is a third of the prism which has the same base, the triangle *ABC*, and *DEF* as its opposite.

<div align="right">Q.E.D.</div>

PORISM. From this it is manifest that any pyramid is a third part of the prism which has the same base with it and equal height.

Proposition 8

Similar pyramids which have triangular bases are in the triplicate ratio of their corresponding sides.

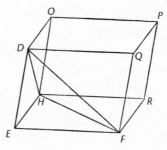

Let there be similar and similarly situated pyramids of which the triangles *ABC*, *DEF* are the bases and the points *G*, *H* the vertices;

 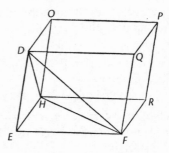

I say that the pyramid *ABCG* has to the pyramid *DEFH* the ratio triplicate of that which *BC* has to *EF*.

For let the parallelepipedal solids *BGML*, *EHQP* be completed.

Now, since the pyramid *ABCG* is similar to the pyramid *DEFH*,
therefore the angle *ABC* is equal to the angle *DEF*,
the angle *GBC* to the angle *HEF*,
and the angle *ABG* to the angle *DEH*;
and, as *AB* is to *DE*, so is *BC* to *EF*, and *BG* to *EH*.

And since, as *AB* is to *DE*, so is *BC* to *EF*,
and the sides are proportional about equal angles,
therefore the parallelogram *BM* is similar to the parallelogram *EQ*.

For the same reason
BN is also similar to *ER*, and *BK* to *EO*;
therefore the three parallelograms *MB*, *BK*, *BN* are similar to the three *EQ*, *EO*, *ER*.

But the three parallelograms *MB*, *BK*, *BN* are equal and similar to their three opposites,
and the three *EQ*, *EO*, *ER* are equal and similar to their three opposites. [XI. 24]

Therefore the solids *BGML*, *EHQP* are contained by similar planes equal in multitude.

Therefore the solid *BGML* is similar to the solid *EHQP*.

But similar parallelepipedal solids are in the triplicate ratio of their corresponding sides. [XI. 33]

Therefore the solid *BGML* has to the solid *EHQP* the ratio triplicate of that which the corresponding side *BC* has to the corresponding side *EF*.

But, as the solid *BGML* is to the solid *EHQP*, so is the pyramid *ABCG* to the pyramid *DEFH*,
inasmuch as the pyramid is a sixth part of the solid, because the prism which is half of the parallelepipedal solid [XI. 28] is also triple of the pyramid. [XII. 7]

Therefore the pyramid *ABCG* also has to the pyramid *DEFH* the ratio triplicate of that which *BC* has to *EF*.

Q.E.D.

PORISM. From this it is manifest that similar pyramids which have polygonal bases are also to one another in the triplicate ratio of their corresponding sides.

For, if they are divided into the pyramids contained in them which have triangular bases, by virtue of the fact that the similar polygons forming their

bases are also divided into similar triangles equal in multitude and corresponding to the wholes, [VI. 20]
then, as the one pyramid which has a triangular base in the one complete pyramid is to the one pyramid which has a triangular base in the other complete pyramid, so also will all the pyramids which have triangular bases contained in the one pyramid be to all the pyramids which have triangular bases contained in the other pyramid [v. 12], that is, the pyramid itself which has a polygonal base, to the pyramid which has a polygonal base.

But the pyramid which has a triangular base is to the pyramid which has a triangular base in the triplicate ratio of the corresponding sides;
therefore also the pyramid which has a polygonal base has to the pyramid which has a similar base the ratio triplicate of that which the side has to the side.

Proposition 9

In equal pyramids which have triangular bases the bases are reciprocally proportional to the heights; and those pyramids in which the bases are reciprocally proportional to the heights are equal.

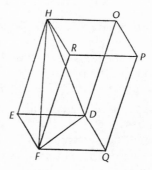

For let there be equal pyramids which have the triangular bases *ABC*, *DEF* and vertices the points *G*, *H*;
I say that in the pyramids *ABCG*, *DEFH* the bases are reciprocally proportional to the heights, that is, as the base *ABC* is to the base *DEF*, so is the height of the pyramid *DEFH* to the height of the pyramid *ABCG*.

For let the parallelepipedal solids *BGML*, *EHQP* be completed.

Now, since the pyramid *ABCG* is equal to the pyramid *DEFH*,
and the solid *BGML* is six times the pyramid *ABCG*,
and the solid *EHQP* six times the pyramid *DEFH*,
 therefore the solid *BGML* is equal to the solid *EHQP*.

But in equal parallelepipedal solids the bases are reciprocally proportional to the heights; [XI. 34]
therefore, as the base *BM* is to the base *EQ*, so is the height of the solid *EHQP* to the height of the solid *BGML*.

But, as the base *BM* is to *EQ*, so is the triangle *ABC* to the triangle *DEF*. [I. 34]

Therefore also, as the triangle *ABC* is to the triangle *DEF*, so is the height of the solid *EHQP* to the height of the solid *BGML*. [V. 11]

But the height of the solid *EHQP* is the same with the height of the pyramid *DEFH*,
and the height of the solid *BGML* is the same with the height of the pyramid *ABCG*,
therefore, as the base *ABC* is to the base *DEF*, so is the height of the pyramid *DEFH* to the height of the pyramid *ABCG*.

Therefore in the pyramids *ABCG*, *DEFH* the bases are reciprocally proportional to the heights.

Next, in the pyramids *ABCG*, *DEFH* let the bases be reciprocally proportional to the heights;
that is, as the base *ABC* is to the base *DEF*, so let the height of the pyramid *DEFH* be to the height of the pyramid *ABCG*;
I say that the pyramid *ABCG* is equal to the pyramid *DEFH*.

For, with the same construction,
since, as the base *ABC* is to the base *DEF*, so is the height of the pyramid *DEFH* to the height of the pyramid *ABCG*,
while, as the base *ABC* is to the base *DEF*, so is the parallelogram *BM* to the parallelogram *EQ*,
therefore also, as the parallelogram *BM* is to the parallelogram *EQ*, so is the height of the pyramid *DEFH* to the height of the pyramid *ABCG*. [V. 11]

But the height of the pyramid *DEFH* is the same with the height of the parallelepiped *EHQP*,
and the height of the pyramid *ABCG* is the same with the height of the parallelepiped *BGML*;
therefore, as the base *BM* is to the base *EQ*, so is the height of the parallelepiped *EHQP* to the height of the parallelepiped *BGML*.

But those parallelepipedal solids in which the bases are reciprocally proportional to the heights are equal; [XI. 34]
therefore the parallelepipedal solid *BGML* is equal to the parallelepipedal solid *EHQP*.

And the pyramid *ABCG* is a sixth part of *BGML*, and the pyramid *DEFH* a sixth part of the parallelepiped *EHQP*;
 therefore the pyramid *ABCG* is equal to the pyramid *DEFH*.

Therefore etc.

 Q.E.D.

Proposition 10

Any cone is a third part of the cylinder which has the same base with it and equal height.

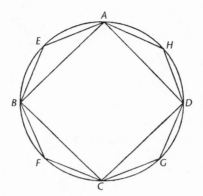

For let a cone have the same base, namely the circle *ABCD*, with a cylinder and equal height;
I say that the cone is a third part of the cylinder, that is, that the cylinder is triple of the cone.

For if the cylinder is not triple of the cone, the cylinder will be either greater than triple or less than triple of the cone.

First let it be greater than triple,
and let the square *ABCD* be inscribed in the circle *ABCD*; [IV. 6]
then the square *ABCD* is greater than the half of the circle *ABCD*.

From the square *ABCD* let there be set up a prism of equal height with the cylinder.

Then the prism so set up is greater than the half of the cylinder, inasmuch as, if we also circumscribe a square about the circle *ABCD*, [IV. 7]
the square inscribed in the circle *ABCD* is half of that circumscribed about it,
and the solids set up from them are parallelepipedal prisms of equal height,
while parallelepipedal solids which are of the same height are to one another as their bases; [XI. 32]
therefore also the prism set up on the square *ABCD* is half of the prism set up from the square circumscribed about the circle *ABCD*;
 [cf. XI. 28, or XII. 6 and 7, Por.]
and the cylinder is less than the prism set up from the square circumscribed about the circle *ABCD*;
therefore the prism set up from the square *ABCD* and of equal height with the cylinder is greater than the half of the cylinder.

Let the circumferences *AB, BC, CD, DA* be bisected at the points *E, F, G, H*,
and let *AE, EB, BF, FC, CG, GD, DH, HA* be joined;
then each of the triangles *AEB, BFC, CGD, DHA* is greater than the half of that segment of the circle *ABCD* which is about it, as we proved before. [XII. 2]

On each of the triangles *AEB, BFC, CGD, DHA* let prisms be set up of equal height with the cylinder;

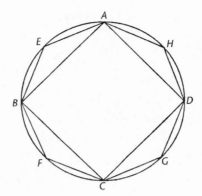

then each of the prisms so set up is greater than the half part of that segment of the cylinder which is about it,
inasmuch as, if we draw through the points *E, F, G, H* parallels to *AB, BC, CD, DA*, complete the parallelograms on *AB, BC, CD, DA*, and set up from them parallelepipedal solids of equal height with the cylinder, the prisms on the triangles *AEB, BFC, CGD, DHA* are halves of the several solids set up;
and the segments of the cylinder are less than the parallelepipedal solids set up; hence also the prisms on the triangles *AEB, BFC, CGD, DHA* are greater than the half of the segments of the cylinder about them.

Thus, bisecting the circumferences that are left, joining straight lines, setting up on each of the triangles prisms of equal height with the cylinder,
and doing this continually,
we shall leave some segments of the cylinder which will be less than the excess by which the cylinder exceeds the triple of the cone. [X. 1]

Let such segments be left, and let them be *AE, EB, BF, FC, CG, GD, DH, HA*;
therefore the remainder, the prism of which the polygon *AEBFCGDH* is the base and the height is the same as that of the cylinder, is greater than triple of the cone.

But the prism of which the polygon *AEBFCGDH* is the base and the height the same as that of the cylinder is triple of the pyramid of which the polygon *AEBFCGDH* is the base and the vertex is the same as that of the cone;

[XII. 7, Por.]

therefore also the pyramid of which the polygon *AEBFCGDH* is the base and the vertex is the same as that of the cone is greater than the cone which has the circle *ABCD* as base.

But it is also less, for it is enclosed by it:
which is impossible.

Therefore the cylinder is not greater than triple of the cone.

I say next that neither is the cylinder less than triple of the cone,
for, if possible, let the cylinder be less than triple of the cone,
 therefore, inversely, the cone is greater than a third part of the cylinder.

Let the square *ABCD* be inscribed in the circle *ABCD*;
therefore the square *ABCD* is greater than the half of the circle *ABCD*.

Now let there be set up from the square *ABCD* a pyramid having the same vertex with the cone;

therefore the pyramid so set up is greater than the half part of the cone,
seeing that, as we proved before, if we circumscribe a square about the circle,
the square ABCD will be half of the square circumscribed about the circle,
and if we set up from the squares parallelepipedal solids of equal height with
the cone, which are also called prisms, the solid set up from the square ABCD
will be half of that set up from the square circumscribed about the circle,
for they are to one another as their bases. [XI. 32]

Hence also the thirds of them are in that ratio;
therefore also the pyramid of which the square ABCD is the base is half of the
pyramid set up from the square circumscribed about the circle.

And the pyramid set up from the square about the circle is greater than the
cone,
for it encloses it.

Therefore the pyramid of which the square ABCD is the base and the vertex is
the same with that of the cone is greater than the half of the cone.

Let the circumferences AB, BC, CD, DA be bisected at the points E, F, G, H,
and let AE, EB, BF, FC, CG, GD, DH, HA be joined;
therefore also each of the triangles AEB, BFC, CGD, DHA is greater than the half
part of that segment of the circle ABCD which is about it.

Now, on each of the triangles AEB, BFC, CGD, DHA let pyramids be set up
which have the same vertex as the cone;
therefore also each of the pyramids so set up is, in the same manner, greater than
the half part of that segment of the cone which is about it.

Thus, by bisecting the circumferences that are left, joining straight lines, setting
up on each of the triangles a pyramid which has the same vertex as the cone,
and doing this continually,
we shall leave some segments of the cone which will be less than the excess by
which the cone exceeds the third part of the cylinder. [X. 1]

Let such be left, and let them be the segments on AE, EB, BF, FC, CG, GD, DH, HA;
therefore the remainder, the pyramid of which the polygon AEBFCGDH is the
base and the vertex the same with that of the cone, is greater than a third part of
the cylinder.

But the pyramid of which the polygon AEBFCGDH is the base and the vertex
the same with that of the cone is a third part of the prism of which the polygon
AEBFCGDH is the base and the height is the same with that of the cylinder;
therefore the prism of which the polygon AEBFCGDH is the base and the height
is the same with that of the cylinder is greater than the cylinder of which the
circle ABCD is the base.

But it is also less, for it is enclosed by it:
which is impossible.

Therefore the cylinder is not less than triple of the cone.

But it was proved that neither is it greater than triple;
therefore the cylinder is triple of the cone;
hence the cone is a third part of the cylinder.

Therefore etc.

Q.E.D.

Proposition 11

Cones and cylinders which are of the same height are to one another as their bases.

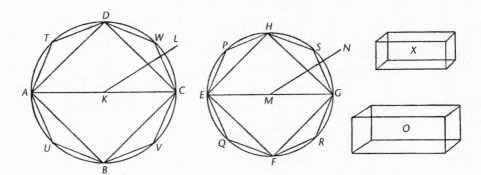

Let there be cones and cylinders of the same height,
let the circles *ABCD*, *EFGH* be their bases, *KL*, *MN* their axes and *AC*, *EG* the diameters of their bases;
I say that, as the circle *ABCD* is to the circle *EFGH*, so is the cone *AL* to the cone *EN*.

For, if not, then, as the circle *ABCD* is to the circle *EFGH*, so will the cone *AL* be either to some solid less than the cone *EN* or to a greater.

First, let it be in that ratio to a less solid *O*,
and let the solid *X* be equal to that by which the solid *O* is less than the cone *EN*;
therefore the cone *EN* is equal to the solids *O*, *X*.

Let the square *EFGH* be inscribed in the circle *EFGH*;
therefore the square is greater than the half of the circle.

Let there be set up from the square *EFGH* a pyramid of equal height with the cone;
therefore the pyramid so set up is greater than the half of the cone,
inasmuch as, if we circumscribe a square about the circle, and set up from it a pyramid of equal height with the cone, the inscribed pyramid is half of the circumscribed pyramid,
for they are to one another as their bases, [XII. 6]
while the cone is less than the circumscribed pyramid.

Let the circumferences *EF*, *FG*, *GH*, *HE* be bisected at the points *P*, *Q*, *R*, *S*, and let *HP*, *PE*, *EQ*, *QF*, *FR*, *RG*, *GS*, *SH* be joined.

Therefore each of the triangles *HPE*, *EQF*, *FRG*, *GSH* is greater than the half of that segment of the circle which is about it.

On each of the triangles *HPE*, *EQF*, *FRG*, *GSH* let there be set up a pyramid of equal height with the cone;
therefore, also, each of the pyramids so set up is greater than the half of that segment of the cone which is about it.

Thus, bisecting the circumferences which are left, joining straight lines, setting up on each of the triangles pyramids of equal height with the cone,
and doing this continually,
we shall leave some segments of the cone which will be less than the solid *X*. [X. 1]

Let such be left, and let them be the segments on *HP, PE, EQ, QF, FR, RG, GS, SH*; therefore the remainder, the pyramid of which the polygon *HPEQFRGS* is the base and the height the same with that of the cone, is greater than the solid *O*.

Let there also be inscribed in the circle *ABCD* the polygon *DTAUBVCW* similar and similarly situated to the polygon *HPEQFRGS*,
and on it let a pyramid be set up of equal height with the cone *AL*.

Since then, as the square on *AC* is to the square on *EG*, so is the polygon *DTAUB-VCW* to the polygon *HPEQFRGIS*, [XII. 1]
while, as the square on *AC* is to the square on *EG*, so is the circle *ABCD* to the circle *EFGH*, [XII. 2]
therefore also, as the circle *ABCD* is to the circle *EEGH*, so is the polygon *DTAUBVCW* to the polygon *HPEQFRGS*.

But, as the circle *ABCD* is to the circle *EFGH*, so is the cone *AL* to the solid *O*,
and, as the polygon *DTAUBVCW* is to the polygon *HPEQFRGS*, so is the pyramid of which the polygon *DTAUBVCW* is the base and the point *L* the vertex to the pyramid of which the polygon *HPEQFRGS* is the base and the point *N* the vertex. [XII. 6]

Therefore also, as the cone *AL* is to the solid *O*, so is the pyramid of which the polygon *DTAUBVCW* is the base and the point *L* the vertex to the pyramid of which the polygon *HPEQFRGS* is the base and the point *N* the vertex; [V. 11]
therefore, alternately, as the cone *AL* is to the pyramid in it, so is the solid *O* to the pyramid in the cone *EN*. [V. 16]

But the cone *AL* is greater than the pyramid in it;
 therefore the solid *O* is also greater than the pyramid in the cone *EN*.

But it is also less:
which is absurd.

Therefore the cone *AL* is not to any solid less than the cone *EN* as the circle *ABCD* is to the circle *EFGH*.

Similarly we can prove that neither is the cone *EN* to any solid less than the cone *AL* as the circle *EFGH* is to the circle *ABCD*.

I say next that neither is the cone *AL* to any solid greater than the cone *EN* as the circle *ABCD* is to the circle *EFGH*.

For, if possible, let it be in that ratio to a greater solid *O*;
therefore, inversely, as the circle *EFGH* is to the circle *ABCD*, so is the solid *O* to the cone *AL*.

But, as the solid *O* is to the cone *AL*, so is the cone *EN* to some solid less than the cone *AL*;
therefore also, as the circle *EFGH* is to the circle *ABCD*, so is the cone *EN* to some solid less than the cone *AL*:
which was proved impossible.

Therefore the cone *AL* is not to any solid greater than the cone *EN* as the circle *ABCD* is to the circle *EFGH*.

But it was proved that neither is it in this ratio to a less solid;
therefore, as the circle *ABCD* is to the circle *EFGH*, so is the cone *AL* to the cone *EN*.

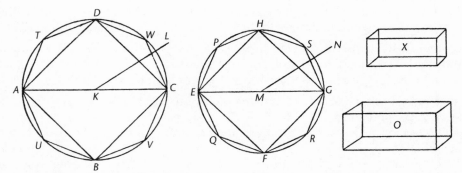

But, as the cone is to the cone, so is the cylinder to the cylinder,
for each is triple of each. [XII. 10]

Therefore also, as the circle *ABCD* is to the circle *EFGH*, so are the cylinders on them which are of equal height.

Therefore etc.

Q.E.D.

Proposition 12

Similar cones and cylinders are to one another in the triplicate ratio of the diameters in their bases.

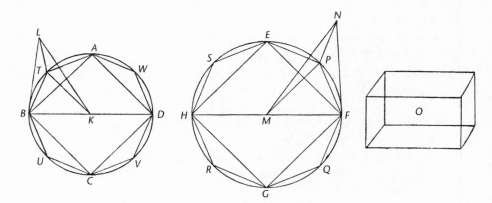

Let there be similar cones and cylinders,
let the circles *ABCD*, *EFGH* be their bases, *BD*, *FH* the diameters of the bases, and *KL*, *MN* the axes of the cones and cylinders;
I say that the cone of which the circle *ABCD* is the base and the point *L* the vertex has to the cone of which the circle *EFGH* is the base and the point *N* the vertex the ratio triplicate of that which *BD* has to *FH*.

For, if the cone *ABCDL* has not to the cone *EFGHN* the ratio triplicate of that which *BD* has to *FH*,
the cone *ABCDL* will have that triplicate ratio either to some solid less than the cone *EFGHN* or to a greater.

First, let it have that triplicate ratio to a less solid O.

Let the square $EFGH$ be inscribed in the circle $EFGH$; [IV. 6]
 therefore the square $EFGH$ is greater than the half of the circle $EFGH$.

Now let there be set up on the square $EFGH$ a pyramid having the same vertex with the cone;
 therefore the pyramid so set up is greater than the half part of the cone.

Let the circumferences EF, FG, GH, HE be bisected at the points P, Q, R, S, and let EP, PF, FQ, QG, GR, RH, HS, SE be joined.

Therefore each of the triangles EPF, FQG, GRH, HSE is also greater than the half part of that segment of the circle $EFGH$ which is about it.

Now on each of the triangles EPF, FQG, GRH, HSE let a pyramid be set up having the same vertex with the cone;
therefore each of the pyramids so set up is also greater than the half part of that segment of the cone which is about it.

Thus, bisecting the circumferences so left, joining straight lines, setting up on each of the triangles pyramids having the same vertex with the cone,
and doing this continually,
we shall leave some segments of the cone which will be less than the excess by which the cone $EFGHN$ exceeds the solid O. [X. 1]

Let such be left, and let them be the segments on EP, PF, FQ, QG, GR, RH, HS, SE;
therefore the remainder, the pyramid of which the polygon $EPFQGRHS$ is the base and the point N the vertex, is greater than the solid O.

Let there be also inscribed in the circle $ABCD$ the polygon $ATBUCVDW$ similar and similarly situated to the polygon $EPFQGRHS$,
and let there be set up on the polygon $ATBUCVDW$ a pyramid having the same vertex with the cone;
of the triangles containing the pyramid of which the polygon $ATBUCVDW$ is the base and the point L the vertex let LBT be one,
and of the triangles containing the pyramid of which the polygon $EPFQGRHS$ is the base and the point N the vertex let NFP be one;
and let KT, MP be joined.

Now, since the cone $ABCDL$ is similar to the cone $EFGHN$,
 therefore, as BD is to FH, so is the axis KL to the axis MN. [XI. Def. 24]

But, as BD is to FH, so is BK to FM;
 therefore also, as BK is to FM, so is KL to MN.

And, alternately, as BK is to KL, so is FM to MN. [V. 16]

And the sides are proportional about equal angles, namely the angles BKL, FMN;
 therefore the triangle BKL is similar to the triangle FMN. [VI. 6]

Again, since, as BK is to KT, so is FM to MP,
and they are about equal angles, namely the angles BKT, FMP,
inasmuch as, whatever part the angle BKT is of the four right angles at the centre K, the same part also is the angle FMP of the four right angles at the centre M;
since then the sides are proportional about equal angles,
 therefore the triangle BKT is similar to the triangle FMP. [VI. 6]

Again, since it was proved that, as BK is to KL, so is FM to MN,
while BK is equal to KT, and FM to PM,

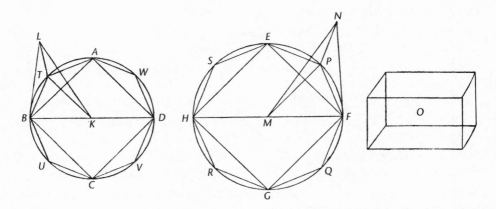

therefore, as *TK* is to *KL*, so is *PM* to *MN*;
and the sides are proportional about equal angles, namely the angles *TKL*, *PMN*, for they are right;

therefore the triangle *LKT* is similar to the triangle *NMP*. [VI. 6]

And since, owing to the similarity of the triangles *LKB*, *NMF*,
as *LB* is to *BK*, so is *NF* to *FM*,
and, owing to the similarity of the triangles *BKT*, *FMP*,
as *KB* is to *BT*, so is *MF* to *FP*,
therefore, *ex aequali*, as *LB* is to *BT*, so is *NF* to *FP*. [V. 22]

Again, since, owing to the similarity of the triangles *LTK*, *NPM*,
as *LT* is to *TK*, so is *NP* to *PM*,
and, owing to the similarity of the triangles *TKB*, *PMF*,
as *KT* is to *TB*, so is *MP* to *PF*;
therefore, *ex aequali*, as *LT* is to *TB*, so is *NP* to *PF*. [V. 22]

But it was also proved that, as *TB* is to *BL*, so is *PF* to *FN*.

Therefore, *ex aequali*, as *TL* is to *LB*, so is *PN* to *NF*. [V. 22]

Therefore in the triangles *LTB*, *NPF* the sides are proportional;
therefore the triangles *LTB*, *NPF* are equiangular; [VI. 5]
hence they are also similar. [VI. Def. 1]

Therefore the pyramid of which the triangle *BKT* is the base and the point *L* the vertex is also similar to the pyramid of which the triangle *FMP* is the base and the point *N* the vertex,
for they are contained by similar planes equal in multitude. [XI. Def. 9]

But similar pyramids which have triangular bases are to one another in the triplicate ratio of their corresponding sides. [XII. 8]

Therefore the pyramid *BKTL* has to the pyramid *FMPN* the ratio triplicate of that which *BK* has to *FM*.

Similarly, by joining straight lines from *A*, *W*, *D*, *V*, *C*, *U* to *K*, and from *E*, *S*, *H*, *R*, *G*, *Q* to *M*, and setting up on each of the triangles pyramids which have the same vertex with the cones,
we can prove that each of the similarly arranged pyramids will also have to each similarly arranged pyramid the ratio triplicate of that which the corresponding side *BK* has to the corresponding side *FM*, that is, which *BD* has to *FH*.

And, as one of the antecedents is to one of the consequents, so are all the antecedents to all the consequents; [V. 12]
therefore also, as the pyramid *BKTL* is to the pyramid *FMPN*, so is the whole pyramid of which the polygon *ATBUCVDW* is the base and the point *L* the vertex to the whole pyramid of which the polygon *EPFQGRHS* is the base and the point *N* the vertex;
hence also the pyramid of which *ATBUCVDW* is the base and the point *L* the vertex has to the pyramid of which the polygon *EPFQGRHS* is the base and the point *N* the vertex the ratio triplicate of that which *BD* has to *FH*.

But, by hypothesis, the cone of which the circle *ABCD* is the base and the point *L* the vertex has also to the solid *O* the ratio triplicate of that which *BD* has to *FH*;
therefore, as the cone of which the circle *ABCD* is the base and the point *L* the vertex is to the solid *O*, so is the pyramid of which the polygon *ATBUCVDW* is the base and *L* the vertex to the pyramid of which the polygon *EPFQGRHS* is the base and the point *N* the vertex;
therefore, alternately, as the cone of which the circle *ABCD* is the base and *L* the vertex is to the pyramid contained in it of which the polygon *ATBUCVDW* is the base and *L* the vertex, so is the solid *O* to the pyramid of which the polygon *EPFQGRHS* is the base and *N* the vertex. [V. 16]

But the said cone is greater than the pyramid in it;
for it encloses it.

Therefore the solid *O* is also greater than the pyramid of which the polygon *EPFQGRHS* is the base and *N* the vertex.

But it is also less:
which is impossible.

Therefore the cone of which the circle *ABCD* is the base and *L* the vertex has not to any solid less than the cone of which the circle *EFGH* is the base and the point *N* the vertex the ratio triplicate of that which *BD* has to *FH*.

Similarly we can prove that neither has the cone *EFGHN* to any solid less than the cone *ABCDL* the ratio triplicate of that which *FH* has to *BD*.

I say next that neither has the cone *ABCDL* to any solid greater than the cone *EFGHN* the ratio triplicate of that which *BD* has to *FH*.

For, if possible, let it have that ratio to a greater solid *O*.

Therefore, inversely, the solid *O* has to the cone *ABCDL* the ratio triplicate of that which *FH* has to *BD*.

But, as the solid *O* is to the cone *ABCDL*, so is the cone *EFGHN* to some solid less than the cone *ABCDL*.

Therefore the cone *EFGHN* also has to some solid less than the cone *ABCDL* the ratio triplicate of that which *FH* has to *BD*:
which was proved impossible.

Therefore the cone *ABCDL* has not to any solid greater than the cone *EFGHN* the ratio triplicate of that which *BD* has to *FH*.

But it was proved that neither has it this ratio to a less solid than the cone *EFGHN*.

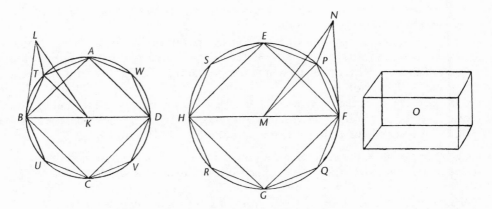

Therefore the cone *ABCDL* has to the cone *EFGHN* the ratio triplicate of that which *BD* has to *FH*.

But, as the cone is to the cone, so is the cylinder to the cylinder,
for the cylinder which is on the same base as the cone and of equal height with it is triple of the cone; [XII. 10]
therefore the cylinder also has to the cylinder the ratio triplicate of that which *BD* has to *FH*.

Therefore etc.

Q.E.D.

Proposition 13

If a cylinder be cut by a plane which is parallel to its opposite planes, then, as the cylinder is to the cylinder, so will the axis be to the axis.

For let the cylinder *AD* be cut by the plane *GH* which is parallel to the opposite planes *AB*, *CD*,
and let the plane *GH* meet the axis at the point *K*;
I say that, as the cylinder *BG* is to the cylinder *GD*, so is the axis *EK* to the axis *KF*.

For let the axis *EF* be produced in both directions to the points *L*, *M*,
and let there be set out any number whatever of axes *EN*, *NL* equal to the axis *EK*,
and any number whatever *FO*, *OM* equal to *FK*;
and let the cylinder *PW* on the axis *LM* be conceived of which the circles *PQ*, *VW* are the bases.

Let planes be carried through the points *N*, *O* parallel to *AB*, *CD* and to the bases of the cylinder *PW*,
and let them produce the circles *RS*, *TU* about the centres *N*, *O*.

Then, since the axes *LN, NE, EK* are equal to one another,
 therefore the cylinders *QR, RB, BG* are to one another as their bases. [XII. 11]

But the bases are equal;
 therefore the cylinders *QR, RB, BG* are also equal to one another.

Since then the axes *LN, NE, EK* are equal to one another,
and the cylinders *QR, RB, BG* are also equal to one another,
and the multitude of the former is equal to the multitude of the latter,
therefore, whatever multiple the axis *KL* is of the axis *EK*, the same multiple also
will the cylinder *QG* be of the cylinder *GB*.

For the same reason, whatever multiple the axis *MK* is of the axis *KF*, the same
multiple also is the cylinder *WG* of the cylinder *GD*.

And, if the axis *KL* is equal to the axis *KM*, the cylinder *QG* will also be equal to
the cylinder *GW*,
if the axis is greater than the axis, the cylinder will also be greater than the
cylinder,
and if less, less.

Thus, there being four magnitudes, the axes *EK, KF* and the cylinders *BG, GD*,
there have been taken equimultiples of the axis *EK* and of the cylinder *BG*,
namely the axis *LK* and the cylinder *QG*,
and equimultiples of the axis *KF* and of the cylinder *GD*, namely the axis *KM*
and the cylinder *GW*;
and it has been proved that, if the axis *KL* is in excess of the axis *KM*, the
cylinder *QG* is also in excess of the cylinder *GW*,
if equal, equal,
and if less, less.

Therefore, as the axis *EK* is to the axis *KF*, so is the cylinder *BG* to the cylinder *GD*.
[V. Def. 5]

Q.E.D.

Proposition 14

Cones and cylinders which are on equal bases are to one another as their heights.

For let *EB, FD* be cylinders on equal bases, the circles *AB, CD*;
I say that, as the cylinder *EB* is to the cylinder *FD*, so is the axis *GH* to the axis *KL*.

For let the axis *KL* be produced to the point *N*,
let *LN* be made equal to the axis *GH*,
and let the cylinder *CM* be conceived about *LN* as axis.

Since then the cylinders *EB, CM* are of the same height, they are to one another as their bases. [XII. 11]

But the bases are equal to one another:
therefore the cylinders *EB, CM* are also equal.

And, since the cylinder *FM* has been cut by the plane *CD* which is parallel to its opposite planes,
therefore, as the cylinder *CM* is to the cylinder *FD*, so is the axis *LN* to the axis *KL*. [XII. 13]

But the cylinder *CM* is equal to the cylinder *EB*,
and the axis *LN* to the axis *GH*;
therefore, as the cylinder *EB* is to the cylinder *FD*, so is the axis *OH* to the axis *KL*.

But, as the cylinder *EB* is to the cylinder *FD*, so is the cone *ABG* to the cone *CDK*. [XII. 10]

Therefore also, as the axis *GH* is to the axis *KL*, so is the cone *ABG* to the cone *CDK* and the cylinder *EB* to the cylinder *ED*.

Q.E.D.

Proposition 15

In equal cones and cylinders the bases are reciprocally proportional to the heights; and those cones and cylinders in which the bases are reciprocally proportional to the heights are equal.

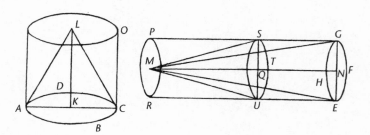

Let there be equal cones and cylinders of which the circles *ABCD, EFGH* are the bases;
let *AC, EG* be the diameters of the bases,
and *KL, MN* the axes, which are also the heights of the cones or cylinders;

let the cylinders *AO, EP* be completed.

I say that in the cylinders *AO, EP* the bases are reciprocally proportional to the heights,
that is, as the base *ABCD* is to the base *EFGH*, so is the height *MN* to the height *KL*.

For the height *LK* is either equal to the height *MN* or not equal.

First, let it be equal.

Now the cylinder *AO* is also equal to the cylinder *EP*.

But cones and cylinders which are of the same height are to one another as their bases; [XII. 11]
 therefore the base *ABCD* is also equal to the base *EFGH*.

Hence also, reciprocally, as the base *ABCD* is to the base *EFGH*, so is the height *MN* to the height *KL*.

Next, let the height *LK* not be equal to *MN*,
but let *MN* be greater;
from the height *MN* let *QN* be cut off equal to *KL*,
through the point *Q* let the cylinder *EP* be cut by the plane *TUS* parallel to the planes of the circles *EFGH, RP,*
and let the cylinder *ES* be conceived erected from the circle *EFGH* as base and with height *NQ*.

Now, since the cylinder *AO* is equal to the cylinder *EP,*
therefore, as the cylinder *AO* is to the cylinder *ES,* so is the cylinder *EP* to the cylinder *ES.* [V. 7]

But, as the cylinder *AO* is to the cylinder *ES,* so is the base *ABCD* to the base *EFGH,*
for the cylinders *AO, ES* are of the same height; [XII. 11]
and, as the cylinder *EP* is to tlie cylinder *ES,* so is the height *MN* to the height *QN,*
for the cylinder *EP* has been cut by a plane which is parallel to its opposite planes. [XII. 13]

Therefore also, as the base *ABCD* is to the base *EFGH,* so is the height *MN* to the height *QN.* [V. 11]

But the height *QN* is equal to the height *KL*;
therefore, as the base *ABCD* is to the base *EFGH,* so is the height *MN* to the height *KL.*

Therefore in the cylinders *AO, EP* the bases are reciprocally proportional to the heights.

Next, in the cylinders *AO, EP* let the bases be reciprocally proportional to the heights,
that is, as the base *ABCD* is to the base *EFGH,* so let the height *MN* be to the height *KL*;
I say that the cylinder *AO* is equal to the cylinder *EP.*

For, with the same construction,
since, as the base *ABCD* is to the base *EFGH,* so is the height *MN* to the height *KL,*
while the height *KL* is equal to the height *QN,*
therefore, as the base *ABCD* is to the base *EFGH,* so is the height *MN* to the height *QN.*

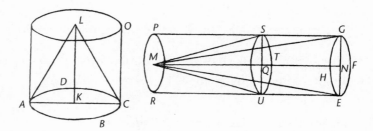

But, as the base *ABCD* is to the base *EFGH*, so is the cylinder *AO* to the cylinder *ES*, for they are of the same height; [XII. 11]
and, as the height *MN* is to *QN*, so is the cylinder *EP* to the cylinder *ES*; [XII. 13]
therefore, as the cylinder *AO* is to the cylinder *ES*, so is the cylinder *EP* to the cylinder *ES*. [V. 11]

Therefore the cylinder *AO* is equal to the cylinder *EP*. [V. 9]

And the same is true for the cones also.

Q.E.D.

Proposition 16

Given two circles about the same centre, to inscribe in the greater circle an equilateral polygon with an even number of sides which does not touch the lesser circle.

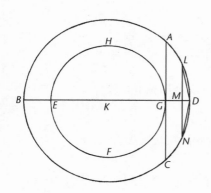

Let *ABCD*, *EFGH* be the two given circles about the same centre *K*;
thus it is required to inscribe in the greater circle *ABCD* an equilateral polygon with an even number of sides which does not touch the circle *EFGH*.

For let the straight line *BKD* be drawn through the centre *K*,
and from the point *G* let *GA* be drawn at right angles to the straight line *BD* and carried through to *C*;
therefore *AC* touches the circle *EFGH*.

[III. 16, Por.]

Then, bisecting the circumference *BAD*, bisecting the half of it, and doing this continually, we shall leave a circumference less than *AD*. [X. 1]

Let such be left, and let it be *LD*;
from *L* let *LM* be drawn perpendicular to *BD* and carried through to *N*,
and let *LD*, *DN* be joined;
therefore *LD* is equal to *DN*. [III. 3, I. 4]

Now, since *LN* is parallel to *AC*,
and *AC* touches the circle *EFGH*,
therefore *LN* does not touch the circle *EFGH*;
therefore *LD*, *DN* are far from touching the circle *EFGH*.

If then we fit into the circle *ABCD* straight lines equal to the straight line *LD* and placed continuously, there will be inscribed in the circle *ABCD* an equilateral polygon with an even number of sides which does not touch the lesser circle *EFGH*.

<div align="right">Q.E.F.</div>

Proposition 17

Given two spheres about the same centre, to inscribe in the greater sphere a polyhedral solid which does not touch the lesser sphere at its surface.

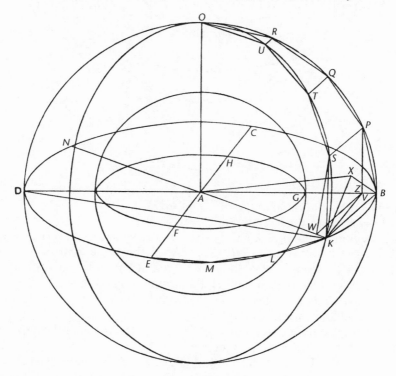

Let two spheres be conceived about the same centre *A*;
thus it is required to inscribe in the greater sphere a polyhedral solid which does not touch the lesser sphere at its surface.

Let the spheres be cut by any plane through the centre;
then the sections will be circles,
inasmuch as the sphere was produced by the diameter remaining fixed and the semicircle being carried round it; [XI. Def. 14]
hence, in whatever position we conceive the semicircle to be, the plane carried through it will produce a circle on the circumference of the sphere.

And it is manifest that this circle is the greatest possible,
inasmuch as the diameter of the sphere, which is of course the diameter both of the semicircle and of the circle, is greater than all the straight lines drawn across in the circle or the sphere.

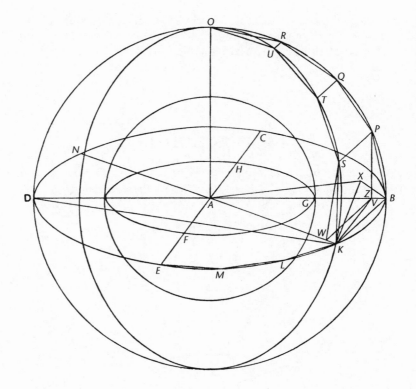

Let then *BCDE* be the circle in the greater sphere,
and *FGH* the circle in the lesser sphere;
let two diameters in them, *BD, CE*, be drawn at right angles to one another;
then, given the two circles *BCDE, FGH* about the same centre, let there be inscribed in the greater circle *BCDE* an equilateral polygon with an even number of sides which does not touch the lesser circle *FGH*,
let *BK, KL, LM, ME* be its sides in the quadrant *BE*,
let *KA* be joined and carried through to *N*,
let *AO* be set up from the point *A* at right angles to the plane of the circle *BCDE*, and let it meet the surface of the sphere at *O*,
and through *AO* and each of the straight lines *BD, KN* let planes be carried;
they will then make greatest circles on the surface of the sphere, for the reason stated.

Let them make such,
and in them let *BOD, KON* be the semicircles on *BD, KN*.

Now, since *OA* is at right angles to the plane of the circle *BCDE*,
therefore all the planes through *OA* are also at right angles to the plane of the circle *BCDE*; [XI. 18]
hence the semicircles *BOD, KON* are also at right angles to the plane of the circle *BCDE*.

And, since the semicircles *BED, BOD, KON* are equal,
for they are on the equal diameters *BD, KN*,
 therefore the quadrants *BE, BO, KO* are also equal to one another.

Therefore there are as many straight lines in the quadrants *BO, KO* equal to the straight lines *BK, KL, LM, ME* as there are sides of the polygon in the quadrant *BE*.

Let them be inscribed, and let them be *BP, PQ, QR, RO* and *KS, ST, TU, UO,*
let *SP, TQ, UR* be joined,
and from *P, S* let perpendiculars be drawn to the plane of the circle *BCDE;* [XI. 11]
　　　these will fall on *BD, KN,* the common sections of the planes,
inasmuch as the planes of *BOD, KON* are also at right angles to the plane of the
circle *BCDE.*
 [cf. XI. Def. 4]

Let them so fall, and let them be *PV, SW,*
and let *WV* be joined.

Now since, in the equal semicircles *BOD, KON,* equal straight lines *BP, KS* have
been cut off,
and the perpendiculars *PV, SW* have been drawn,
　　　　　　therefore *PV* is equal to *SW,* and *BV* to *KW.* [III. 27, I. 26]

But the whole *BA* is also equal to the whole *KA;*
　　　therefore the remainder *VA* is also equal to the remainder *WA;*
　　　　　therefore, as *BV* is to *VA,* so is *KW* to *WA;*
　　　　　　　therefore *WV* is parallel to *KB.* [VI. 2]

And, since each of the straight lines *PV, SW* is at right angles to the plane of the
circle *BCDE,*
　　　　　　therefore *PV* is parallel to *SW.* [XI. 6]

But it was also proved equal to it;
　　　　　therefore *WV, SP* are also equal and parallel. [I. 33]

And, since *WV* is parallel to *SP,*
while *WV* is parallel to *KB,*
　　　　　therefore *SP* is also parallel to *KB.* [XI. 9]

And *BP, KS* join their extremities;
　　　　　therefore the quadrilateral *KBPS* is in one plane,
inasmuch as, if two straight. lines be parallel, and points be taken at random on
each of them, the straight line joining the points is in the same plane with the
parallels. [XI. 7]

For the same reason
　　　　each of the quadrilaterals *SPQT, TQRU* is also in one plane.

But the triangle *URO* is also in one plane. [XI. 2]

If then we conceive straight lines joined from the points *P, S, Q, T, R, U* to *A,*
there will be constructed a certain polyhedral solid figure between the circum-
ferences *BO, KO,* consisting of pyramids of which the quadrilaterals *KBPS,
SPQT, TQRU* and the triangle *URO* are the bases and the point *A* the vertex.

And, if we make the same construction in the case of each of the sides *KL, LM, ME*
as in the case of *BK,* and further, in the case of the remaining three quadrants,
there will be constructed a certain polyhedral figure inscribed in the sphere
and contained by pyramids, of which the said quadrilaterals and the triangle
URO, and the others corresponding to them, are the bases and the point *A* the
vertex.

I say that the said polyhedron will not touch the lesser sphere at the surface on
which the circle *FGH* is.

Let *AX* be drawn from the point *A* perpendicular to the plane of the quadrilateral
KBPS, and let it meet the plane at the point *X;* [XI. 11]
let *XB, XK* be joined.

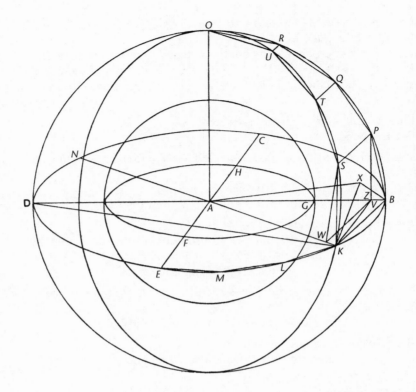

Then, since AX is at right angles to the plane of the quadrilateral $KBPS$, therefore it is also at right angles to all the straight lines which meet it and are in the plane of the quadrilateral. [XI. Def. 3]

Therefore AX is at right angles to each of the straight lines BX, XK.

And, since AB is equal to AK,
 the square on AB is also equal to the square on AK.

And the squares on AX, XB are equal to the square on AB,
for the angle at X is right; [I. 47]
 and the squares on AX, XK are equal to the square on AK. [I. 47]

Therefore the squares on AX, XB are equal to the squares on AX, XK.

Let the square on AX be subtracted from each;
therefore the remainder, the square on BX, is equal to the remainder, the square on XK;

 therefore BX is equal to XK.

Similarly we can prove that the straight lines joined from X to P, S are equal to each of the straight lines BX, XK.

Therefore the circle described with centre X and distance one of the straight lines XB, XK will pass through P, S also,
 and $KBPS$ will be a quadrilateral in a circle.

Now, since KB is greater than WV,
while WV is equal to SP,
 therefore KB is greater than SP.

But KB is equal to each of the straight lines KS, BP;

therefore each of the straight lines KS, BP is greater than SP.

And, since KBPS is a quadrilateral in a circle,
and KB, BP, KS are equal, and PS less,
and BX is the radius of the circle,
 therefore the square on KB is greater than double of the square on BX.

Let KZ be drawn from K perpendicular to BV.[1]

Then, since BD is less than double of DZ,
and, as BD is to DZ, so is the rectangle DB, BZ to the rectangle DZ, ZB,
if a square be described upon BZ and the parallelogram on ZD be completed,
then the rectangle DB, BZ is also less than double of the rectangle DZ, ZB.

And, if KD be joined,
 the rectangle DB, BZ is equal to the square on BK,
and the rectangle DZ, ZB equal to the square on KZ; [III. 31, VI. 8 and Por.]
 therefore the square on KB is less than double of the square on KZ.

But the square on KB is greater than double of the square on BX;
 therefore the square on KZ is greater than the square on BX.

And, since BA is equal to KA,
 the square on BA is equal to the square on AK.

And the squares on BX, XA are equal to the square on BA,
 and the squares on KZ, ZA equal to the square on KA; [I. 47]
 therefore the squares on BX, XA are equal to the squares on KZ, ZA,
and of these the square on KZ is greater than the square on BX;
 therefore the remainder, the square on ZA, is less than the square on XA.

Therefore AX is greater than AZ;
 therefore AX is much greater than AG.

And AX is the perpendicular on one base of the polyhedron,
and AG on the surface of the lesser sphere;
 hence the polyhedron will not touch the lesser sphere on its surface.

Therefore, given two spheres about the same centre, a polyhedral solid has been inscribed in the greater sphere which does not touch the lesser sphere at its surface.

Q.E.F.

PORISM. But if in another sphere also a polyhedral solid be inscribed similar to the solid in the sphere BCDE,
the polyhedral solid in the sphere BCDE has to the polyhedral solid in the other sphere the ratio triplicate of that which the diameter of the sphere BCDE has to the diameter of the other sphere.

For, the solids being divided into their pyramids similar in multitude and arrangement, the pyramids will be similar.

1. Heath comments that Z and V denote the same point, as indeed the figure indicates. For draw KZ perpendicular to diameter BD. Then the rectangle DB, BZ will equal the square on BK. Similarly, because PV is perpendicular to BD, the rectangle DB, BV is equal to the square on BP. But BK is equal to BP. Hence the rectangle DB, BZ equals the rectangle DB, BV; thus BV equals BZ and Z, V coincide. —Ed.

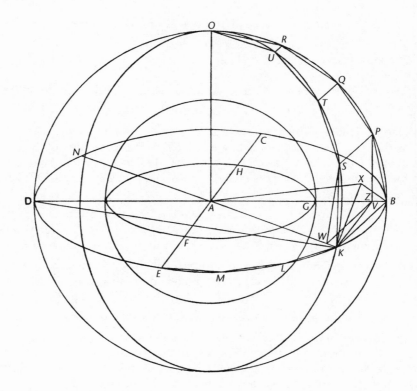

But similar pyramids are to one another in the triplicate ratio of their corresponding sides; [XII. 8, Por.]
therefore the pyramid of which the quadrilateral *KBPS* is the base, and the point *A* the vertex, has to the similarly arranged pyramid in the other sphere the ratio triplicate of that which the corresponding side has to the corresponding side, that is, of that which the radius *AB* of the sphere about *A* as centre has to the radius of the other sphere.

Similarly also each pyramid of those in the sphere about *A* as centre has to each similarly arranged pyramid of those in the other sphere the ratio triplicate of that which *AB* has to the radius of the other sphere.

And, as one of the antecedents is to one of the consequents, so are all the antecedents to all the consequents; [V. 12]
hence the whole polyhedral solid in the sphere about *A* as centre has to the whole polyhedral solid in the other sphere the ratio triplicate of that which *AB* has to the radius of the other sphere, that is, of that which the diameter *BD* has to the diameter of the other sphere.

Q.E.D.

Proposition 18

Spheres are to one another in the triplicate ratio of their respective diameters.

Let the spheres *ABC*, *DEF* be conceived,
and let *BC*, *EF* be their diameters;

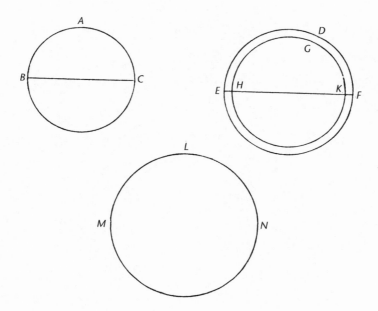

I say that the sphere *ABC* has to the sphere *DEF* the ratio triplicate of that which *BC* has to *EF*.

For, if the sphere *ABC* has not to the sphere *DEF* the ratio triplicate of that which *BC* has to *EF*,
then the sphere *ABC* will have either to some less sphere than the sphere *DEF*, or to a greater, the ratio triplicate of that which *BC* has to *EF*.

First, let it have that ratio to a less sphere *GHK*,
let *DEF* be conceived about the same centre with *GHK*,
let there be inscribed in the greater sphere *DEF* a polyhedral solid which does not touch the lesser sphere *GHK* at its surface, [XII. 17]
and let there also be inscribed in the sphere *ABC* a polyhedral solid similar to the polyhedral solid in the sphere *DEF*;
therefore the polyhedral solid in *ABC* has to the polyhedral solid in *DEF* the ratio triplicate of that which *BC* has to *EF*. [XII. 17, Por.]

But the sphere *ABC* also has to the sphere *GHK* the ratio triplicate of that which *BC* has to *EF*;
therefore, as the sphere *ABC* is to the sphere *GHK*, so is the polyhedral solid in the sphere *ABC* to the polyhedral solid in the sphere *DEF*;
and, alternately, as the sphere *ABC* is to the polyhedron in it, so is the sphere *GHK* to the polyhedral solid in the sphere *DEF*. [V. 16]

But the sphere *ABC* is greater than the polyhedron in it;
therefore the sphere *GHK* is also greater than the polyhedron in the sphere *DEF*.

But it is also less,
for it is enclosed by it.

Therefore the sphere *ABC* has not to a less sphere than the sphere *DEF* the ratio triplicate of that which the diameter *BC* has to *EF*.

Similarly we can prove that neither has the sphere *DEF* to a less sphere than the sphere *ABC* the ratio triplicate of that which *EF* has to *BC*.

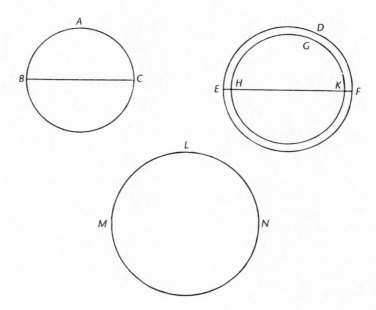

I say next that neither has the sphere *ABC* to any greater sphere than the sphere *DEF* the ratio triplicate of that which *BC* has to *EF*.

For, if possible, let it have that ratio to a greater, *LMN*;
therefore, inversely, the sphere *LMN* has to the sphere *ABC* the ratio triplicate of that which the diameter *EF* has to the diameter *BC*.

But, inasmuch as *LMN* is greater than *DEF*,
therefore, as the sphere *LMN* is to the sphere *ABC*, so is the sphere *DEF* to some less sphere than the sphere *ABC*, as was before proved. [XII. 2, Lemma]

Therefore the sphere *DEF* also has to some less sphere than the sphere *ABC* the ratio triplicate of that which *EF* has to *BC*:
which was proved impossible.

Therefore the sphere *ABC* has not to any sphere greater than the sphere *DEF* the ratio triplicate of that which *BC* has to *EF*.

But it was proved that neither has it that ratio to a less sphere.

Therefore the sphere *ABC* has to the sphere *DEF* the ratio triplicate of that which *BC* has to *EF*.

Q.E.D.

Book XIII

Proposition 1

If a straight line be cut in extreme and mean ratio, the square on the greater segment added to the half of the whole is five times the square on the half.

For let the straight line AB be cut in extreme and mean ratio at the point C, and let AC be the greater segment; let the straight line AD be produced in a straight line with CA, and let AD be made half of AB; I say that the square on CD is five times the square on AD.

For let the squares AE, DF be described on AB, DC, and let the figure in DF be drawn; let FC be carried through to G.

Now, since AB has been cut in extreme and mean ratio at C, therefore the rectangle AB, BC is equal to the square on AC.　　　**[VI. Def. 3, VI. 17]**

And CE is the rectangle AB, BC, and FH the square on AC;
　　　　　therefore CE is equal to FH.

And, since BA is double of AD, while BA is equal to KA, and AD to AH,
　　　　　therefore KA is also double of AH.

But, as KA is to AH, so is CK to CH;　　　　　**[VI. 1]**
　　　　　therefore CK is double of CH.

But LH, HC are also double of CH.

　　　　Therefore KC is equal to LH, HC.

But CE was also proved equal to HF;
　　　therefore the whole square AE is equal to the gnomon MNO.

And, since BA is double of AD,
　　　the square on BA is quadruple of the square on AD,
　　　　　that is, AE is quadruple of DH.

But AE is equal to the gnomon MNO;
　　　　therefore the gnomon MNO is also quadruple of AP;
　　　　　therefore the whole DF is five times AP.

And DF is the square on DC, and AP the square on DA;
　　　therefore the square on CD is five times the square on DA.

　Therefore etc.

　　　　　　　　　　　　　　　　　　　　　Q.E.D.

Proposition 2

If the square on a straight line be five times the square on a segment of it, then, when the double of the said segment is cut in extreme and mean ratio, the greater segment is the remaining part of the original straight line.

For let the square on the straight line *AB* be five times the square on the segment *AC* of it,
and let *CD* be double of *AC*;
I say that, when *CD* is cut in extreme and mean ratio, the greater segment is *CB*.

Let the squares *AF*, *CG* be described on *AB*, *CD* respectively,
let the figure in *AF* be drawn,
and let *BE* be drawn through.

Now, since the square on *BA* is five times the square on *AC*,

AF is five times *AH*.

Therefore the gnomon *MNO* is quadruple of *AH*.

And, since *DC* is double of *CA*,
therefore the square on *DC* is quadruple of the square on *CA*,

that is, *CG* is quadruple of *AH*.

But the gnomon *MNO* was also proved quadruple of *AH*;
therefore the gnomon *MNO* is equal to *CG*.

And, since *DC* is double of *CA*,
while *DC* is equal to *CK*, and *AC* to *CH*,
therefore *KB* is also double of *BH*. [VI. 1]

But *LH*, *HB* are also double of *HB*;
therefore *KB* is equal to *LH*, *HB*.

But the whole gnomon *MNO* was also proved equal to the whole *CG*;
therefore the remainder *HF* is equal to *BG*.

And *BG* is the rectangle *CD*, *DB*,
for *CD* is equal to *DG*;
and *HF* is the square on *CB*;
therefore the rectangle *CD*, *DB* is equal to the square on *CB*.

Therefore, as *DC* is to *CB*, so is *CB* to *BD*.

But *DC* is greater than *CB*;
therefore *CB* is also greater than *BD*.

Therefore, when the straight line *CD* is cut in extreme and mean ratio, *CB* is the greater segment.

Therefore etc.

Q.E.D.

LEMMA. That the double of *AC* is greater than *BC* is to be proved thus.

If not, let *BC* be, if possible, double of *CA*.

Therefore the square on *BC* is quadruple of the square on *CA*;
 therefore the squares on *BC, CA* are five times the square on *CA*.

But, by hypothesis, the square on *BA* is also five times the square on *CA*;
 therefore the square on *BA* is equal to the squares on *BC, CA*:
which is impossible.

 [II. 4]

Therefore *CB* is not double of *AC*.

Similarly we can prove that neither is a straight line less than *CB* double of *CA*;
for the absurdity is much greater.

Therefore the double of *AC* is greater than *CB*.

 Q.E.D.

Proposition 3

*If a straight line be cut in extreme and mean ratio, the square on the lesser
segment added to the half of the greater segment is five times the square on the
half of the greater segment.*

For let any straight line *AB* be cut in
extreme and mean ratio at the point *C*,
let *AC* be the greater segment,
and let *AC* be bisected at *D*;
I say that the square on *BD* is five times the
square on *DC*.

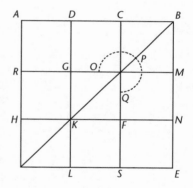

For let the square *AE* be described on *AB*,
and let the figure be drawn double.

Since *AC* is double of *DC*,
therefore the square on *AC* is quadruple of
the square on *DC*,
 that is, *RS* is quadruple of *FG*.

And, since the rectangle *AB, BC* is equal to
the square on *AC*,
and *CE* is the rectangle *AB, BC*,
 therefore *CE* is equal to *RS*.

But *RS* is quadruple of *FG*;
 therefore *CE* is also quadruple of *FG*.

Again, since *AD* is equal to *DC*,
 HK is also equal to *KF*.

Hence the square *GF* is also equal to the square *HL*.

Therefore *GK* is equal to *KL*, that is *MN* to *NE*;
 hence *MF* is also equal to *FE*.

But *MF* is equal to *CG*;
 therefore *CG* is also equal to *FE*.

Let *CN* be added to each;
 therefore the gnomon *OPQ* is equal to *CE*.

But *CE* was proved quadruple of *GF*;

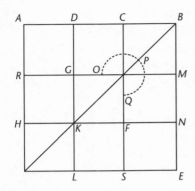

therefore the gnomon *OPQ* is also quadruple of the square *FG*.

Therefore the gnomon *OPQ* and the square *FG* are five times *FG*.

But the gnomon *OPQ* and the square *FG* are the square *DN*.

And *DN* is the square on *DB*, and *GF* the square on *DC*.

Therefore the square on *DB* is five times the square on *DC*.

Q.E.D.

Proposition 4

If a straight line be cut in extreme and mean ratio, the square on the whole and the square on the lesser segment together are triple of the square on the greater segment.

Let *AB* be a straight line,
let it be cut in extreme and mean ratio at *C*,
and let *AC* be the greater segment;
I say that the squares on *AB*, *BC* are triple of the square on *CA*.

For let the square *ADEB* be described on *AB*,
and let the figure be drawn.

Since, then, *AB* has been cut in extreme and mean ratio at *C*,
and *AC* is the greater segment,
therefore the rectangle *AB*, *BC* is equal to the square on *AC*.

[VI. Def. 3, VI. 17]

And *AK* is the rectangle *AB*, *BC*, and *HG* the square on *AC*;
therefore *AK* is equal to *HG*.

And, since *AF* is equal to *FE*,
let *CK* be added to each;
therefore the whole *AK* is equal to the whole *CE*;
therefore *AK*, *CE* are double of *AK*.

But *AK*, *CE* are the gnomon *LMN* and the square *CK*;
therefore the gnomon *LMN* and the square *CK* are double of *AK*.

But, further, *AK* was also proved equal to *HG*;
therefore the gnomon *LMN* and the squares *CK*, *HG* are triple of the square *HG*.

And the gnomon *LMN* and the squares *CK*, *HG* are the whole square *AE* and *CK*,
which are the squares on *AB*, *BC*,
while *HG* is the square on *AC*.

Therefore the squares on *AB*, *BC* are triple of the square on *AC*.

Q.E.D.

Proposition 5

If a straight line be cut in extreme and mean ratio, and there be added to it a straight line equal to the greater segment, the whole straight line has been cut in extreme and mean ratio, and the original straight line is the greater segment.

For let the straight line *AB* be cut in extreme and mean ratio at the point *C*, let *AC* be the greater segment, and let *AD* be equal to *AC*.

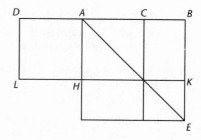

I say that the straight line *DB* has been cut in extreme and mean ratio at *A*, and the original straight line *AB* is the greater segment.

For let the square *AE* be described on *AB*, and let the figure be drawn.

Since *AB* has been cut in extreme and mean ratio at *C*,
 therefore the rectangle *AB, BC* is equal to the square on *AC*.

[VI. Def. 3, VI. 17]

And *CE* is the rectangle *AB, BC,* and *CH* the square on *AC*;
 therefore *CE* is equal to *HC*.

But *HE* is equal to *CE*,
and *DH* is equal to *HC*;
 therefore *DH* is also equal to *HE*.

Therefore the whole *DK* is equal to the whole *AE*.

And *DK* is the rectangle *BD, DA,* for *AD* is equal to *DL*;
and *AE* is the square on *AB*;
 therefore the rectangle *BD, DA* is equal to the square on *AB*.

Therefore, as *DB* is to *BA*, so is *BA* to *AD*. [VI. 17]

And *DB* is greater than *BA*;
 therefore *BA* is also greater than *AD*. [V. 14]

Therefore *DB* has been cut in extreme and mean ratio at *A*, and *AB* is the greater segment.

Q.E.D.

⟨Proposition 6

If a rational straight line be cut in extreme and mean ratio, each of the segments is the irrational straight line called apotome.

Let *AB* be a rational straight line,
let it be cut in extreme and mean ratio at *C*,
and let *AC* be the greater segment;

D A C B

I say that each of the straight lines AC, CB is the irrational straight line called apotome.

For let BA be produced, and let AD be made half of BA.

Since, then, the straight line AB has been cut in extreme and mean ratio,
and to the greater segment AC is added AD which is half of AB,
 therefore the square on CD is five times the square on DA. [XIII. 1]

Therefore the square on CD has to the square on DA the ratio which a number has to a number;
 therefore the square on CD is commensurable with the square on DA. [X. 6]

But the square on DA is rational,
for DA is rational, being half of AB which is rational;
 therefore the square on CD is also rational; [X. Def. 4]
 therefore CD is also rational.

And, since the square on CD has not to the square on DA the ratio which a square number has to a square number,
 therefore CD is incommensurable in length with DA; [X. 9]
therefore CD, DA are rational straight lines commensurable in square only;
 therefore AC is an apotome. [X. 73]

Again, since AB has been cut in extreme and mean ratio,
and AC is the greater segment,
 therefore the rectangle AB, BC is equal to the square on AC.
 [VI. Def. 3, VI. 17]

Therefore the square on the apotome AC, if applied to the rational straight line AB, produces BC as breadth.

But the square on an apotome, if applied to a rational straight line, produces as breadth a first apotome; [X. 97]
 therefore CB is a first apotome.

And CA was also proved to be an apotome.

Therefore etc.

Q.E.D.⟩

Proposition 7

If three angles of an equilateral pentagon, taken either in order or not in order, be equal, the pentagon will be equiangular.

For in the equilateral pentagon ABCDE let, first, three angles taken in order, those at A, B, C, be equal to one another;
I say that the pentagon ABCDE is equiangular.

For let AC, BE, FD be joined.

Now, since the two sides *CB, BA* are equal to
the two sides *BA, AE* respectively,
and the angle *CBA* is equal to the angle *BAE,*
therefore the base *AC* is equal to the base *BE,*
the triangle *ABC* is equal to the triangle *ABE,*
and the remaining angles will be equal to the
remaining angles,
namely those which the equal sides subtend,

[I. 4]

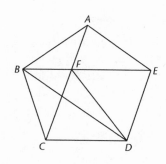

that is, the angle *BCA* to the angle *BEA,* and the
angle *ABE* to the angle *CAB;*
hence the side *AF* is also equal to the side *BF.*

[I. 6]

But the whole *AC* was also proved equal to the whole *BE;*
therefore the remainder *FC* is also equal to the remainder *FE.*

But *CD* is also equal to *DE.*

Therefore the two sides *FC, CD* are equal to the two sides *FE, ED;*
and the base *FD* is common to them;
therefore the angle *FCD* is equal to the angle *FED.* [I. 8]

But the angle *BCA* was also proved equal to the angle *AEB;*
therefore the whole angle *BCD* is also equal to the whole angle *AED.*

But, by hypothesis, the angle *BCD* is equal to the angles at *A, B;*
therefore the angle *AED* is also equal to the angles at *A, B.*
Similarly we can prove that the angle *CDE* is also equal to the angles at *A, B, C;*
therefore the pentagon *ABCDE* is equiangular.

Next, let the given equal angles not be angles taken in order, but let the angles
at the points *A, C, D* be equal;
I say that in this case too the pentagon *ABCDE* is equiangular.

For let *BD* be joined.

Then, since the two sides *BA, AE* are equal to the two sides *BC, CD,*
and they contain equal angles,
therefore the base *BE* is equal to the base *BD,*
the triangle *ABE* is equal to the triangle *BCD,*
and the remaining angles will be equal to the remaining angles,
namely those which the equal sides subtend; [I. 4]
therefore the angle *AEB* is equal to the angle *CDB.*

But the angle *BED* is also equal to the angle *BDE,*
since the side *BE* is also equal to the side *BD.* [I. 5]

Therefore the whole angle *AED* is equal to the whole angle *CDE.*

But the angle *CDE* is, by hypothesis, equal to the angles at *A, C;*
therefore the angle *AED* is also equal to the angles at *A, C.*

For the same reason
the angle *ABC* is also equal to the angles at *A, C, D.*

Therefore the pentagon *ABCDE* is equiangular.

Q.E.D.

Proposition 8

If in an equilateral and equiangular pentagon straight lines subtend two angles taken in order, they cut one another in extreme and mean ratio, and their greater segments are equal to the side of the pentagon.

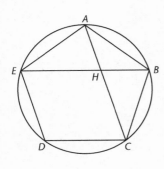

For in the equilateral and equiangular pentagon *ABCDE* let the straight lines *AC, BE,* cutting one another at the point *H,* subtend two angles taken in order, the angles at *A, B*;

I say that each of them has been cut in extreme and mean ratio at the point *H,* and their greater segments are equal to the side of the pentagon.

For let the circle *ABCDE* be circumscribed about the pentagon *ABCDE.* [IV. 14]

Then, since the two straight lines *EA, AB* are equal to the two *AB, BC,*
and they contain equal angles,
therefore the base *BE* is equal to the base *AC,*
the triangle *ABE* is equal to the triangle *ABC,*
and the remaining angles will be equal to the remaining angles respectively,
namely those which the equal sides subtend. [I. 4]

Therefore the angle *BAC* is equal to the angle *ABE*;
therefore the angle *AHE* is double of the angle *BAH.* [I. 32]

But the angle *EAC* is also double of the angle *BAC,*
inasmuch as the circumference *EDC* is also double of the circumference *CB*;
[III. 28, VI. 33]

therefore the angle *HAE* is equal to the angle *AHE*;
hence the straight line *HE* is also equal to *EA,* that is, to *AB.* [I. 6]

And, since the straight line *BA* is equal to *AE,*
the angle *ABE* is also equal to the angle *AEB.* [I. 5]

But the angle *ABE* was proved equal to the angle *BAH*;
therefore the angle *BEA* is also equal to the angle *BAH.*

And the angle *ABE* is common to the two triangles *ABE* and *ABH*;
therefore the remaining angle *BAE* is equal to the remaining angle *AHB*;
[I. 32]

therefore the triangle *ABE* is equiangular with the triangle *ABH*;
therefore, proportionally, as *EB* is to *BA,* so is *AB* to *BH.* [VI. 4]

But *BA* is equal to *EH*;
therefore, as *BE* is to *EH,* so is *EH* to *HB.*

And *BE* is greater than *EH*;
therefore *EH* is also greater than *HB.* [V. 14]

Therefore *BE* has been cut in extreme and mean ratio at *H,* and the greater segment *HE* is equal to the side of the pentagon.

Similarly we can prove that *AC* has also been cut in extreme and mean ratio at *H,* and its greater segment *CH* is equal to the side of the pentagon.

Q.E.D.

Proposition 9

If the side of the hexagon and that of the decagon inscribed in the same circle be added together, the whole straight line has been cut in extreme and mean ratio, and its greater segment is the side of the hexagon.

Let *ABC* be a circle;
of the figures inscribed in the circle *ABC* let *BC* be the side of a decagon, *CD* that of a hexagon, and let them be in a straight line;
I say that the whole straight line *BD* has been cut in extreme and mean ratio, and *CD* is its greater segment.

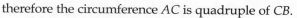

For let the centre of the circle, the point *E*, be taken,
let *EB, EC, ED* be joined,
and let *BE* be carried through to *A*.

Since *BC* is the side of an equilateral decagon, therefore the circumference *ACB* is five times the circumference *BC*;

therefore the circumference *AC* is quadruple of *CB*.

But, as the circumference *AC* is to *CB*, so is the angle *AEC* to the angle *CEB*;

[VI. 33]

therefore the angle *AEC* is quadruple of the angle *CEB*.

And, since the angle *EBC* is equal to the angle *ECB*, [I. 5]

therefore the angle *AEC* is double of the angle *ECB*. [I. 32]

And, since the straight line *EC* is equal to *CD*, for each of them is equal to the side of the hexagon inscribed in the circle *ABC*,

[IV. 15, Por.]

the angle *CED* is also equal to the angle *CDE*; [I. 5]

therefore the angle *ECB* is double of the angle *EDC*. [I. 32]

But the angle *AEC* was proved double of the angle *ECB*;

therefore the angle *AEC* is quadruple of the angle *EDC*.

But the angle *AEC* was also proved quadruple of the angle *BEC*;

therefore the angle *EDC* is equal to the angle *BEC*.

But the angle *EBD* is common to the two triangles *BEC* and *BED*;

therefore the remaining angle *BED* is also equal to the remaining angle *ECB*;

[I. 32]

therefore the triangle *EBD* is equiangular with the triangle *EBC*.

Therefore, proportionally, as *DB* is to *BE*, so is *EB* to *BC*. [VI. 4]

But *EB* is equal to *CD*.

Therefore, as *BD* is to *DC*, so is *DC* to *CB*.

And *BD* is greater than *DC*;

therefore *DC* is also greater than *CB*.

Therefore the straight line *BD* has been cut in extreme and mean ratio, and *DC* is its greater segment.

Q.E.D.

Proposition 10

If an equilateral pentagon be inscribed in a circle, the square on the side of the pentagon is equal to the squares on the side of the hexagon and on that of the decagon inscribed in the same circle.

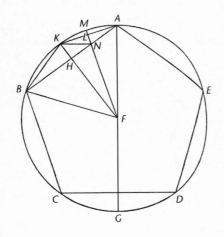

Let *ABCDE* be a circle, and let the equilateral pentagon *ABCDE* be inscribed in the circle *ABCDE*.

I say that the square on the side of the pentagon *ABCDE* is equal to the squares on the side of the hexagon and on that of the decagon inscribed in the circle *ABCDE*.

For let the centre of the circle, the point *F*, be taken, let *AF* be joined and carried through to the point *G*, let *FB* be joined, let *FH* be drawn from *F* perpendicular to *AB* and be carried through to *K*, let *AK*, *KB* be joined,

let *FL* be again drawn from *F* perpendicular to *AK*, and be carried through to *M*, and let *KN* be joined.

Since the circumference *ABCG* is equal to the circumference *AEDG*, and in them *ABC* is equal to *AED*, therefore the remainder, the circumference *CG*, is equal to the remainder *GD*.

But *CD* belongs to a pentagon; therefore *CG* belongs to a decagon.

And, since *FA* is equal to *FB*, and *FH* is perpendicular, therefore the angle *AFK* is also equal to the angle *KFB*. [I. 5, I. 26]

Hence the circumference *AK* is also equal to *KB*; [III. 26] therefore the circumference *AB* is double of the circumference *BK*; therefore the straight line *AK* is a side of a decagon.

For the same reason
AK is also double of *KM*.

Now, since the circumference *AB* is double of the circumference *BK*, while the circumference *CD* is equal to the circumference *AB*, therefore the circumference *CD* is also double of the circumference *BK*.

But the circumference *CD* is also double of *CG*; therefore the circumference *CG* is equal to the circumference *BK*.

But *BK* is double of *KM*, since *KA* is so also; therefore *CG* is also double of *KM*.

But, further, the circumference *CB* is also double of the circumference *BK*, for the circumference *CB* is equal to *BA*.

Therefore the whole circumference *GB* is also double of *BM*; hence the angle *GFB* is also double of the angle *BFM*. [VI. 33]

But the angle *GFB* is also double of the angle *FAB*,
for the angle *FAB* is equal to the angle *ABF*.

Therefore the angle *BFN* is also equal to the angle *FAB*.

But the angle *ABF* is common to the two triangles *ABF* and *BFN*;
therefore the remaining angle *AFB* is equal to the remaining angle *BNF*; [I. 32]
therefore the triangle *ABF* is equiangular with the triangle *BFN*.

Therefore, proportionally, as the straight line *AB* is to *BF*, so is *FB* to *BN*; [VI. 4]
therefore the rectangle *AB*, *BN* is equal to the square on *BF*. [VI. 17]

Again, since *AL* is equal to *LK*,
while *LN* is common and at right angles,
therefore the base *KN* is equal to the base *AN*; [I. 4]
therefore the angle *LKN* is also equal to the angle *LAN*.

But the angle *LAN* is equal to the angle *KBN*;
therefore the angle *LKN* is also equal to the angle *KBN*.

And the angle at *A* is common to the two triangles *AKB* and *AKN*.

Therefore the remaining angle *AKB* is equal to the remaining angle *KNA*; [I. 32]
therefore the triangle *KBA* is equiangular with the triangle *KNA*.

Therefore, proportionally, as the straight line *BA* is to *AK*, so is *KA* to *AN*; [VI. 4]
therefore the rectangle *BA*, *AN* is equal to the square on *AK*. [VI. 17]

But the rectangle *AB*, *BN* was also proved equal to the square on *BF*;
therefore the rectangle *AB*, *BN* together with the rectangle *BA*, *AN*, that is, the
square on *BA* [II. 2], is equal to the square on *BF* together with the square on *AK*.

And *BA* is a side of the pentagon, *BF* of the hexagon [IV. 15, Por.], and *AK* of the
decagon.

Therefore etc.

Q.E.D.

Proposition 11

*If in a circle which has its diameter rational an equilateral pentagon be
inscribed, the side of the pentagon is the irrational straight line called minor.*

For in the circle *ABCDE* which has its
diameter rational let the equilateral
pentagon *ABCDE* be inscribed;
I say that the side of the pentagon is the
irrational straight line called minor.

For let the centre of the circle, the point *F*,
be taken,
let *AF*, *FB* be joined and carried through
to the points, *G*, *H*,
let *AC* be joined,
and let *FK* be made a fourth part of *AF*.

Now *AF* is rational;
therefore *FK* is also rational.

But *BF* is also rational;

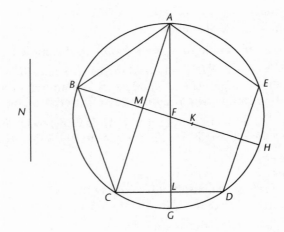

therefore the whole *BK* is rational.

And, since the circumference *ACG* is equal to the circumference *ADG*,
and in them *ABC* is equal to *AED*,

therefore the remainder *CG* is equal to the remainder *GD*.

And, if we join *AD*, we conclude that

the angles at *L* are right,
and *CD* is double of *CL*.

For the same reason

the angles at *M* are also right,
and *AC* is double of *CM*.

Since then the angle *ALC* is equal to the angle *AMF*,
and the angle *LAC* is common to the two triangles *ACL* and *AMF*,
therefore the remaining angle *ACL* is equal to the remaining angle *MFA*; **[I. 32]**
therefore the triangle *ACL* is equiangular with the triangle *AMF*;
therefore, proportionally, as *LC* is to *CA*, so is *MF* to *FA*.

And the doubles of the antecedents may be taken;
therefore, as the double of *LC* is to *CA*, so is the double of *MF* to *FA*.

But, as the double of *MF* is to *FA*, so is *MF* to the half of *FA*;
therefore also, as the double of *LC* is to *CA*, so is *MF* to the half of *FA*.

And the halves of the consequents may be taken;
therefore, as the double of *LC* is to the half of *CA*, so is *MF* to the fourth of *FA*.

And *DC* is double of *LC*, *CM* is half of *CA*, and *FK* a fourth part of *FA*;
therefore, as *DC* is to *CM*, so is *MF* to *FK*.

Componendo also, as the sum of *DC*, *CM* is to *CM*, so is *MK* to *KF*; **[V. 18]**
therefore also, as the square on the sum of *DC*, *CM* is to the square on *CM*, so is
the square on *MK* to the square on *KF*.

And since, when the straight line subtending two sides of the pentagon, as *AC*,
is cut in extreme and mean ratio, the greater segment is equal to the side of the
pentagon, that is, to *DC*, **[XIII. 8]**
while the square on the greater segment added to the half of the whole is five
times the square on the half of the whole, **[XIII. 1]**
and *CM* is half of the whole *AC*,

therefore the square on *DC, CM* taken as one straight line is five times the square on *CM*.

But it was proved that, as the square on *DC, CM* taken as one straight line is to the square on *CM*, so is the square on *MK* to the square on *KF*;

therefore the square on *MK* is five times the square on *KF*.

But the square on *KF* is rational,
for the diameter is rational;

therefore the square on *MK* is also rational;

therefore *MK* is rational.

And, since *BF* is quadruple of *FK*,

therefore *BK* is five times *KF*;

therefore the square on *BK* is twenty-five times the square on *KF*.

But the square on *MK* is five times the square on *KF*;

therefore the square on *BK* is five times the square on *KM*;
therefore the square on *BK* has not to the square on *KM* the ratio which a square number has to a square number;

therefore *BK* is incommensurable in length with *KM*. [X. 9]

And each of them is rational.

Therefore *BK, KM* are rational straight lines commensurable in square only.

But, if from a rational straight line there be subtracted a rational straight line which is commensurable with the whole in square only, the remainder is irrational, namely an apotome;

therefore *MB* is an apotome and *MK* the annex to it. [X. 73]

I say next that *MB* is also a fourth apotome.

Let the square on *N* be equal to that by which the square on *BK* is greater than the square on *KM*;

therefore the square on *BK* is greater than the square on *KM* by the square on *N*. And, since *KF* is commensurable with *FB*,

componendo also, *KB* is commensurable with *FB*. [X. 15]

But *BF* is commensurable with *BH*;

therefore *BK* is also commensurable with *BH*. [X. 12]

And, since the square on *BK* is five times the square on *KM*,

therefore the square on *BK* has to the square on *KM* the ratio which 5 has to 1.

Therefore, *convertendo*,

the square on *BK* has to the square on *N* the ratio which 5 has to 4,
[V. 19, Por.]
and this is not the ratio which a square number has to a square number;

therefore *BK* is incommensurable with *N*; [X. 9]
therefore the square on *BK* is greater than the square on *KM* by the square on a straight line incommensurable with *BK*.

Since then the square on the whole *BK* is greater than the square on the annex *KM* by the square on a straight line incommensurable with *BK*,
and the whole *BK* is commensurable with the rational straight line, *BH*, set out,

therefore *MB* is a fourth apotome. [X. Deff. III. 4]

But the rectangle contained by a rational straight line and a fourth apotome is irrational,

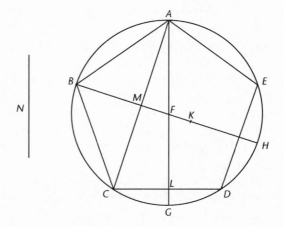

and its square root is irrational, and is called minor. [X. 94]

But the square on *AB* is equal to the rectangle *HB, BM,*
because, when *AH* is joined, the triangle *ABH* is equiangular with the triangle *ABM,* and, as *HB* is to *BA,* so is *AB* to *BM.*

Therefore the side *AB* of the pentagon is the irrational straight line called minor.

Q.E.D.

Proposition 12

If an equilateral triangle be inscribed in a circle, the square on the side of the triangle is triple of the square on the radius of the circle.

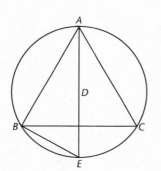

Let *ABC* be a circle,
and let the equilateral triangle *ABC* be inscribed in it;
I say that the square on one side of the triangle *ABC* is triple of the square on the radius of the circle.

For let the centre *D* of the circle *ABC* be taken,
let *AD* be joined and carried through to *E,*
and let *BE* be joined.

Then, since the triangle *ABC* is equilateral,
therefore the circumference *BEC* is a third part of the circumference of the circle *ABC.*

Therefore the circumference *BE* is a sixth part of the circumference of the circle;
therefore the straight line *BE* belongs to a hexagon;
therefore it is equal to the radius *DE*. [IV. 15, Par.]

And, since *AE* is double of *DE,*
the square on *AE* is quadruple of the square on *ED,* that is, of the square on *BE.*

But the square on *AE* is equal to the squares on *AB, BE*; [III. 31, I. 47]
therefore the squares on *AB, BE* are quadruple of the square on *BE.*

Therefore, *separando*, the square on *AB* is triple of the square on *BE*.

But *BE* is equal to *DE*;
therefore the square on *AB* is triple of the square on *DE*.

Therefore the square on the side of the triangle is triple of the square on the radius.

Q.E.D.

Proposition 13

To construct a pyramid, to comprehend it in a given sphere, and to prove that the square on the diameter of the sphere is one and a half times the square on the side of the pyramid.

 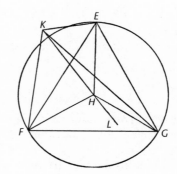

Let the diameter *AB* of the given sphere be set out,
and let it be cut at the point *C* so that *A C* is double of *CB*;
let the semicircle *ADB* be described on *AB*,
let *CD* be drawn from the point *C* at right angles to *AB*,
and let *DA* be joined;
let the circle *EFG* which has its radius equal to *DC* be set out,
let the equilateral triangle *EFG* be inscribed in the circle *EFG*, [IV. 2]
let the centre of the circle, the point *H*, be taken, [III. 1]
let *EH, HF, HG* be joined;
from the point *H* let *HK* be set up at right angles to the plane of the circle *EFG*,
 [XI. 12]
let *HK* equal to the straight line *AC* be cut off from *HK*,
and let *KE, KF, KG* be joined.

Now, since *KH* is at right angles to the plane of the circle *EFG*,
therefore it will also make right angles with all the straight lines which meet it
and are in the plane of the circle *EFG*. [XI. Def. 3]

But each of the straight lines *HE, HF, HG* meets it:
therefore *HK* is at right angles to each of the straight lines *HE, HF, HG*.

And, since *AC* is equal to *HK*, and *CD* to *HE*,
and they contain right angles,
therefore the base *DA* is equal to the base *KE*. [I. 4]

For the same reason
each of the straight lines *KF, KG* is also equal to *DA*;
therefore the three straight lines *KE, KF, KG* are equal to one another.

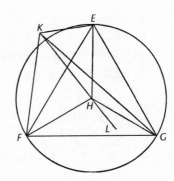

And, since AC is double of CB,

therefore AB is triple of BC.

But, as AB is to BC, so is the square on AD to the square on DC, as will be proved afterwards.

Therefore the square on AD is triple of the square on DC.

But the square on FE is also triple of the square on EH, [XIII. 12]
and DC is equal to EH;

therefore DA is also equal to EF.

But DA was proved equal to each of the straight lines KE, KF, KG;
therefore each of the straight lines EF, FG, GE is also equal to each of the straight lines KE, KF, KG;

therefore the four triangles EFG, KEF, KFG, KEG are equilateral.

Therefore a pyramid has been constructed out of four equilateral triangles, the triangle EFG being its base and the point K its vertex.

It is next required to comprehend it in the given sphere and to prove that the square on the diameter of the sphere is one and a half times the square on the side of the pyramid.

For let the straight line HL be produced in a straight line with KH,
and let HL be made equal to CB.

Now, since, as AC is to CD, so is CD to CB, [VI. 8, Por.]
while AC is equal to KH, CD to HE, and CB to HL,

therefore, as KH is to HE, so is EH to HL;

therefore the rectangle KH, HL is equal to the square on EH. [VI. 17]

And each of the angles KHE, EHL is right;

therefore the semicircle described on KL will pass through E also.
[cf. VI. 8, III. 31]

If then, KL remaining fixed, the semicircle be carried round and restored to the same position from which it began to be moved, it will also pass through the points F, G,
since, if FL, LG be joined, the angles at F, G similarly become right angles;
and the pyramid will be comprehended in the given sphere.

For KL, the diameter of the sphere, is equal to the diameter AB of the given sphere, inasmuch as KH was made equal to AC, and HL to CB.

I say next that the square on the diameter of the sphere is one and a half times the square on the side of the pyramid.

For, since *AC* is double of *CB*,

<div align="center">therefore *AB* is triple of *BC*;</div>

<div align="center">and, *convertendo*, *BA* is one and a half times *AC*.</div>

But, as *BA* is to *AC*, so is the square on *BA* to the square on *AD*.

Therefore the square on *BA* is also one and a half times the square on *AD*.

And *BA* is the diameter of the given sphere, and *AD* is equal to the side of the pyramid.

Therefore the square on the diameter of the sphere is one and a half times the square on the side of the pyramid.

<div align="right">Q.E.D.</div>

LEMMA. It is to be proved that, as *AB* is to *BC*, so is the square on *AD* to the square on *DC*.

For let the figure of the semicircle be set out,
let *DB* be joined,
let the square *EC* be described on *AC*,
and let the parallelogram *FB* be completed.

Since then, because the triangle *DAB* is equiangular with the triangle *DAC*,
as *BA* is to *AD*, so is *DA* to *AC*, [VI. 8, VI. 4]
therefore the rectangle *BA*, *AC* is equal to the square on *AD*. [VI. 17]

And since, as *AB* is to *BC*, so is *EB* to *BF*, [VI. 1]
and *EB* is the rectangle *BA*, *AC*, for *EA* is equal to *AC*,
and *BF* is the rectangle *AC*, *CB*,

 therefore, as *AB* is to *BC*, so is the rectangle *BA*, *AC* to the rectangle *AC*, *CB*.

And the rectangle *BA*, *AC* is equal to the square on *AD*,
and the rectangle *AC*, *CB* to the square on *DC*,
for the perpendicular *DC* is mean proportional between the segments *AC*, *CB* of the base, because the angle *ADB* is right. [VI. 8, Por.]

Therefore, as *AB* is to *BC*, so is the square on *AD* to the square on *DC*.

<div align="right">Q.E.D.</div>

Proposition 14

To construct an octahedron and comprehend it in a sphere, as in the preceding case; and to prove that the square on the diameter of the sphere is double of the square on the side of the octahedron.

Let the diameter *AB* of the given sphere be set out,
and let it be bisected at *C*;
let the semicircle *ADB* be described on *AB*,
let *CD* be drawn from *C* at right angles to *AB*,

let *DB* be joined;
let the square *EFGH*, having each of its sides equal to *DB*, be set out,
let *HF, EG* be joined,
from the point *K* let the straight line *KL* be set up at right angles to the plane of
the square *EFGH*, [XI. 12]
and let it be carried through to the other side of the plane, as *KM*;
from the straight lines *KL, KM* let *KL, KM* be respectively cut off equal to one of
the straight lines *EK, FK, GK, HK*,
and let *LE, LF, LG, LH, ME, MF, MG, MH* be joined.

Then, since *KE* is equal to KH, and the angle *EKH* is right,
therefore the square on *HE* is double of the square on *EK*. [I. 47]

Again, since *LK* is equal to *KE*,
and the angle *LKE* is right,
therefore the square on *EL* is double of the square on *EK*. [I. 47]

But the square on *HE* was also proved double of the square on *EK*;
therefore the square on *LE* is equal to the square on *EH*;
therefore *LE* is equal to *EH*.

For the same reason
LH is also equal to *HE*;
therefore the triangle *LEH* is equilateral.

Similarly we can prove that each of the remaining triangles of which the sides
of the square *EFGH* are the bases, and the points *L, M* the vertices, is equilateral;
therefore an octahedron has been constructed which is contained by eight
equilateral triangles.

It is next required to comprehend it in the given sphere, and to prove that the
square on the diameter of the sphere is double of the square on the side of the
octahedron.

For, since the three straight lines *LK, KM, KE* are equal to one another,
therefore the semicircle described on *LM* will also pass through *E*.

And for the same reason,
if, *LM* remaining fixed, the semicircle be carried round and restored to the same
position from which it began to be moved,
it will also pass through the points *F, G, H*,
and the octahedron will have been comprehended in a sphere.

I say next that it is also comprehended in the given sphere.

For, since *LK* is equal to *KM*,
while *KE* is common,
and they contain right angles,
therefore the base *LE* is equal to the base *EM*. [I. 4]

And, since the angle *LEM* is right, for it is in a semicircle, [III. 31]
therefore the square on *LM* is double of the square on *LE*. [I. 47]

Again, since *AC* is equal to *CB*,
AB is double of *BC*.

But, as *AB* is to *BC*, so is the square on *AB* to the square on *BD*;
therefore the square on *AB* is double of the square on *BD*.

But the square on *LM* was also proved double of the square on *LE*.

And the square on *DB* is equal to the square on *LE*,
for *EH* was made equal to *DB*.

Therefore the square on *AB* is also equal to the square on *LM*;
therefore *AB* is equal to *LM*.

And *AB* is the diameter of the given sphere;
therefore *LM* is equal to the diameter of the given sphere.

Therefore the octahedron has been comprehended in the given sphere, and it has been demonstrated at the same time that the square on the diameter of the sphere is double of the square on the. side of the octahedron.

Q.E.D.

Proposition 15

To construct a cube and comprehend it in a sphere, like the pyramid; and to prove that the square on the diameter of the sphere is triple of the square on the side of the cube.

Let the diameter *AB* of the given sphere be set out,
and let it be cut at *C* so that *AC* is double of *CB*;
let the semicircle *ADB* be described on *AB*,
let *CD* be drawn from *C* at right angles to *AB*,
and let *DB* be joined;
let the square *EFGH* having its side equal to *DB* be set out,
from *E*, *F*, *G*, *H* let *EK*, *FL*, *GM*, *HN* be drawn at right angles to the plane of the square *EFGH*,
from *EK*, *FL*, *GM*, *HN* let *EK*, *FL*, *GM*, *HN* respectively be cut off equal to one of the straight lines *EF*, *FG*, *GH*, *HE*,
and let *KL*, *LM*, *MN*, *NK* be joined;

therefore the cube *FN* has been constructed which is contained by six equal squares.

It is then required to comprehend it in the given sphere, and to prove that the square on the diameter of the sphere is triple of the square on the side of the cube.

For let *KG, EG* be joined.

Then, since the angle *KEG* is right, because *KE* is also at right angles to the plane *EG* and of course to the straight line *EG* also, [XI. Def. 3]
 therefore the semicircle described on *KG* will also pass through the point *E*.

Again, since *GF* is at right angles to each of the straight lines *FL, FE*,
 GF is also at right angles to the plane *FK*;
hence also, if we join *FK, GF* will be at right angles to *FK*;
and for this reason again the semicircle described on *GK* will also pass through *F*.

Similarly it will also pass through the remaining angular points of the cube.

If then, *KG* remaining fixed, the semicircle be carried round and restored to the same position from which it began to be moved,
 the cube will be comprehended in a sphere.

I say next that it is also comprehended in the given sphere.

For, since *GF* is equal to *FE*,
and the angle at *F* is right,
 therefore the square on *EG* is double of the square on *EF*.

But *EF* is equal to *EK*;
 therefore the square on *EG* is double of the square on *EK*;
hence the squares on *GE, EK*, that is the square on *GK* [I. 47], is triple of the square on *EK*.

And, since *AB* is triple of *BC*,
while, as *AB* is to *BC*, so is the square on *AB* to the square on *BD*,
 therefore the square on *AB* is triple of the square on *BD*.

But the square on *GK* was also proved triple of the square on *KE*.

And *KE* was made equal to *DB*;
 therefore *KG* is also equal to *AB*.

And *AB* is the diameter of the given sphere;
 therefore *KG* is also equal to the diameter of the given sphere.

Therefore the cube has been comprehended in the given sphere; and it has been demonstrated at the same time that the square on the diameter of the sphere is triple of the square on the side of the cube.

<div align="right">Q.E.D.</div>

Proposition 16

To construct an icosahedron and comprehend it in a sphere, like the aforesaid figures; and to prove that the side of the icosahedron is the irrational straight line called minor.

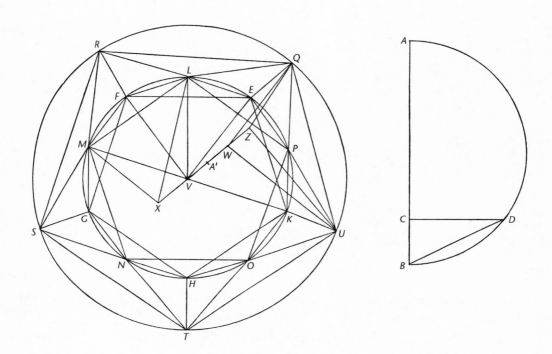

Let the diameter *AB* of the given sphere be set out,
and let it be cut at *C* so that *AC* is quadruple of *CB*,
let the semicircle *ADB* be described on *AB*,
let the straight line *CD* be drawn from *C* at right angles to *AB*,
and let *DB* be joined;
let the circle *EFGHK* be set out and let its radius be equal to *DB*,
let the equilateral and equiangular pentagon *EFGHK* be inscribed in the circle *EFGHK*,
let the circumferences *EF*, *FG*, *GH*, *HK*, *KE* be bisected at the points *L*, *M*, *N*, *O*, *P*,
and let *LM*, *MN*, *NO*, *OP*, *PL*, *EP* be joined.

Therefore the pentagon *LMNOP* is also equilateral,
and the straight line *EP* belongs to a decagon.

Now from the points *E*, *F*, *G*, *H*, *K* let the straight lines *EQ*, *FR*, *GS*, *HT*, *KU* be set up at right angles to the plane of the circle,
and let them be equal to the radius of the circle *EFGHK*,
let *QR*, *RS*, *ST*, *TU*, *UQ*, *QL*, *LR*, *RM*, *MS*, *SN*, *NT*, *TO*, *OU*, *UP*, *PQ* be joined.

Now, since each of the straight lines *EQ*, *KU* is at right angles to the same plane,
therefore *EQ* is parallel to *KU*. [XI. 6]

But it is also equal to it;

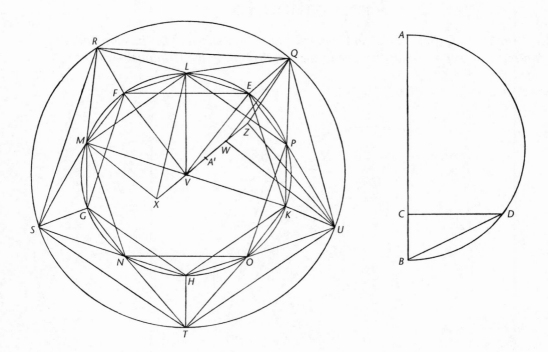

and the straight lines joining those extremities of equal and parallel straight lines which are in the same direction are equal and parallel. [I. 33]

Therefore *QU* is equal and parallel to *EK*.

But *EK* belongs to an equilateral pentagon;
therefore *QU* also belongs to the equilateral pentagon inscribed in the circle *EFGHK*.

For the same reason
each of the straight lines *QR, RS, ST, TU* also belongs to the equilateral pentagon inscribed in the circle *EFGHK*;
therefore the pentagon *QRSTU* is equilateral.

And, since *QE* belongs to a hexagon,
and *EP* to a decagon,
and the angle *QEP* is right,
therefore *QP* belongs to a pentagon;
for the square on the side of the pentagon is equal to the square on the side of the hexagon and the square on the side of the decagon inscribed in the same circle. [XIII. 10]

For the same reason
PU is also a side of a pentagon.
But *QU* also belongs to a pentagon;
therefore the triangle *QPU* is equilateral.

For the same reason
each of the triangles *QLR, RMS, SNT, TOU* is also equilateral.

And, since each of the straight lines QL, QP was proved to belong to a pentagon, and LP also belongs to a pentagon,

therefore the triangle QLP is equilateral.

For the same reason

each of the triangles LRM, MSN, NTO, OUP is also equilateral.

Let the centre of the circle EFGHK, the point V, be taken;
from V let VZ be set up at right angles to the plane of the circle,
let it be produced in the other direction, as VX,
let there be cut off VW, the side of a hexagon, and each of the straight lines VX, WZ, being sides of a decagon,
and let QZ, QW, UZ, EV, LV, LX, XM be joined.

Now, since each of the straight lines VW, QE is at right angles to the plane of the circle,

therefore VW is parallel to QE. [XI. 6]

But they are also equal;

therefore EV, QW are also equal and parallel. [I. 33]

But EV belongs to a hexagon;

therefore QW also belongs to a hexagon.

And, since QW belongs to a hexagon,
and WZ to a decagon,
and the angle QWZ is right,

therefore QZ belongs to a pentagon. [XIII. 10]

For the same reason

UZ also belongs to a pentagon,

inasmuch as, if we join VK, WU, they will be equal and opposite, and VK, being a radius, belongs to a hexagon; [IV. 15, Por.]

therefore WU also belongs to a hexagon.

But WZ belongs to a decagon,
and the angle UWZ is right;

therefore UZ belongs to a pentagon. [XIII. 10]

But QU also belongs to a pentagon;

therefore the triangle QUZ is equilateral.

For the same reason
each of the remaining triangles of which the straight lines QR, RS, ST, TU are the bases, and the point Z the vertex, is also equilateral.

Again, since VL belongs to a hexagon,
and VX to a decagon,
and the angle LVX is right,

therefore LX belongs to a pentagon. [XIII. 10]

For the same reason,
if we join MV, which belongs to a hexagon,

MX is also inferred to belong to a pentagon.

But LM also belongs to a pentagon;

therefore the triangle LMX is equilateral.

Similarly it can be proved that each of the remaining triangles of which MN, NO, OP, PL are the bases, and the point X the vertex, is also equilateral.

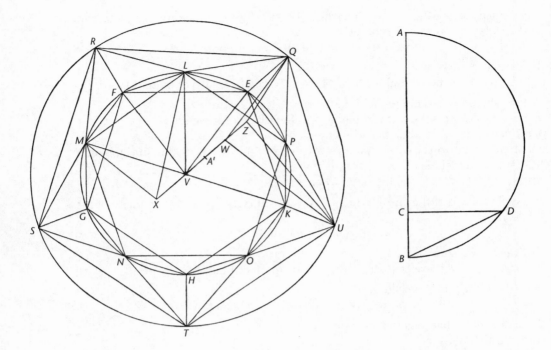

Therefore an icosahedron has been constructed which is contained by twenty equilateral triangles.

It is next required to comprehend it in the given sphere, and to prove that the side of the icosahedron is the irrational straight line called minor.

For, since VW belongs to a hexagon,
and WZ to a decagon,
therefore VZ has been cut in extreme and mean ratio at W,
and VW is its greater segment; [XIII. 9]
therefore as ZV is to VW, so is VW to WZ.

But VW is equal to VE, and WZ to VX;
therefore, as ZV is to VE, so is EV to VX.

And the angles ZVE, EVX are right;
therefore, if we join the straight line EZ, the angle XEZ will be right because of the similarity of the triangles XEZ, VEZ.

For the same reason,
since, as ZV is to VW, so is VW to WZ,
and ZV is equal to XW, and VW to WQ,
therefore, as XW is to WQ, so is QW to WZ.

And for this reason again,
if we join QX, the angle at Q will be right; [VI. 8]
therefore the semicircle described on XZ will also pass through Q. [III. 31]

And if, XZ remaining fixed, the semicircle be carried round and restored to the same position from which it began to be moved, it will also pass through Q and the remaining angular points of the icosahedron,
and the icosahedron will have been comprehended in a sphere.

I say next that it is also comprehended in the given sphere.

For let VW be bisected at A'.

Then, since the straight line VZ has been cut in extreme and mean ratio at W and ZW is its lesser segment,
therefore the square on ZW added to the half of the greater segment, that is WA', is five times the square on the half of the greater segment; [XIII. 3]
therefore the square on ZA' is five times the square on A'W.

And ZX is double of ZA', and VW double of A'W;
therefore the square on ZX is five times the square on WV.

And, since AC is quadruple of CB,
therefore AB is five times BC.
But, as AB is to BC, so is the square on AB to the square on BD; [VI. 8, V. Def. 9]
therefore the square on AB is five times the square on BD.

But the square on ZX was also proved to be five times the square on VW.

And DB is equal to VW,
for each of them is equal to the radius of the circle EFGHK;
therefore AB is also equal to XZ.

And AB is the diameter of the given sphere;
therefore XZ is also equal to the diameter of the given sphere.

Therefore the icosahedron has been comprehended in the given sphere.

I say next that the side of the icosahedron is the irrational straight line called minor.

For, since the diameter of the sphere is rational,
and the square on it is five times the square on the radius of the circle EFGHK,
therefore the radius of the circle EFGHK is also rational;
hence its diameter is also rational.

But, if an equilateral pentagon be inscribed in a circle which has its diameter rational, the side of the pentagon is the irrational straight line called minor.
 [XIII. 11]

And the side of the pentagon EFGHK is the side of the icosahedron.

Therefore the side of the icosahedron is the irrational straight line called minor.

Q.E.D.

PORISM. From this it is manifest that the square on the diameter of the sphere is five times the square on the radius of the circle from which the icosahedron has been described, and that the diameter of the sphere is composed of the side of the hexagon and two of the sides of the decagon inscribed in the same circle.

Proposition 17

To construct a dodecahedron and comprehend it in a sphere, like the aforesaid figures, and to prove that the side of the dodecahedron is the irrational straight line called apotome.

Let *ABCD*, *CBEF*, two planes of the aforesaid cube at right angles to one another, be set out,

let the sides *AB*, *BC*, *CD*, *DA*, *EF*, *EB*, *FC* be bisected at *G*, *H*, *K*, *L*, *M*, *N*, *O* respectively,

let *GK*, *HL*, *MH*, *NO* be joined,

let the straight lines *NP*, *PO*, *HQ* be cut in extreme and mean ratio at the points *R*, *S*, *T* respectively,

and let *RP*, *PS*, *TQ* be their greater segments;

from the points *R*, *S*, *T* let *RU*, *SV*, *TW* be set up at right angles to the planes of the cube towards the outside of the cube,

let them be made equal to *RP*, *PS*, *TQ*, and let *UB*, *BW*, *WC*, *CV*, *VU* be joined.

I say that the pentagon *UBWCV* is equilateral, and in one plane, and is further equiangular.

For let *RB*, *SB*, *VB* be joined.

Then, since the straight line *NP* has been cut in extreme and mean ratio at *R*, and *RP* is the greater segment,

therefore the squares on *PN*, *NR* are triple of the square on *RP*. [XIII. 4]

But *PN* is equal to *NB*, and *PR* to *RU*;

therefore the squares on *BN*, *NR* are triple of the square on *RU*.

But the square on *BR* is equal to the squares on *BN*, *NR*; [I. 47]

therefore the square on *BR* is triple of the square on *RU*;

hence the squares on *BR*, *RU* are quadruple of the square on *RU*.

But the square on *BU* is equal to the squares on *BR*, *RU*;

therefore the square on *BU* is quadruple of the square on *RU*;

therefore *BU* is double of *RU*.

But *VU* is also double of *UR*,

inasmuch as *SR* is also double of *PR*, that is, of *RU*;

therefore *BU* is equal to *UV*.

Similarly it can be proved that each of the straight lines *BW*, *WC*, *CV* is also equal to each of the straight lines *BU*, *UV*.

Therefore the pentagon *BUVCW* is equilateral.

I say next that it is also in one plane.

For let *PX* be drawn from *P* parallel to each of the straight lines *RU*, *SV* and towards the outside of the cube, and let *XH*, *HW* be joined;
I say that *XHW* is a straight line.

For, since *HQ* has been cut in extreme and mean ratio at *T*, and *QT* is its greater segment,

therefore, as *HQ* is to *QT*, so is *QT* to *TH*.

But *HQ* is equal to *HP*, and *QT* to each of the straight lines *TW*, *PX*;

therefore, as *HP* is to *PX*, so is *WT* to *TH*.

And *HP* is parallel to *TW*,
for each of them is at right angles to the plane *BD*; [XI. 6]
and *TH* is parallel to *PX*,
for each of them is at right angles to the plane *BF*. [XI. 6]

But if two triangles, as *XPH*, *HTW*, which have two sides proportional to two sides be placed together at one angle so that their corresponding sides are also parallel, the remaining straight lines will be in a straight line; [VI. 32]
therefore *XH* is in a straight line with *HW*.

But every straight line is in one plane; [XI. 1]
therefore the pentagon *UBWCV* is in one plane.

I say next that it is also equiangular.

For, since the straight line *NP* has been cut in extreme and mean ratio at *R*, and *PR* is the greater segment,
while *PR* is equal to *PS*,
therefore *NS* has also been cut in extreme and mean ratio at *P*,
and *NP* is the greater segment; [XIII. 5]
therefore the squares on *NS*, *SP* are triple of the square on *NP*. [XIII. 4]

But *NP* is equal to *NB*, and *PS* to *SV*;
therefore the squares on *NS*, *SV* are triple of the square on *NB*;
hence the squares on *VS*, *SN*, *NB* are quadruple of the square on *NB*.

But the square on *SB* is equal to the squares on *SN*, *NB*;
therefore the squares on *BS*, *SV*, that is, the square on *BV*—
for the angle *VSB* is right—
is quadruple of the square on *NB*;
therefore *VB* is double of *BN*.

But *BC* is also double of *BN*;
therefore *BV* is equal to *BC*.

And, since the two sides *BU*, *UV* are equal to the two sides *BW*, *WC*,
and the base *BV* is equal to the base *BC*,
therefore the angle *BUV* is equal to the angle *BWC*. [III. 8]

Similarly we can prove that the angle *UVC* is also equal to the angle *BWC*;
therefore the three angles *BWC*, *BUV*, *UVC* are equal to one another.

But if in an equilateral pentagon three angles are equal to one another, the pentagon will be equiangular, [XIII. 7]
therefore the pentagon *BUVCW* is equiangular.

And it was also proved equilateral;
therefore the pentagon *BUVCW* is equilateral and equiangular, and it is on one side *BC* of the cube.

Therefore, if we make the same construction in the case of each of the twelve sides of the cube, a solid figure will have been constructed which is contained by twelve equilateral and equiangular pentagons, and which is called a dodecahedron.

It is then required to comprehend it in the given sphere, and to prove that the side of the dodecahedron is the irrational straight line called apotome.

For let *XP* be produced, and let the produced straight line be *XZ*;

therefore *PZ* meets the diameter of the cube, and they bisect one another, for this has been proved in the last theorem but one of the eleventh book.

[XI. 38]

Let them cut at *Z;*
therefore *Z* is the centre of the sphere which comprehends the cube, and *ZP* is half of the side of the cube.

Let *UZ* be joined.

Now, since the straight line *NS* has been cut in extreme and mean ratio at *P,* and *NP* is its greater segment, therefore the squares on *NS, SP* are triple of the square on *NP.* [XIII. 4]

But *NS* is equal to *XZ,* inasmuch as *NP* is also equal to *PZ,* and *XP* to *PS.*

But further, *PS* is also equal to *XU,* since it is also equal to *RP;*
therefore the squares on *ZX, XU* are triple of the square on *NP.*

But the square on *UZ* is equal to the squares on *ZX, XU;*
therefore the square on *UZ* is triple of the square on *NP.*

But the square on the radius of the sphere which comprehends the cube is also triple of the square on the half of the side of the cube, for it has previously been shown how to construct a cube and comprehend it in a sphere, and to prove that the square on the diameter of the sphere is triple of the square on the side of the cube. [XIII. 15]

But, if whole is so related to whole, so is half to half also; and *NP* is half of the side of the cube; therefore *UZ* is equal to the radius of the sphere which comprehends the cube.

And *Z* is the centre of the sphere which comprehends the cube; therefore the point *U* is on the surface of the sphere.

Similarly we can prove that each of the remaining angles of the dodecahedron is also on the surface of the sphere; therefore the dodecahedron has been comprehended in the given sphere.

I say next that the side of the dodecahedron is the irrational straight line called apotome.

For since, when *NP* has been cut in extreme and mean ratio, *RP* is the greater segment, and, when *PO* has been cut in extreme and mean ratio, *PS* is the greater segment, therefore, when the whole *NO* is cut in extreme and mean ratio, *RS* is the greater segment.

{Thus, since, as *NP* is to *PR,* so is *PR* to *RN,* the same is true of the doubles also,

for parts have the same ratio as their equimultiples; [V. 15]
 therefore as *NO* is to *RS*, so is *RS* to the sum of *NR, SO.*

But *NO* is greater than *RS*;
 therefore *RS* is also greater than the sum of *NR, SO*;
therefore *NO* has been cut in extreme and mean ratio, and *RS* is its greater segment.}[1]

But *RS* is equal to *UV*;
therefore, when NO is cut in extreme and mean ratio, UV is the greater segment.

And, since the diameter of the sphere is rational,
and the square on it is triple of the square on the side of the cube,
 therefore *NO*, being a side of the cube, is rational.

⟨But if a rational line be cut in extreme and mean ratio, each of the segments is an irrational apotome.⟩

Therefore UV, being a side of the dodecahedron, is an irrational apotome. [XIII. 6]

 Q.E.D.

PORISM. From this it is manifest that, when the side of the cube is cut in extreme and mean ratio, the greater segment is the side of the dodecahedron.

Proposition 18

To set out the sides of the five figures and to compare them with one another.

Let *AB*, the diameter of the given sphere, be set out,
and let it be cut at *C* so that *AC* is equal to *CB*,
and at *D* so that *AD* is double of *DB*;
let the semicircle *AEB* be described on *AB*,
from *C, D* let *CE, DF* be drawn at right angles to *AB*,
and let *AF, FB, EB* be joined.

Then, since *AD* is double of *DB*,
 therefore *AB* is triple of *BD*.

Convertendo, therefore, *BA* is one and a half times *AD*.

But, as *BA* is to *AD*, so is the square on *BA* to the square on *AF*, [V. Def. 9, VI. 8]
for the triangle *AFB* is equiangular with the triangle *AFD*;
 therefore the square on *BA* is one and a half times the square on *AF*.

But the square on the diameter of the sphere is also one and a half times the square on the side of the pyramid. [XIII. 13]

1. Heath brackets this passage, perhaps because it seems redundant. —Ed.

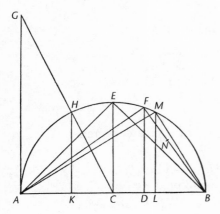

And *AB* is the diameter of the sphere; therefore *AF* is equal to the side of the pyramid.

Again, since *AD* is double of *DB*,
therefore *AB* is triple of *BD*.

But, as *AB* is to *BD*, so is the square on *AB* to the square on *BF*; **[VI. 8, V. Def. 9]** therefore the square on *AB* is triple of the square on *BF*.

But the square on the diameter of the sphere is also triple of the square on the side of the cube. **[XIII. 15]**

And *AB* is the diameter of the sphere;
therefore *BF* is the side of the cube.

And, since *AC* is equal to *CB*,
therefore *AB* is double of *BC*.

But, as *AB* is to *BC*, so is the square on *AB* to the square on *BE*;
therefore the square on *AB* is double of the square on *BE*.

But the square on the diameter of the sphere is also double of the square on the side of the octahedron. **[XIII. 14]**

And *AB* is the diameter of the given sphere;
therefore *BE* is the side of the octahedron.

Next, let *AG* be drawn from the point *A* at right angles to the straight line *AB*, let *AG* be made equal to *AB*,
let *GC* be joined,
and from *H* let *HK* be drawn perpendicular to *AB*.

Then, since *GA* is double of *AC*,
for *GA* is equal to *AB*,
and, as *GA* is to *AC*, so is *HK* to *KC*,
therefore *HK* is also double of *KC*.

Therefore the square on *HK* is quadruple of the square on *KC*;
therefore the squares on *HK*, *KC*, that is, the square on *HC*, is five times the square on *KC*.

But *HC* is equal to *CB*;
therefore the square on *BC* is five times the square on *CK*.

And, since *AB* is double of *CB*,
and, in them, *AD* is double of *DB*,
therefore the remainder *BD* is double of the remainder *DC*.

Therefore *BC* is triple of *CD*;
therefore the square on *BC* is nine times the square on *CD*.

But the square on *BC* is five times the square on *CK*;
therefore the square on *CK* is greater than the square on *CD*;
therefore *CK* is greater than *CD*.

Let *CL* be made equal to *CK*,
from *L* let *LM* be drawn at right angles to *AB*,
and let *MB* be joined.

Now, since the square on *BC* is five times the square on *CK*,
and *AB* is double of *BC*, and *KL* double of *CK*,
> therefore the square on *AB* is five times the square on *KL*.

But the square on the diameter of the sphere is also five times the square on the radius of the circle from which the icosahedron has been described. [XIII. 16, Por.]

And *AB* is the diameter of the sphere;
therefore *KL* is the radius of the circle from which the icosahedron has been described;
> therefore *KL* is a side of the hexagon in the said circle. [IV. 15, Por.]

And, since the diameter of the sphere is made up of the side of the hexagon and two of the sides of the decagon inscribed in the same circle, [XIII. 16, Por.]
and *AB* is the diameter of the sphere,
while *KL* is a side of the hexagon,
and *AK* is equal to *LB*,
therefore each of the straight lines *AK*, *LB* is a side of the decagon inscribed in the circle from which the icosahedron has been described.

And, since *LB* belongs to a decagon, and *ML* to a hexagon,
for *ML* is equal to *KL*, since it is also equal to *HK*, being the same distance from the centre, and each of the straight lines *HK*, *KL* is double of *KC*,
> therefore *MB* belongs to a pentagon. [XIII. 10]

But the side of the pentagon is the side of the icosahedron; [XIII. 16]
> therefore *MB* belongs to the icosahedron.

Now, since *FB* is a side of the cube,
let it be cut in extreme and mean ratio at *N*,
and let *NB* be the greater segment;
> therefore *NB* is a side of the dodecahedron. [XIII. 17, Por.]

And, since the square on the diameter of the sphere was proved to be one and a half times the square on the side *AF* of the pyramid, double of the square on the side *BE* of the octahedron and triple of the side *FB* of the cube,
therefore, of parts of which the square on the diameter of the sphere contains six, the square on the side of the pyramid contains four, the square on the side of the octahedron three, and the square on the side of the cube two.

Therefore the square on the side of the pyramid is four-thirds of the square on the side of the octahedron, and double of the square on the side of the cube;
and the square on the side of the octahedron is one and a half times the square on the side of the cube.

The said sides, therefore, of the three figures, I mean the pyramid, the octahedron and the cube, are to one another in rational ratios.

But the remaining two, I mean the side of the icosahedron and the side of the dodecahedron, are not in rational ratios either to one another or to the aforesaid sides;
for they are irrational, the one being minor [XIII. 16] and the other an apotome [XIII. 17].

That the side *MB* of the icosahedron is greater than the side *NB* of the dodecahedron we can prove thus.

For, since the triangle *FDB* is equiangular with the triangle *FAB*, [VI. 8]
proportionally, as *DB* is to *BF*, so is *BF* to *BA*. [VI. 4]

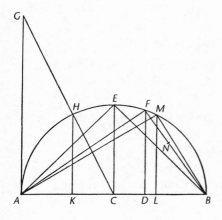

And, since the three straight lines are proportional,

as the first is to the third, so is the square on the first to the square on the second;

[V. Def. 9, VI. 20, Por.]

therefore, as *DB* is to *BA*, so is the square on *DB* to the square on *BF*;

therefore, inversely, as *AB* is to *BD*, so is the square on *FB* to the square on *BD*.

But *AB* is triple of *BD*;

therefore the square on *FB* is triple of the square on *BD*.

But the square on *AD* is also quadruple of the square on *DB*,

for *AD* is double of *DB*;

therefore the square on *AD* is greater than the square on *FB*;

therefore *AD* is greater than *FB*;

therefore *AL* is by far greater than *FB*.

And, when *AL* is cut in extreme and mean ratio, *KL* is the greater segment, inasmuch as *LK* belongs to a hexagon, and *KA* to a decagon; [XIII. 9]

and, when *FB* is cut in extreme and mean ratio, *NB* is the greater segment;

therefore *KL* is greater than *NB*.

But *KL* is equal to *LM*;

therefore *LM* is greater than *NB*.

Therefore *MB*, which is a side of the icosahedron, is by far greater than *NB* which is a side of the dodecahedron.

Q.E.D.

I say next that *no other figure, besides the said five figures, can be constructed which is contained by equilateral and equiangular figures equal to one another.*

For a solid angle cannot be constructed with two triangles, or indeed planes.

With three triangles the angle of the pyramid is constructed, with four the angle of the octahedron, and with five the angle of the icosahedron;

but a solid angle cannot be formed by six equilateral triangles placed together at one point,

for, the angle of the equilateral triangle being two-thirds of a right angle, the six will be equal to four right angles:

which is impossible, for any solid angle is contained by angles less than four right angles.

For the same reason, neither can a solid angle be constructed by more than six plane angles.

By three squares the angle of the cube is contained, but by four it is impossible for a solid angle to be contained,

for they will again be four right angles.

By three equilateral and equiangular pentagons the angle of the dodecahedron is contained;

but by four such it is impossible for any solid angle to be contained,

for, the angle of the equilateral pentagon being a right angle and a fifth, the four angles will be greater than four right angles:
which is impossible.

Neither again will a solid angle be contained by other polygonal figures by reason of the same absurdity.

Therefore etc.

Q.E.D.

LEMMA. But that *the angle of the equilateral and equiangular pentagon is a right angle and a fifth* we must prove thus.

Let *ABCDE* be an equilateral and equiangular pentagon,
let the circle *ABCDE* be circumscribed about it,
let its centre *F* be taken,
and let *FA, FB, FC, FD, FE* be joined.

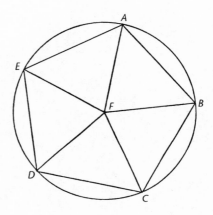

Therefore they bisect the angles of the pentagon at *A, B, C, D, E*.

And, since the angles at *F* are equal to four right angles and are equal,
therefore one of them, as the angle *AFB*, is one right less a fifth;
therefore the remaining angles *FAB, ABF* consist of one right angle and a fifth.

But the angle *FAB* is equal to the angle *FBC*;
therefore the whole angle *ABC* of the pentagon consists of one right angle and a fifth.

Q.E.D.

Bibliography

Editions and translations of *Elements*

Euclidis Opera Omnia, J. L. Heiberg and H. Menge, eds., 8 vols. (Leipzig: Teubner, 1883–1916). Greek with Latin translation.

Euclid, *Elementa*, J. L. Heiberg and E. S. Stamatis, eds., 6 vols. (Leipzig: Teubner, 1969–1977). The standard edition of the Greek text and scholia.

Euclid, *The thirteen books of Euclid's Elements,* translation, introduction, and commentary by Sir Thomas L. Heath, 3 vols. (Cambridge: Cambridge University Press, 1903). The first edition of the standard translation.

Euclid, *The thirteen books of Euclid's Elements*, tr. Heath, 2nd edition, revised with additions, 3 vols. (Cambridge: Cambridge University Press, 1926). The basis for the Dover edition.

Euclid, *The thirteen books of Euclid's Elements*, tr. Heath, 3 vols. (New York: Dover Publications, 1956). Facsimile reprint of the 1926 Cambridge edition.

Euclid, *The thirteen books of Euclid's Elements*, tr. Heath, in *Great Books of the Western World*, Vol. 11 (Chicago, Encyclopaedea Britannica, 1952). A new version of the 1926 edition, with some changes.

Euclid in Greek, Book I. Sir Thomas L. Heath, ed., (Cambridge: Cambridge University Press, 1920). The Greek text of Book I with many useful notes on Euclid's usage.

Other works of Euclid

"Euclid," Chapter 15 of *Greek Mathematical Works, I: Thales to Euclid*, ed. and trs. by Ivor Thomas (Cambridge, Massachusetts and London: Loeb Classical Library, 1957). Selections from *Elements, Data, Porisms, Conics, Surface Loci,* and *Optics,* in Greek with English translation.

Euclid, *The Data of Euclid*, George L. McDowell and Merle A. Sokolik, trs. (Baltimore: Union Square Press, 1993). A different approach to geometry, aimed at analysis and the art of solving problems.

Euclid, *Recipients, Commonly Called the Data* (Golden Hind, 1989). Another translation of the *Data*.

Euclid, *Phaenomena: a translation and study of a hellenistic treatise in spherical astronomy,* trs., ed., and commentary by J.L. Berggren and R.S.D. Thomas. (New York : Garland Pub., 1996).

Euclid, *Optics* and *Catoptrics,* Evangelos Spandagos, ed. (Athens: Aithra, 2000). Greek text with a modern Greek translation.

Euclid, "The Optics of Euclid," trs. Harry E. Burton, *Journal of the American Optical Society* **35** (1945), pp. 357–372.

Smith, A. Mark, *Foundations of Ancient Mathematical Optics: A Source Based Guided Study* (Philadelphia: American Philosophical Society, 1999). Contains selections from Euclid's *Optics* and *Catoptrics*.

Euclid, *The Arabic Version of Euclid's Optics: Edited and Translated with Historical Introduction and Commentary*, Elaheh Kheirandish, trs. (Springer Verlag, 1998).

Euclid's book on divisions of figures, Raymond Clare Archibald, ed. (Cambridge: Cambridge University Press, 1915).

The Lost Treatises of Euclid, Evangelos Spandagos, ed. (Athens: Aithra, 2000). Greek text with modern Greek translation. (Euclid's Fragmenta)

Meibom, Marcus, *Antiquae musicae auctores septem: Graece et Latine* (New York: Broude Brothers, 1977). A facsimile of the 1652 edition; includes Euclid's *Introductio harmonica* in Greek with a Latin translation.

Other ancient authors

Proclus, *A Commentary on the First Book of Euclid's* Elements, trs. Glenn R. Morrow (Princeton: Princeton University Press, 1970, 1992).

Secondary materials

Artmann, Benno, *Euclid : The Creation of Mathematics* (Springer Verlag, 1999). An overview of *Elements* with commentary.

Boyer, Carl B., *A History of Mathematics* (Princeton: Princeton University Press, 1985). General information on Euclid and Greek mathematics.

Fowler, David, *The Mathematics of Plato's Academy* (Oxford: Oxford University Press, 1999). An original and controversial interpretation of Greek mathematics, with particular attention to Books II, IV, X, and XIII of *Elements*. Includes plates of many early Euclid manuscripts.

Hartshorne, Robin, *Companion to Euclid* (Providence: American Mathematical Society, 1997). Begins with Euclid and contrasts his work with classical and modern mathematical approaches.

Heath, Thomas L., *A History of Greek Mathematics,* 2 vols. (Oxford, 1921; reprinted: New York: Dover Publications, 1981).

Heath, Thomas L., *A Manual of Greek Mathematics* (Oxford, 1931; reprinted New York: Dover Publications, 1963).

Klein, Jacob, *Greek Mathematical Thought and the Origin of Algebra* (New York: Dover Publications, 1992). Contains a perceptive account of ancient number theory.

Netz, Reviel, *The Shaping of Deduction in Greek Mathematics* (Cambridge: Cambridge University Press, 1999). Concentrates especially on diagrams and the way their use affected the structure of mathematical arguments.

Internet

Joyce, David E., *Euclid's Elements.*
Elements in modernized language by a present-day mathematician.
http://aleph0.clarku.edu/~djoyce/java/elements/elements.html.
(Internet resources are, of course, ephemeral. This address may have changed, or the website may have been discontinued, by the time you read this.)

Index (with Glossary)